AF869577

Lecture Notes in Chemistry

Volume 108

Series Editors

Barry Carpenter, Cardiff, UK

Paola Ceroni, Bologna, Italy

Katharina Landfester, Mainz, Germany

Jerzy Leszczynski, Jackson, USA

Tien-Yau Luh, Taipei, Taiwan

Eva Perlt, Bonn, Germany

Nicolas C. Polfer, Gainesville, USA

Reiner Salzer, Dresden, Germany

Kazuya Saito, Department of Chemistry, University of Tsukuba, Tsukuba, Japan

The series Lecture Notes in Chemistry (LNC), reports new developments in chemistry and molecular science - quickly and informally, but with a high quality and the explicit aim to summarize and communicate current knowledge for teaching and training purposes. Books published in this series are conceived as bridging material between advanced graduate textbooks and the forefront of research. They will serve the following purposes:

- provide an accessible introduction to the field to postgraduate students and nonspecialist researchers from related areas,
- provide a source of advanced teaching material for specialized seminars, courses and schools, and
- be readily accessible in print and online.

The series covers all established fields of chemistry such as analytical chemistry, organic chemistry, inorganic chemistry, physical chemistry including electrochemistry, theoretical and computational chemistry, industrial chemistry, and catalysis. It is also a particularly suitable forum for volumes addressing the interfaces of chemistry with other disciplines, such as biology, medicine, physics, engineering, materials science including polymer and nanoscience, or earth and environmental science.

Both authored and edited volumes will be considered for publication. Edited volumes should however consist of a very limited number of contributions only. Proceedings will not be considered for LNC.

The year 2010 marked the relaunch of LNC.

Junpei Yamanaka · Tohru Okuzono ·
Akiko Toyotama

Colloidal Self-Assembly

Springer

Junpei Yamanaka
Nagoya City University
Nagoya, Japan

Tohru Okuzono
Nagoya City University
Nagoya, Japan

Akiko Toyotama
Nagoya City University
Nagoya, Japan

ISSN 0342-4901 ISSN 2192-6603 (electronic)
Lecture Notes in Chemistry
ISBN 978-981-99-5051-5 ISBN 978-981-99-5052-2 (eBook)
https://doi.org/10.1007/978-981-99-5052-2

© The Editor(s) (if applicable) and The Author(s), under exclusive license to Springer Nature Singapore Pte Ltd. 2023

This work is subject to copyright. All rights are solely and exclusively licensed by the Publisher, whether the whole or part of the material is concerned, specifically the rights of translation, reprinting, reuse of illustrations, recitation, broadcasting, reproduction on microfilms or in any other physical way, and transmission or information storage and retrieval, electronic adaptation, computer software, or by similar or dissimilar methodology now known or hereafter developed.
The use of general descriptive names, registered names, trademarks, service marks, etc. in this publication does not imply, even in the absence of a specific statement, that such names are exempt from the relevant protective laws and regulations and therefore free for general use.
The publisher, the authors, and the editors are safe to assume that the advice and information in this book are believed to be true and accurate at the date of publication. Neither the publisher nor the authors or the editors give a warranty, expressed or implied, with respect to the material contained herein or for any errors or omissions that may have been made. The publisher remains neutral with regard to jurisdictional claims in published maps and institutional affiliations.

This Springer imprint is published by the registered company Springer Nature Singapore Pte Ltd.
The registered company address is: 152 Beach Road, #21-01/04 Gateway East, Singapore 189721, Singapore

Preface

This book is intended to introduce the basics of the study of structure formation by self-assembly of colloidal particles. The authors have studied colloidal self-assembly for many years through experiments (JY, AT) and computer simulations (TO). This book is a concise compilation of information necessary for research in this field. The main subject of the book is the self-assembly of mostly spherical colloid particles.

Colloids have attracted attention as model experimental systems in soft matter physics and as useful materials in chemistry and pharmacology. This book will be useful to researchers in these fields as well. In addition, many examples in this book relate to other scientific and technological fields, such as optical materials, sensors, and cosmetics, and are expected to have applications in a variety of scientific and technological fields.

First, the basic knowledge of colloid science necessary to understand this publication is explained, followed by a description of the various experimental techniques, from particle synthesis to structural analysis. We also explain basic computer simulation methods with examples.

Based on the above, the authors will then introduce the various self-assembly studies of colloidal systems that we have carried out, particularly the self-assembly of colloidal crystals. The main topics are (1) crystallization of colloidal dispersions focusing on the role of surface charge, (2) preparation of large and high-quality colloidal crystals by controlled growth methods, (3) aggregation and crystallization using void attraction in the presence of polymers, and (4) applications of colloidal crystals from cosmetics to sensing materials. We also present (5) space experiments on colloidal self-assembly on the International Space Station. In the last chapter, we described (6) a brief review and perspective in this research field.

This book is targeted at students and researchers in the fields of chemistry, pharmacy, and engineering. Basic knowledge of electromagnetism and the physical chemistry of electrolytes, which is necessary for understanding this book, is explained in the Appendices.

We would like to express our sincere gratitude to the many students who have collaborated with us on the works presented in this publication. We would also like to thank the students who were enrolled in our laboratory at the time of writing this book for their valuable cooperation in the preparation of this book.

The two space experiments in which we participated were conducted in collaboration with researchers from JAXA and many other institutions, as listed below, for which we hereby express our sincere appreciation.

Three-Dimensional Photonic Crystal Project. Tsuyoshi Kano, Tomoyuki Kobayashi, Yoshimasa Ohki, Toshitami Ikeda, and Teruhiko Tabuchi (JAXA); Kyoichi Arakane, Yuki Watanabe, and Erika Yoda (AES Co., Ltd.); Kensaku Ito (Toyama University); Sachiko Onda and Masako Murai (Nagoya City University); Yoshihiro Takiguchi and Shoichi Uchiyama (Hamamatsu Photonics Co., Ltd.); Tsutomu Sawada (National Institute for Materials Science); Fumio Uchida, Hiroshi Yamada, and Hiroshi Ozaki (Fuji Chemicals Co., Ltd.); Shigeru Okamoto (Nagoya Institute of Technology) and Yoichi Oba (Interface Technology Labs., Co., Ltd.).

Colloidal Clusters Project. Satoshi Adachi and Tetsuya Sakashita (JAXA); Taro Shimaoka and Masae Nagai (Japan Space Forum); Yuki Watanabe, Seijiro Fukuyama, and Yui Nakata (AES Co., Ltd.), Tsunehiko Higuchi (Nagoya City University); Jitendra Mata (Australian Nuclear Science and Technology Organisation).

The senior author (JY) would like to take this opportunity to sincerely thank Prof. Norio Ise, Professor Emeritus of Kyoto University, who gave me the opportunity to start research on colloidal crystals. Sincere thanks are due to Prof. Takeji Hashimoto, Professor Emeritus of Kyoto University and Leader of the ERATO Polymer Phasing Project, and Prof. Masakatsu Yonese, Professor Emeritus of Nagoya City University.

We would like to sincerely thank many researchers with whom I have collaborated. In particular, we are grateful to Prof. Satoshi Uda and Dr. Jun Nozawa of the Institute for Materials Research, Tohoku University, with whom we had discussions through joint seminars for many years. Sincere thanks are due to Prof. B. V. R. Tata of Hyderabad University, Prof. Priti S. Mohanty of Kalinga Institute of Industrial Technology, India, and also many students in their laboratories, for their collaborative research on colloidal crystallization and microgel colloidal crystals.

Finally, we would like to express our deepest gratitude to Shinichi Koizumi and the staff of Springer Nature for giving us the opportunity to write this book and for waiting so long for the manuscript to be completed.

A video describing the contents of this book has been published as part of the Springer Nature Video Series. This video will include many movies and simulation results on the studies of the self-assembly process of colloidal systems, which we believe will further deepen your understanding of this book.

We hope that this book will be helpful to readers in their future research on colloidal self-assembly.

Nagoya, Japan
June 2023

Junpei Yamanaka
Tohru Okuzono
Akiko Toyotama

Contents

Chapter 1
An Introduction to Colloid Science and Colloidal Self-Assembly

Abstract We wrote this book for those who wish to begin the study of colloidal self-assembly. Colloids are model systems for soft matter physics and other basic sciences and are also useful materials in engineering, chemistry, and pharmacology. The main subject of the book is the self-assembly of spherical colloid particles. In this chapter, we begin with a definition and examples of colloids and discuss the interest and significance of studying the self-assembly and structure formation of colloids. Typical self-assembled structures, including colloidal crystals, are then introduced. We will also give an overview of this book.

Keywords Colloids · Dispersions · Colloidal crystals · Self-assembly · Photonic materials

The main subject of this book is the self-assembly of mostly spherical colloid particles. In the first chapter, we explain the definition of colloids and describe the significance of self-assembly and structure formation of colloids. We then explain the relationship between colloid science and various other scientific and technological fields. An overview of this book will also be given.

1.1 What is a Colloid?

Let us start with the definition of colloid. Scottish Scientist Thomas Graham proposed the idea of "colloid" in 1861 [1]. He classified substances into those that diffuse quickly, such as inorganic salts in water, and those that diffuse slowly, such as gelatin. He named the former *crystalloid* because it often crystallizes. Gelatin, the main ingredient of animal glue, is an example of the latter, and it was classified as *colloid* from Greek $\kappa o\lambda\lambda\alpha$, which means bond. Therefore, the original meaning of the word "colloid" is "a substance that diffuses slowly like gelatin."

While this definition has historical significance, colloids are defined from a different perspective today. The definition of colloid currently adopted is that recommended by the International Union of Pure and Applied Chemistry (IUPAC), the

© The Author(s), under exclusive license to Springer Nature Singapore Pte Ltd. 2023

J. Yamanaka et al., *Colloidal Self-Assembly*, Lecture Notes in Chemistry 108,
https://doi.org/10.1007/978-981-99-5052-2_1

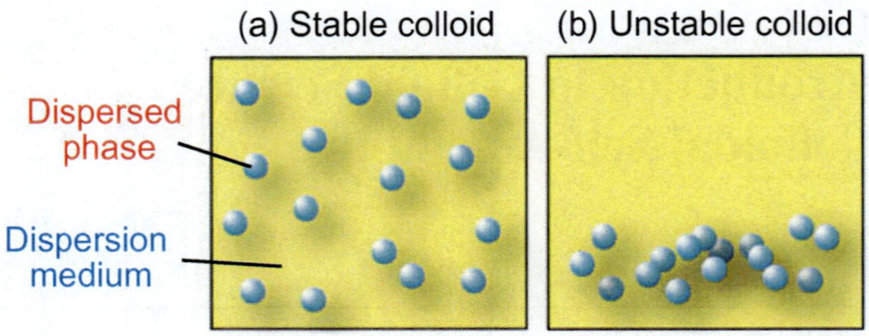

Fig. 1.1 Illustrations of **a** stable and **b** unstable colloids

organization that defines terminology in chemistry, in 1972 [2]. The term refers to a state of subdivision, implying that the molecules or polymolecular particles dispersed in a medium have at least in one direction a dimension roughly between 1 nm and 1 μm, or that in a system discontinuities are found at distances of that order. Figure 1.1a shows the schematic diagram of the colloid. This definition is based on the ideas of a Dutch physical chemist, Wolfgang Ostwald. The "phase" refers to the part with uniform material properties, and "dispersion" means the state in which other substances are scattered in one phase. According to the definition of IUPAC, a phase is "an entity of a material system which is uniform in chemical composition and physical state." Some points to keep in mind are summarized below.

(i) "Dispersion" and "solution" are different. For example, in an aqueous solution of sodium chloride NaCl, NaCl dissociates into Na^+ and Cl^- ions and presents almost uniformly in water. On the other hand, NaCl is insoluble in oil, but if NaCl powder was added to the oil and stirred vigorously with a mixer, NaCl would be crushed and present in pieces in the oil as fine particles. These NaCl particles are small in size, but their physicochemical properties, *e.g.*, density and melting point, are the same as macroscopic solid NaCl. Such a state is referred to as a "dispersed" state, and the entire system is called a dispersion.

(ii) In the definition of colloid science given above, "colloid" refers to the entire system consisting of a dispersed phase and a dispersion medium (dispersant). In other words, a colloid is synonymous with a "colloidal dispersion system." In soft matter physics, however, the dispersed phase is sometimes called colloid. In this book, we follow the definition of colloid science. When it is necessary to identify the dispersed phase, it is denoted as a "colloidal particle," etc.

(iii) The boundary where two phases are in contact is called the "interface." Colloidal systems have an internal interface between the two phases: the dispersed phase and the dispersion media phase. The interface is also called the "surface" of the phase. Usually, the dense side is the standard phase in such cases. For example, if a solid and a liquid are in contact, the interface is called the "surface of the solid" and not the "surface of the liquid."

The total interface area inside the colloid is much larger than the system's size. Consider the case where the dispersed phase is a spherical particle of radius R, and the volume fraction of the dispersed phase is φ. Here φ is defined as $\varphi = V_2/V_1$, where V_1 is the volume of the entire system and V_2 is the volume of the dispersed phase. For a single particle, the volume $V = (4/3)\pi R^3$ and the surface area $S = 4\pi R^2$. Since the number of particles in 1 mL of dispersion is $N = \varphi/V$, the total surface area S_{tot} is given by $S_{tot} = NS = 3\varphi/R$. For example, when $\varphi = 0.1$ and $2R = 1$ nm, we have $N = 1.9 \times 10^{20}$ and $S_{tot} = 6 \times 10^6$ cm^2. The surface area of a cube with a volume of 1 mL is 6 cm^2, and S_{tot} is one million times larger. This S_{tot} value also corresponds to the area of a square with one side approximately 25 m.

The smaller the particle size, the greater the percentage of atoms on the surface. For a gold particle of 1 nm diameter, the atoms on the surface rate are about 50% of the atoms that make up the particle. Because surface atoms are more thermally mobile than internal atoms, metal nanoparticles are known to have interesting features, such as a lower melting point than macroscopic metals. Nanoparticles are not discussed in detail in this book, so please refer to other books listed in the "Further Reading."

(iv) The definition of "colloid" relates to size and state, and any system that satisfies these conditions is considered a colloid; i.e., it does not depend on the type of matter. There are three phases of matter: gas phase, liquid phase, and solid phase; therefore, there are nine different combinations of dispersed steps and media. However, gas–gas colloids do not seem to have been reported because the two gases mix and no interface is formed. Any other combination will result in an interface. Examples of typical colloids are compiled in Fig. 1.2. Note that milk is a colloid of solid protein and liquid oil droplets dispersed in water. Polymer gels, such as jelly, are listed in the table as examples of colloids because they can be regarded as structures in which water is dispersed in a solid polymer. In reality, however, polymer gels are two-component systems in which the polymer is swollen with water, and the polymer and water phases often have a continuous structure.

(v) The wavelength of visible light is about 400–800 nm, and the size of the dispersed phase of colloid is within this range. In general, objects scatter electromagnetic waves of wavelengths close to their size. Thus, colloidal systems with particles of a few hundred nm scatter visible light and appear colored (See Column 1.1). In particular, when all wavelengths of visible light are scattered to the same degree, they appear white to the human eye. Clouds and fog appear white because the tiny colloidal-sized droplets in them scatter sunlight. Note that such coloration of colloids occurs by a mechanism quite different from that of coloration by dyes. Pigment molecules absorb light at wavelengths specific to the molecule and thus color it. For example, chloroplasts absorb light "other than" green, thus making plants appear green. Water, a component of clouds and fog, is transparent to visible light. Milk and various other suspensions also appear opaque. Muddy water and ink appear turbid due to the substance's absorption and scattering of light.

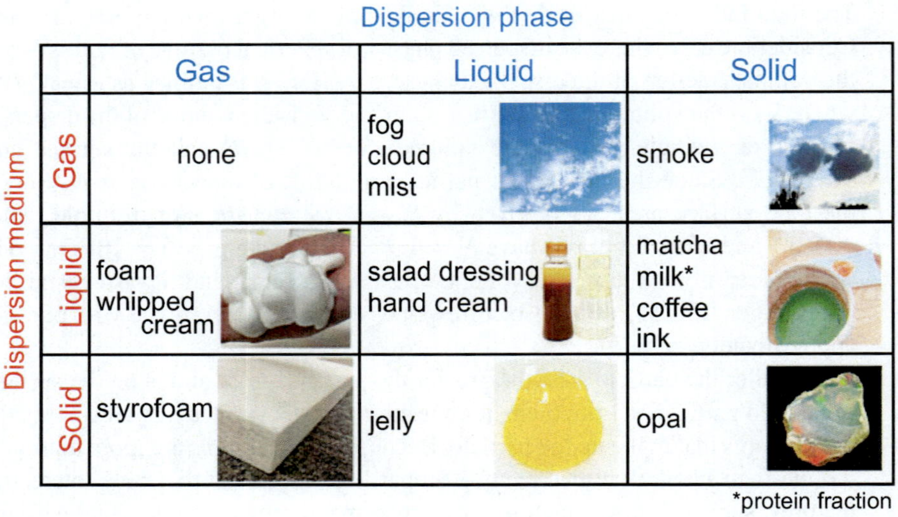

Fig. 1.2 Examples of colloids

The gemstone opal also exhibits a variety of colors, which is due to the phenomenon of reflection of light at specific wavelengths due to the Bragg diffraction (See Column 1.2). The colors derived from the structure are called structural colors. In addition to opal, many other examples of structural color are known in nature [3]. For instance, the beautiful coloration of the shell (Fig. 1.3) is due to the alternating layers of inorganic crystalline materials and proteins.

Column 1.1

Rayleigh scattering and Mie scattering

The scattering of light by a substance depends on the ratio of the size of the substance to the wavelength of light. Light is an electromagnetic wave, i.e., a wave in which an electric field and a magnetic field are propagating through space, oscillating.

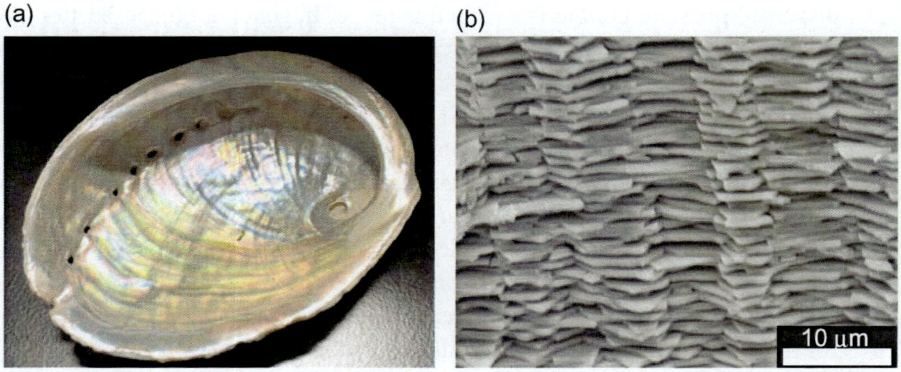

Fig. 1.3 **a** Structural color from the shell and **b** SEM image of the cross-section of the shell

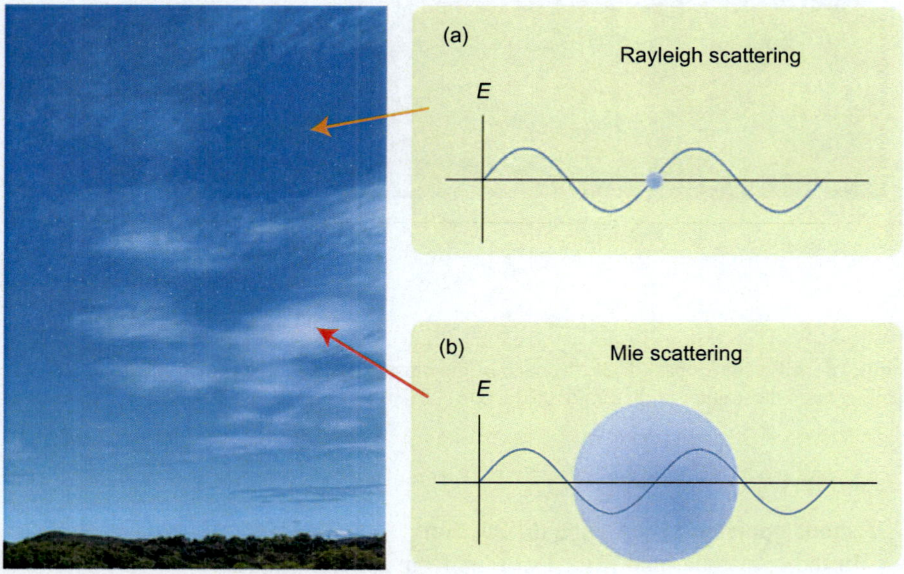

Fig. 1.4 Relationship between particle size and wavelength of light for **a** Rayleigh scattering and **b** Mie scattering theory. Rayleigh scattering can explain the blue color of the sky. Mie scattering theory is needed to explain the white color of clouds

Although the magnetic field is in a plane orthogonal to the electric field, its effect on light scattering is small. Therefore, here we consider only the electric field. When the particle size is sufficiently smaller than the wavelength of the incident light, we can safely assume that the electric field strength is the same at any location on the particle (Fig 1.4a). Lord Rayleigh studied light scattering in this case. Rayleigh's theory explains experimental results well for scattering from particles down to about one-tenth the size of the wavelength. It was derived that the scattering intensity is inversely proportional to the fourth power of the wavelength of the incident light. The shorter the wavelength, the stronger the light is scattered. The sky appears blue because of Rayleigh scattering by fine particles in the atmosphere.

On the other hand, as the size of particles increases further, the same particle cannot be regarded as having the same electric field strength at different locations, and the assumption used to derive Rayleigh's equation no longer holds (Fig. 1.4b). G. Mie has reported a theory that considers the difference in the phase of light within the particle [4]. In this case, the scattering is called Mie scattering. If the particle size is sufficiently larger than the wavelength of light, the wavelength dependence of the diffuse light intensity becomes small. Therefore, the scattered light becomes white. Clouds and fog appear white due to Mie scattering by water droplets in the atmosphere (the size of the droplets may exceed the upper limit of the colloidal definition).

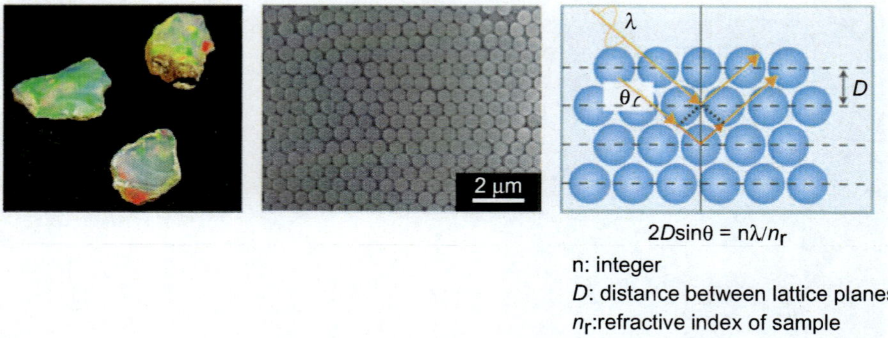

Fig. 1.5 **a** Precious opals, **b** SEM image of the closest-packed structure of silica particles (d = 500), and **c** illustration of the Bragg diffraction

Column 1.2

Precious opals and the Bragg diffraction

In the gemstone opal (Fig. 1.5a), uniform spherical particles of silicon dioxide (silica) of several hundred nanometers in size form densely packed colloidal crystals. Opal is formed when rainwater or underground thermal water containing silicic acid fills the gaps in geological strata, from which uniformly sized silica particles precipitate over a long period and accumulate in a regular crystal structure. Figure 1.5b is a scanning electron microscope (SEM) image of the crystalline structure of silica particles prepared in our laboratory. The particle arrangement in Fig. 1.5b is called the opal structure because it is identical to the arrangement of silica particles in precious opal.

The beautiful coloring of opal is due to the Bragg diffraction caused by this crystal structure (Fig. 1.5c). Colloidal crystals can be considered two-dimensional planar particle layers stacked at equal intervals. The planes created by these particle layers are called "lattice planes." Reflected light from neighboring lattice planes travels along light paths of different lengths. For example, if the wavelength in a vacuum is λ, the wavelength in a material is λ/n_r, where n_r is the refractive index of the sample. When light with an optical path difference of an integer multiple of the wavelength overlaps, intensifying interference occurs, and intense light is observed in that direction. This phenomenon is called Bragg diffraction. Light is amplified when the angle θ between the crystal surface and the material satisfies Bragg's condition $2D\sin\theta = n\lambda/n_r$ (n is a positive integer, and D is the distance between lattice planes). The angle θ is called the Bragg angle.

(vi) From the above definition of colloid, it can be seen that the dispersed phase must be stably dispersed in the medium. However, colloidal particles often aggregate due to van der Waals (vdW) attraction between particles (Fig. 1.1b). Therefore, maintaining a stable colloid (stabilization) is one of the central problems in colloid science, and much knowledge has been accumulated. Two stabilization methods are well known. One is steric stabilization, in which particles are covered with a polymer or other materials to keep them away from the range of vdW attraction. The other is electrostatic stabilization, in which an electric charge is introduced into the particles to repel them electrically. The stabilization of colloids is discussed in detail in Chap. 2.

1.2 Why Do We Study the Self-Assembly of Colloidal Particles?

As mentioned above, "colloid" is defined in terms of the way things are viewed, and as can be seen from Fig. 1.2, a wide variety of colloids exist in nature. You will also see that colloidal dispersions are used in various industrial fields, including paints, detergents, foods, and pharmaceuticals. Thus, colloid science is an interdisciplinary study and is associated with many areas of science and technology, both basic and applied. Therefore, colloidal self-assembly has been studied from many different perspectives.

Recent studies on colloidal self-assembly are described in Sect. 7.2. Also, please refer to recent review articles [5–8]. A number of studies have been reported focusing on the following three respects.

1.2.1 To Elucidate Colloidal Assembly in Nature

Colloids and assembled structures are abundant in nature, as mentioned in Sect. 1.1. Therefore, studies using colloidal particles with well-defined properties can lead to a better understanding of association phenomena.

It has also recently been shown that chameleon skin is covered with a two-dimensional (2D) regular array of biomolecular colloidal particles related to body color variation [9]. Because the lattice spacings of these assembled structures are in the visible light regime, they exhibit structural color due to the Bragg diffraction of visible light. The phenomenon of colloidal aggregation is also important in biology [10]. Most globular protein molecules are nm in size and can be classified as colloidal particles. The shell structure of viruses is also formed by the aggregation of one or several types of protein molecules. Because of their small size, viruses cannot store genes for many types of proteins. For this reason, one or a few kinds of proteins assemble to form the body of the virus. Studying the organization of colloidal systems is also helpful in understanding the structure formation mechanism.

1.2.2 As Models to Study the Phase Behavior of Atomic and Molecular Systems

Colloidal dispersions are interesting model systems for studying the phase behavior of atoms and molecules, including crystallization and phase separation. As discussed in more detail in Chap. 2 and beyond, colloidal particles arrange themselves regularly in dispersion to form structures called "colloidal crystals," similar to atomic or molecular crystals. For example, in electrostatically stabilized colloids, crystals are formed when the electrostatic repulsion between particles is strong enough. Figure 1.6a is an appearance of the charged colloidal crystals, which shows structural color due to the Bragg diffraction. Figure 1.6b is a three-dimensional (3D) microscope image of the crystal structure taken with a confocal laser scanning microscope (see Chap. 3). Colloidal and atomic crystals have various similarities and have been studied as model systems. Colloidal systems are considered sound model systems for atoms and molecules for the following reasons.

(i) Micron- and submicron-sized particles can be observed in situ and in real time using optical microscopy. This allows us to keep the association structure of particles at the single-particle level.
(ii) A variety of interactions can be made to act between particles, including excluded volume interactions and electrostatic forces. Furthermore, the magnitude of these interactions can be easily varied over a wide range. For example, the electrostatic repulsive forces between charged particles can be easily tuned over several orders of magnitude with the concentration of salt added. Using

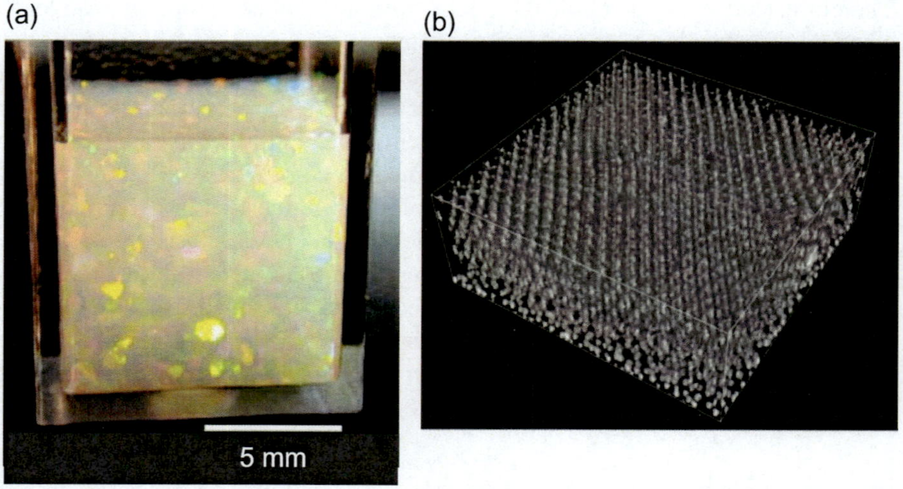

Fig. 1.6 Charged colloidal crystals. **a** Overview (colloidal silica/water diameter $d = 100$ nm, 3.2vol%). **b** 3D micrograph (polystyrene particle/water, $d = 482$ nm; image size, $7.62 \times 27.62 \times 11.55$ µm)

colloidal systems, experimental conditions corresponding to ultra-high temperature and ultra-high pressure in atomic systems can be easily achieved. It is also possible to create a state corresponding to an ideal gas without interaction between the particles.

(iii) Since colloidal particles have slow motion, i.e., long characteristic time, one can easily observe non-equilibrium and dynamic processes such as crystal growth. This makes it possible to study the methods of particle association and phase transitions at the single-particle level.

1.2.3 As Novel Materials

Many studies have been reported on the application of colloidal self-assembled structures as novel materials. Of note are applications in the field of optics, as will be discussed in detail in Chaps. 6 and 7. The diffraction wavelength of colloidal crystals can be set from visible light to the near-infrared region. Structures in which the refractive index changes periodically are called "photonic crystals" (derived from photons) [5, 11, 12]. The way light propagates can be controlled by such structures. 1D and 2D photonic crystals can be constructed by bottom-up methods such as thin-film lamination and lithographic techniques. Colloidal crystals are three-dimensional photonic crystals, which are not always easy to fabricate by other methods. In addition to face-centered-cubic (FCC) and body-centered-cubic (BCC) structures, recent interest has focused on constructing complex structures with low filling factors, such as diamond structures. Diamond lattice structures are theoretically known to reflect incident light at any angle. The construction of diamond lattices using colloidal crystals and their application to photonic crystals has attracted much attention [13, 14]. In Chap. 6, we also describe applied studies of colloidal self-assembly for sensing, decoration, and cosmetics.

1.3 Outline of the Book

Chapter 2 and subsequent chapters of this book are organized to provide a detailed description of the study of the self-assembly of colloidal systems; from the basics to applications, Fig. 1.7 presents examples of the topics covered in Chap. 2 onward.

Chapter 2 describes the fundamentals necessary to understand the contents of the subsequent chapters. As discussed in this chapter, in a colloidal dispersion, the colloidal particles are stably dispersed in a medium. Colloidal particles aggregate when the interaction between the particles is attractive in total and is stronger than the effect of random thermal motion of the particles. In Chap. 2, we explain the concept of stability and stabilization of such colloidal dispersion systems. Next, typical interactions of colloidal systems, such as hard-sphere repulsion, vdW forces, electrostatic interactions, and depletion attraction, are described. Furthermore, basic knowledge of

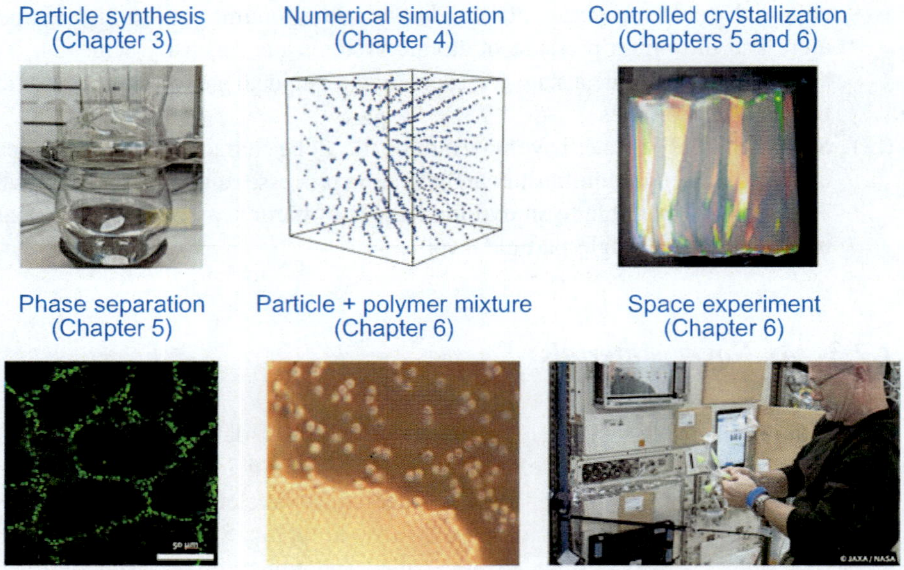

Fig. 1.7 Examples of topics from Chap. 2 onward

the phenomenon of colloidal crystallization, in which colloidal particles are regularly arranged in a crystal lattice due to these interactions, will be explained.

Chapter 3 describes the basic knowledge and methods necessary for the experimental study of the self-assembly of colloidal systems. First, we describe the synthesis of colloidal samples and techniques for characterizing particle concentration, particle size, and surface charge number. Next, we will show how to prepare various colloidal crystals described in Chap. 2. Some practical tips will also be presented. Furthermore, the principles and measurement examples of crystal structure analysis by spectroscopy, optical microscopy, and scattering methods are described using charged colloidal crystals as an example.

Chapter 4 explains the basic concepts of numerical modeling and algorithms and methods for analyzing numerical data obtained by computer simulation of colloidal systems. First, molecular dynamics methods are briefly introduced, and then some ways that are also useful for simulating colloidal systems are presented. Next, Brownian dynamics and Monte Carlo methods will be described for manufacturing colloidal systems. Finally, pseudocodes for the main algorithms and some images obtained from the simulations are presented.

Chapter 5 presents some of our previously reported methods for controlled colloidal crystallization. We will first give some methods we have reported for the controlled crystallization of colloids for preparing large, high-quality, and charged colloidal crystals. In particular, we describe unidirectional growth under pH and salt concentration gradients. Then, the temperature dependence of the colloidal system will be described, and controlled crystallization using temperature variation will be discussed. Note that small amounts of different types of particles added to a single-component colloidal system behave as "impurities." The processes involved in eliminating various impurities from colloidal crystals will also be presented.

Chapter 6 will introduce applied research on colloidal crystals and self-assembly. First, we present research on the preparation of self-supporting materials by fixing the crystal structure with a polymer gel. We also report on the crystallization of microgels using depletion attraction and the application of gold colloidal crystals.

Our laboratory participated in a space experiment project on colloidal self-assembly in the International Space Station microgravity environment in collaboration with the Japan Aerospace Exploration Agency (JAXA) and many other organizations. Finally, we also report on our space experiments on colloidal systems.

In Chap. 7, a summary of the previous chapters is given. We also discuss recent developments in colloidal self-assembly, such as anisotropically interacting particles and multicomponent systems, as well as future prospects in this field.

To understand this book, the reader should have some knowledge of elementary mechanics, electromagnetism, and mathematics on partial derivatives. Appendices to this book include A. electromagnetism, B. the physical chemistry of electrolyte solutions (Debye–Hückel theory), C. the basics of titration experiments, D. the derivation of electrostatic potentials between particles, and E. the diffusion equation, which should be referred to as needed. Columns are also provided throughout the text to summarize the necessary related knowledge and information. The topics covered in this book also include phase transitions, which are studied in polymer science and statistical physics. Many textbooks and review articles are included in each chapter as Further Readings and references for readers who wish to study the subject in more depth.

A movie of the contents of this book was published under the title "Visual Guide to Study Colloidal Self-Assembly" by Springer Nature in 2022. We hope that the readers will also watch this movie for a better understanding of the contents of this book.

References

1. Graham T (1861) Phil Trans Roy Soc London 151:183–224
2. https://goldbook.iupac.org/terms/view/C01172
3. Heepe L, Xue L, Gorb SN (eds) (2017) Bio-inspired structured adhesives: biological prototypes, fabrication, tribological properties, contact mechanics, and novel concepts. Springer International, Cham
4. van de Hulst HC (1957) Light scattering by small particles. Wiley, New York
5. Li F, Josephson DP, Stein A (2011) Angew Chem Int Ed 50:360–388
6. Vogel N et al (2015) Chem Rev 115:6265–6311
7. Li B, Zhou D, Han Y (2016) Nat Rev Mater 1:1–13
8. Cai Z et al (2021) Chem Soc Rev 50:5898–5951
9. Teyssier J, Saenko SV, Van Der Marel D, Milinkovitch MC (2015) Nat Commun 6:6368
10. Tyedmers J, Mogk A, Bukau B (2010) Nat Rev Mol Cell Biol 11:777–788
11. Joannopoulos JD, Winn JN (2008) Photonic crystals: molding the flow of light, 2nd edn. Princeton University Press, New Jersey

12. Moon JH, Yang S (2010) Chem Rev 110:547–574
13. Ducrot É, He M, Yi GR, Pine DJ (2017) Nat Mater 16:652–657
14. Wang Y, Jenkins IC, McGinley JT, Sinno T, Crocker JC (2017) Nat Commun 8:1–8

Further Reading

15. Textbooks on colloid and interface science.
16. Everett DH (1988) Basic principles of colloid science. The Royal Society of Chemistry, London
17. Russel WB, Saville DA, Schowalter WR (1989) Colloidal dispersions. Cambridge University Press, Cambridge
18. Hunter RJ (2001) Foundations of colloid science, 2nd edn. Oxford University Press, Oxford
19. Hiemenz PC, Rajagopalan R (1997) Principles of colloid and surface chemistry, 3rd edn. CRC Press, Boca Raton
20. Masliyah JH, Bhattacharjee S (2006) Electrokinetic and colloid transport phenomena. Wiley, New Jersey
21. Berg JC (2010) An introduction to interfaces and colloids, the bridge to nanoscience. World Scientific, Singapore
22. Israelachvili JN (2011) Intermolecular and surface forces, 3rd edn. Academic Press, Massachusetts
23. Butt H-J, Kappl M (2018) Surface and interfacial forces, 2nd edn. Wiley-VCH, Weinheim

Chapter 2
Fundamentals of Colloidal Self-Assembly

Abstract This chapter describes the fundamental sciences of colloidal self-assembly, from the basic concepts of colloid science. In dispersions, colloidal particles must be stably dispersed. However, the particles aggregate when the interaction between particles is attractive in total, and its magnitude is much stronger than the effect of random thermal motion of the particles. In this chapter, we first explain the stability of colloidal dispersions and the stabilization concept. Then, we explain important interactions of colloidal systems, including hard-sphere (HS) repulsion, van der Waals (vdW) force, electrostatic interaction, and depletion attraction. We also describe the basis of colloidal crystallization, in which colloidal particles are regularly arranged in a crystal lattice due to these interactions.

Keywords Stabilization · Hard-sphere repulsion · Van der Waals attraction · Electrostatic interaction · Depletion attraction · Colloidal crystals · Effects of external fields

2.1 Stability and Stabilization of Colloids

As described in Chap. 1, colloidal dispersion is a system in which colloidal particles are stably dispersed in a medium. Therefore, the stability of colloidal dispersion has been one of the central issues in colloid science [1–9]. Colloidal particles are composed of a large number of atoms the size of which is approximately 0.1 nm (1 Angstrom). For example, a particle with a diameter of 100 nm consists of 10^3 atoms in the length scale, namely about 10^9 atoms in its volume. The vdW forces between colloidal particles, which are made up of such a large number of atoms, are much stronger and act over a longer distance than for atoms. When the vdW force is sufficiently stronger than the effect of random thermal motion of the particles, the particles aggregate mutually.

As we will explain below, the vdW force is approximately proportional to the difference in refractive index between the particles and the dispersion medium [10, 11]. Therefore, by matching the refractive indices of the particles and the medium, one can obtain a system in which the vdW force is practically ineffective. However,

© The Author(s), under exclusive license to Springer Nature Singapore Pte Ltd. 2023
J. Yamanaka et al., *Colloidal Self-Assembly*, Lecture Notes in Chemistry 108,
https://doi.org/10.1007/978-981-99-5052-2_2

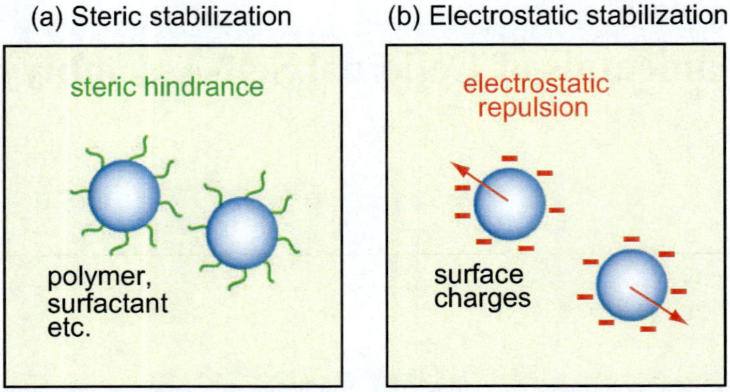

Fig. 2.1 Illustrations of **a** steric and **b** electrostatic stabilizations of colloids

such "refractive index matching" is not always possible for any kind of particle. To overcome the vdW attraction and maintain the colloids stable, two types of stabilization methods—steric and electrostatic stabilizations—have frequently been used [1–9]. Steric stabilization utilizes the adsorption of a layer of polymer or surfactant molecules on the particles, avoiding direct contact with the particles (Fig. 2.1a). Electrostatic stabilization uses the Coulomb repulsion between like-charged particles to prevent the particles from approaching each other (Fig. 2.1b).

In understanding the stabilization of colloidal systems, we must first learn about the interactions between particles. Section 2.2 will introduce the basic concepts of interparticle interactions based on pair potentials. We will describe hard-sphere repulsion, vdW forces, and electrostatic interactions. We will also explain the depletion attraction that occurs when linear polymers are added to colloidal dispersions.

Due to these interactions, colloidal particles of uniform size self-assemble to form regular structures called "colloidal crystals" under appropriate conditions [2, 8, 9, 12–18]. Section 2.3 describes the basics of colloidal crystallization through these various kinds of interactions.

2.2 Interaction Between Two Colloidal Particles

2.2.1 Interaction Pair Potential

Suppose there are two particles and a force F is acting between them, as illustrated in Fig. 2.2a.

Usually, F varies with the distance r between particle centers, i.e., $F = F(r)$. Here we take the center of the left particle as the origin ($r = 0$). If $F(r)$ is repulsive, the particle on the right moves to the right, namely to the direction of increasing r, while it moves to the left (decreasing r) if $F(r)$ is attractive. These interparticle interactions are

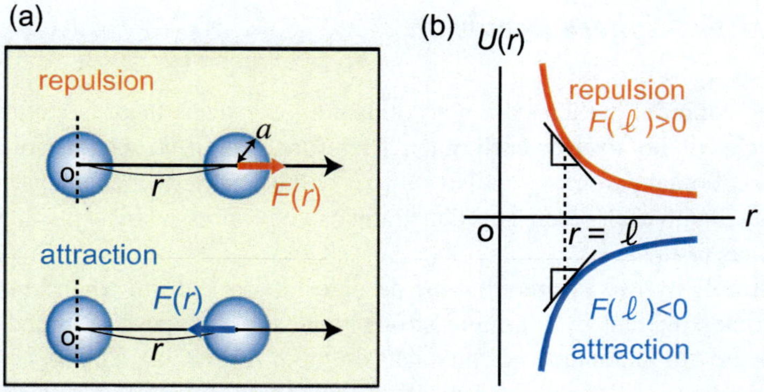

Fig. 2.2 **a** Force F acting between the colloidal particles and **b** the potential curves for attractive and repulsive interactions

usually expressed by an interaction potential $U(r)$, which is the potential energy of the right particle when placed at a distance r from the left particle. When two particles are infinitely far apart, they do not interact. Therefore, the interaction potential is defined so that $U(r) \to 0$ for $r \to \infty$. Then $U(r)$ is defined as the work required to transport the right particle from a distance $= \infty$ to r, against the force F(r) from the left particles. That is, by applying a force $-F(r)$, to the right particle. Thus, we have

$$U(r) = -\int_r^{\infty} F(r')\mathrm{d}r'. \tag{2.1}$$

Conversely, $F(r)$ is given by

$$F(r) = \mathrm{d}U(r)/\mathrm{d}r. \tag{2.2}$$

The potential curves for attractive and repulsive interactions are illustrated in Fig. 2.2b.

We can estimate the average position of the particles from $U(r)$ as follows. According to statistical mechanics, the density of particles $n(r)$ is known to follow a Boltzmann distribution $n(r) = n_0 \exp(-E(r)/k_B T)$. Here $E(r)$ is the potential energy at locations r. n_0 is the particle density at $r \to \infty$. k_B is Boltzmann's constant, and T is the temperature. Therefore, when observing a sufficiently large number of particle pairs interacting via $U(r)$, we find that the relative concentration of the particles $n(r)/n_0$ is equal to $\exp(-U(r)/k_B T)$. When $U(r) = 1\ k_B T$ and $5\ k_B T$, for example, $n(r)/n_0 =$ approximately 0.48 and 0.02, respectively. Thus, we understand that particles move away in repulsive potentials and particles move closer in attractive potentials.

2.2.2 Hard-Sphere Repulsion

Since the colloidal particles are not points but have sizes; there is a constraint that the particles do not overlap each other. Therefore, even if no specific force is acting between colloidal particles, repulsive forces will be exerted when the particles come into contact with each other. This constraint can be expressed by using an interaction potential as follows.

For simplicity, we assume that the particles do not deform when in contact; the two particles interact by infinitely strong repulsion. It is also assumed that when particles are not in contact, no force acts on them (Fig. 2.3a). This system is called "the hard-sphere (HS)" colloid. The HS systems are the most basic colloidal systems.

Note that the HS is an idealized concept and does not exist in reality. However, for many kinds of solid particles, such as polymer and metal oxide particles, the deformation of the particles when they collide in dispersion by thermal motion is usually negligibly small. Thus, the HS catches some aspects of real colloids.

Here we assume that all particles are spheres of the same radius a. U_{HS} gives the interaction

$$U_{\mathrm{HS}}(r) = \begin{cases} 0 & (r > 2a) \\ \infty & (r \leq 2a) \end{cases}. \tag{2.3}$$

Figure 2.3b shows the potential curve of the HS system. At $r = 2a$, $F(r) = -\mathrm{d}U(r)/\mathrm{d}r$ is infinity.

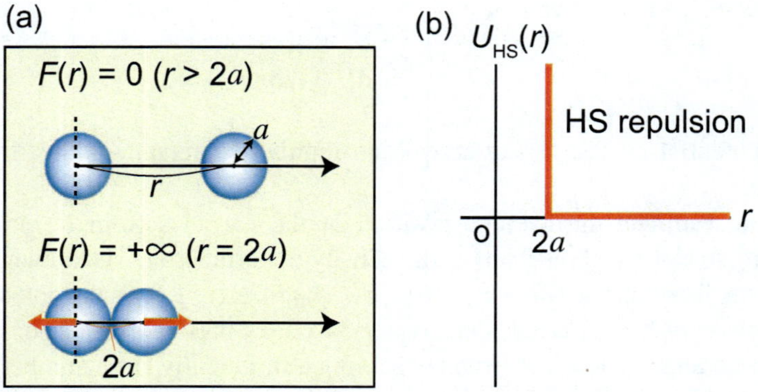

Fig. 2.3 **a** Force and **b** the potential curve for the hard-sphere colloids

2.2.3 *Van der Waals Force*

Origin of the vdW force

Many materials are more or less electrically non-uniform. Molecules made up of different types of atoms bonded together have permanent electric dipoles due to differences in the electronegativity of the atoms. For example, in the keto group C=O, the oxygen atom is slightly negative ($\delta-$), and conversely, the carbon atom is positively ($\delta+$) charged (Fig. 2.4a).

This pair of charges is called an *electric dipole.* In particular, electric dipoles formed by intramolecular polarization, as in this example, are called *permanent dipoles*. Moreover, electrical inhomogeneity can be instantaneously induced by thermal fluctuations or by the action of external fields. If a positive charge $+q$ is generated locally in an electrically neutral material, a charge of the opposite sign with the same magnitude $-q$ should be developed nearby. They are called *induced dipoles.*

These electric dipoles interact via electrostatic forces (Coulomb forces). The interactions are classified into three types; permanent dipole–permanent dipole, permanent dipole–induced dipole, and induced dipole–induced dipole. They are referred to as the Keesom, Debye, and London interactions, respectively (Fig. 2.4b). The vdW interaction is the sum of these three interactions [10].

Generally, the vdW potential $U_{\mathrm{vdW}}(x)$ acting between two objects separated by a distance x can be described in the form of Eq. (2.4).

$$U_{\mathrm{vdW}}(x) = -Af(x). \tag{2.4}$$

Here, A is a constant determined by the dielectric properties of substances of the dispersed phase and the dispersion medium and is called the Hamaker constant. x is a variable convenient for expression that depends on the object's shape. For example,

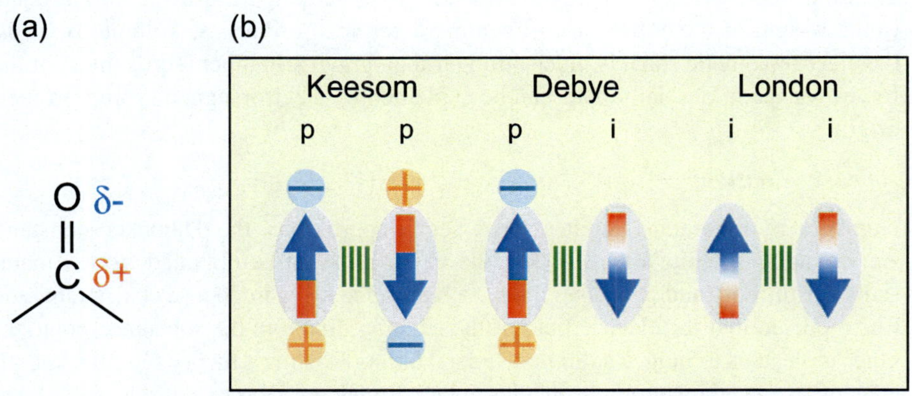

Fig. 2.4 **a** Example of a permanent electric dipole. **b** Three kinds of dipole–dipole interactions. *p* and *i* represent permanent and induced electric dipoles, respectively

for a flat plate, the distance between the planes should be x. $f(x)$ is a function of x and varies with the shape and size of the object. As described below, $f(x)$ takes positive values and increases monotonically as x increases.

Column 2.1

Why the vdW force is explained by electromagnetism?

As described in the text, the vdW force is explained in terms of electromagnetic interactions between electric dipoles in materials. Other than the vdW force, many forces we see in everyday life are attributed to the electromagnetic force if we trace back their principles. The reason for this is as follows.

It is known that all interactions in nature can be classified into four types (see, for example, J. N. Israelachvili, "Intermolecular and Surface Forces," 3rd Edition). They are the "strong interaction," "weak interaction," "electromagnetic interaction," and "gravitational force" (*strong* and *weak* are not mere adjectives, but names for types of forces). The first two of these work only inside the atomic nucleus. As you know, the nucleus is composed of protons and neutrons. Since the protons have a positive charge and are present very close to each other, there should be strong Coulomb repulsion between the protons. Nevertheless, the protons in the nucleus do not break apart. In order for protons to stay together, another kind of attractive force, which is stronger than the Coulomb force, must act between protons. This force is the "strong interaction." However, this force is a short-range force that acts only in the nucleus and does not appear in everyday life, except in nuclear reactions and so on. The "weak interaction" is also a short-range force. Therefore, the phenomena we see in our daily life are explained by electromagnetic and gravitational forces.

In colloid science, gravity is an important force, as it causes particles to settle. But its magnitude is much smaller than that of the electromagnetic force. The famous physicist R.P. Feynman introduces the following calculation. Suppose you stand at a distance of about arm's length from someone, and the number of electrons that make up the two of you is each 1 percent greater than the number of protons. The Coulomb repulsive force acting at this time is large enough to lift a weight equal to the weight of the entire earth (Feynman Lectures on Physics, Volume 3). Thus, the electromagnetic force is much stronger than gravity. In other words, most of the forces we see in our daily lives can be explained by electromagnetic forces in their origin.

Hamaker Constant

Numbers of theoretical studies have been reported on the Hamaker constant. According to Lifshitz's theory [10], the value of A can be obtained from a material's permittivity and refractive index. A substance is regarded as a continuum, and it is modeled that the electric field of the existing dipole in the substance polarizes other molecules to induce a dipole. Understanding Lifshitz's theory requires knowledge of advanced quantum mechanics, which is beyond the scope of this book. Here we discuss only the results from the theory. The Hamaker constant is divided into the contribution of permanent dipole (Keesom and Debye interaction) $A_{\nu=0}$ and the induced dipole (London force) $A_{\nu>0}$, where ν is the frequency of the applied electric field. The value of A is given by $A = A_{\nu=0} + A_{\nu>0}$. When particles 1 and 2 are dispersed in a medium 3, $A_{\nu=0}$ and $A_{\nu>0}$ are expressed as

$$A_{\nu=0} = \frac{3}{4}k_B T\left(\frac{\varepsilon_{r1} - \varepsilon_{r3}}{\varepsilon_{r1} + \varepsilon_{r3}}\right)\left(\frac{\varepsilon_{r2} - \varepsilon_{r3}}{\varepsilon_{r2} + \varepsilon_{r3}}\right) \tag{2.5}$$

$$A_{\nu>0} = \frac{3h\nu_e}{8\sqrt{2}} \frac{(n_1^2 - n_3^2)(n_2^2 - n_3^2)}{(n_1^2 + n_3^2)^{\frac{1}{2}}(n_2^2 + n_3^2)^{\frac{1}{2}}\left[(n_1^2 + n_3^2)^{\frac{1}{2}} + (n_2^2 + n_3^2)^{\frac{1}{2}}\right]}. \tag{2.6}$$

Here h is Planck's constant and ν_e the frequency at which a substance mainly absorbs electromagnetic waves, which is usually in the ultraviolet region and approximately 3×10^{15}/s. ε_{ri} and n_{ri} are the relative permittivities and refractive indexes of the particles and medium. We note that ε_{ri} and n_{ri} are not independent of each other. One can derive a relation $\varepsilon_{ri}{}^2 = n_{ri}$ from basic electromagnetics for materials that do not have permanent electric dipoles. This relationship applies well to non-polar molecules. For example, for normal-dodecane, $\varepsilon_r = 2.01$ and $n_r = 1.41$. From Eq. (2.6), if either or both of n_{r1} and n_{r2} are equal to n_{r3}, then $A_{\nu>0} = 0$. Also, in the ideal case where $\varepsilon_r{}^2 = n_r$, $A_{\nu=0} = 0$ simultaneously. It turns out that $A = 0$, and thus $U(r) = 0$. Therefore, by matching the refractive indexes of the particles and the medium, it is possible to obtain a colloidal system in which the vdW force is not substantial. A mixture of decalin and carbon disulfide is the refractive index matching medium for polymethyl methacrylate particles. An ethylene glycol solution can be used as a medium for silica particles to match the refractive index.

From (2.5) and (2.6), we can see that A is always positive between two particles made of the same kind of matter ($n_{r1} = n_{r2}$ and $\varepsilon_{r1} = \varepsilon_{r2}$) if the refractive index (and permittivity) is not matched with the medium. This means that the vdW force is always attractive for the same kind of particles. On the other hand, when $n_{r1} < n_{r3} < n_{r2}$ or $n_{r2} < n_{r3} < n_{r1}$ hold, A has a negative value. That is, the vdW force is repulsive. However, the refractive index of the medium is usually smaller than those of particles, and in many cases, the vdW force acting between any kind of colloidal particles is attractive. Especially in a vacuum ($n_{r3} = 1$), the vdW force acting between particles of any sort is always attractive.

The value of the Hamaker constant is usually of the order of about 10^{-20} J. For example, for two silica particles interacting in water, calculated from (2.5) and (2.6), $A = 0.57 \times 10^{-20}$ J (=1.4 k_BT at $T = 25$ °C). $A = 1.98 \times 10^{-20}$ J (=4.8 k_BT) for two polystyrene particles in water [19]. The Hamaker constant of metals is an order of magnitude larger than that of dielectric particles. For this reason, metallic particles such as gold particles readily coagulate in dispersions.

Potential Curves

The Lennard–Jones (L-P) potential $u_{LP}(x)$ is a well-known empirical model for the interaction potential between two atoms [10, 20]. Here x is the distance between two atoms. The L-P potential consists of a repulsive force between the atoms (reflecting the fact that the atoms do not overlap, called the Born repulsion) and an attractive term corresponding to the vdW force, expressed in the form, $u_{LP}(x) = A/x^p - B/x^q$. A and B are constants, and p and q are positive integers. A potential of $p = 12$ and $q = 6$ explains the experimental results well. vdW potential between two macroscopic objects can be calculated by assuming that the vdW potential for volume elements of the objects is proportional to $1/x^6$ and by integrating it for the entire volume of the objects. For example, for a pair of semi-infinite half-spaces interacting across a distance h,

$$f(h) = \frac{1}{12\pi h^2}, \tag{2.7}$$

and for a pair of two flat plates of thickness δ,

$$f(h) = \frac{1}{12\pi}\left(\frac{1}{h^2} + \frac{1}{(h+2\delta)^2} - \frac{2}{(h+\delta)^2}\right). \tag{2.8}$$

To calculate the vdW potential between objects other than flat plates, Derjagin used an approximation that considers an object a collection of flat plates (Derjaguin approximation) [2, 6, 10, 11]. Using this method, $f(r)$ between objects of arbitrary shapes, such as spheres, can be calculated from $f(r)$ between flat plates. When a flat plate and a sphere of radius a are separated by the closest distance s, we have

$$f(s) = \frac{1}{6}\left(\frac{a}{s} + \frac{a}{2a+s} + \ln\frac{s}{2a+s}\right), \tag{2.9}$$

which approaches $f(s) \to a/6s$, when $s \to 0$. For two spheres having radii of a_1 and a_2, separated by a center-to-center distance of r,

$$f(r) = \frac{1}{6}\left(\frac{2a_1a_2}{r^2 - (a_1+a_2)^2} + \frac{2a_1a_2}{r^2 - (a_1-a_2)^2} + \ln\frac{r^2 - (a_1+a_2)^2}{r^2 - (a_1-a_2)^2}\right). \tag{2.10}$$

Figure 2.5 shows the potential curves calculated from (2.10) for several colloids (silica particles in ethylene glycol (EG), silica particles in water, and polystyrene (PS) particles in water). Hamaker constants were estimated by (2.5) and (2.6). Here, we calculated assuming particles with a diameter of $2a = 100$ nm in a one-component system ($a_1 = a_2 = a$). The particles come into contact with each other at $r = 2a$.

The vdW between micron-sized objects is as strong as gravity, especially in the air. It has recently been reported that geckos stick to vertical walls due to the vdW force between the micron-sized protrusions of their fingers and the wall [21–23]. We have also reported that colloidal particles adsorb to the surface of gels in water only by vdW forces, even in the absence of other specific interactions [19, 24].

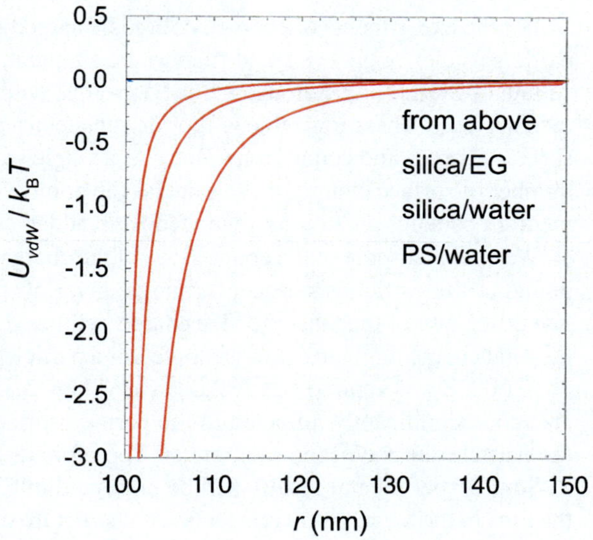

Fig. 2.5 vdW potential curve calculated from (2.10) for several colloids. $2a = 100$ nm

2.2.4 Electrostatic Interaction

Particles with electric charges interact through Coulomb forces. When charges are in motion, an electric current is generated, and the electric current also generates a magnetic field, which very much complicates the interaction between the charges. However, the behavior of charged colloids can usually be explained by electrostatic interactions, except when the effect of the external electric or magnetic fields is significant. Basic electrostatics concepts, including Coulomb interaction, electric potential, and electric field, are summarized in Appendix A.

Figure 2.6 is a schematic diagram of the interaction of two charged colloidal particles.

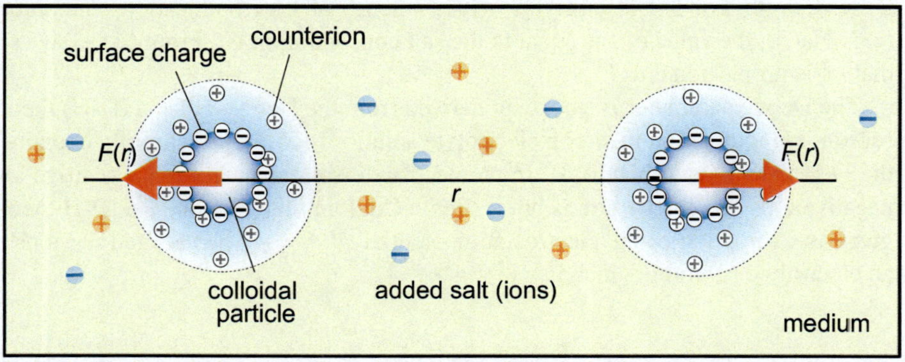

Fig. 2.6 Schematic diagram of the interaction of two charged colloidal particles

For the experiments of charged colloids, charged polystyrene particles and metal oxide particles (e.g., silica and titania) are frequently used. The former often have sulfate ($-SO_4H$) and sulfone ($-SO_3H$) groups, while the latter have OH groups on their surfaces. These ionizable groups dissociate in a polar medium like water, giving surface charges and counterions. Because a single particle is electrically neutral, the number of surface charges is the same as the number of counterions. In addition, the medium contains small ions generated from added salts and ionic impurities.

We note that some of the counterions of the charged colloidal particle are strongly bound to the particle's surface, forming a layer of trapped charges called the Stern layer. Because of this, the effective charge number Z_{eff} of the particle is smaller than the total charge number Z. For example, for a particle of diameter $2a = 100$ nm and Z = 10,000, Z_{eff} is approximately 2,000 (see Chap. 3). As Z increases, the counterions are more significantly attracted to the particle surface because the electric field on the particle surface becomes stronger. Therefore, Z_{eff}/Z decreases with increasing Z. Some experimental results will be presented in Chap. 3. The surface charge and the ions in the Stern layer create a layered structure of positive and negative charges. This structure is called an electric double layer. Counterions that are not trapped in the Stern layer are spread throughout the medium, and their distribution follows a balance between electrostatic attraction from surface charges and thermal motion. This structure is referred to as a diffuse double layer.

For discussion on the spatial distribution of counterions, the Debye parameter κ, defined by (2.11), is important.

$$\kappa^2 = \frac{e_0^2}{\varepsilon k_{\mathrm{B}} T} \sum_{i=1}^{N} c_i z_i^2. \tag{2.11}$$

Here, e_0 is the elementary charge (=1.602 × 10^{-19} C), and ε is the permittivity of the medium. i is the number assigned to each type of ion, and z_i is the valence of the i-th ion ($i = 1, 2, 3, \ldots, N$), which is positive for cations and negative for anions. n_i is the number density of the i-th ion. We note that $I = \frac{1}{2}\sum c_i z_i^2$ is a quantity called the ionic strength. For 1–1 electrolytes, with a valency of 1 for both cations and anions (e.g., NaCl), the value of I is equal to the salt concentration C_{s}. From (2.11), we see that κ^2 is proportional to I.

The Debye parameter is a concept derived from the Debye–Hückel (D-H) theory for ion–ion interaction in strong electrolyte solutions [25, 26]. We briefly explained the D-H theory in Appendix B. In an electrolyte solution with many positive and negative ions, the interaction is not a simple Coulomb interaction. The D-H theory gives an expression of the electrostatic potential $\phi(r) = \phi_{\mathrm{DH}}(r)$ around the ions in an electrolyte solution, which is

$$\phi_{\mathrm{DH}}(r) = \frac{ze_0}{4\pi\varepsilon} \frac{\mathrm{e}^{-\kappa r}}{r}, \tag{2.12}$$

where z is the valency of the central ion. Because exp(-κr) is a decreasing function of r, $\phi_{DH}(r)$ decays faster than the Coulomb potential $\phi_C(r)$,

$$\phi_C(r) = \frac{ze_0}{4\pi\varepsilon}\frac{1}{r}. \tag{2.13}$$

Note that $\phi_{DH}(r)$ approaches $\phi_C(r)$ when $\kappa \to 0$, that is, when no ions are present. The term exp(-κr) indicates that the electrostatic potential is screened due to the presence of ions. This phenomenon is called the electrostatic screening effect. Since the exponent $-\kappa r$ in (2.12) is dimensionless, $1/\kappa$ has a dimension of length and is called the Debye length or Debye screening length. $1/\kappa$ is inversely proportional to $\sqrt{I}$, and for 1–1 electrolytes at temperature = 25 °C, $1/\kappa$ is approximately equal to $0.30/\sqrt{C}$ (nm); for example, $1/\kappa \sim 100$, 10, and 1 nm, when $C_s = 10^{-5}$, 10^{-3}, and 10^{-1} mol/L, respectively. Potential with the form e^{-ar}/r (where a is a constant), as in (2.12), is called a Yukawa-type potential after physicist H. Yukawa, who introduced this form of potential for the force (nuclear force) acting in the atomic nucleus.

Note that $\phi_{DH}(r)$ is the electrostatic work required to carry a unit charge (1C) from infinity to position r when the charge ze_0 is at $r = 0$ (see Appendix A). Thus, the interaction potential $U_{DH}(r)$ between two charges ze_0 is

$$U_{DH}(r) = ze_0\phi_{DH}(r) = \frac{(ze_0)^2}{4\pi\varepsilon}\frac{e^{-\kappa r}}{r}. \tag{2.14}$$

The above discussion concerns the interaction between small ions, which were assumed as point charges. In the case of charged particles, it is necessary to consider that they have a much larger size than ions. The theories describing the interaction between charged colloidal particles were proposed independently by Russian scientists Derjaguin and Landau [27] and Dutch colloid scientists Verwey and Overbeek [28]. Their theory is now called the DLVO theory, after their names [1–6, 13]. In the DLVO theory, the interaction between charged colloidal particles is considered the sum of electrostatic repulsion and vdW attraction. Here we discuss solely electrostatic interactions.

When charged particles approach each other in a medium, an overlap of their diffuse electric double layer occurs. As a result, the counterion concentration increases in the interparticle region, which increases the osmotic pressure between the charged particles. In the DLVO theory, the electrostatic interaction potential is calculated based on this idea. An equation for the electrostatic interaction potential between spheres is derived [2, 13, 28]. The Yukawa-type electrostatic interaction potential $U_Y(r)$ is obtained for two spherical particles, by assuming that the surface potential of the particles is sufficiently small and the distance between the particles is large enough.

$$U_Y(r) = G(\kappa, a)\frac{(Ze_0)^2}{4\pi\varepsilon}\frac{e^{-\kappa r}}{r}. \tag{2.15}$$

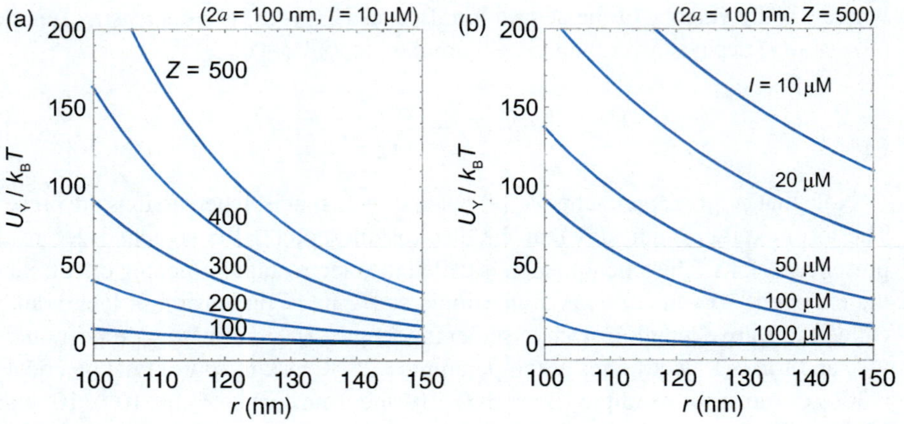

Fig. 2.7 Yukawa potential (2.15), for **a** several Z values at $I = 10\ \mu$M, and **b** several I values at $Z = 500$. $2a = 100$ nm

Here $G(\kappa, a) \equiv (\exp 2\,\kappa a)/(1 + \kappa a)^2$ includes the effect of particle size and is called the geometric factor. At $a \to 0$, Eq. (2.15) agrees with Eq. (2.14).

From (2.15), we can see that the magnitude of the electrostatic interaction depends on Z and κ. When the added salt is a 1–1 electrolyte and the counterion of the colloidal particles is monovalent, $I = C_s + \frac{1}{2}\,C_c$. Figure 2.7a presents the potential curves for several Z_{eff} values calculated using (2.15) ($2a = 100$ nm, $I = 10\ \mu$M). Figure 2.7(b) shows the influence of I ($2a = 100$ nm $Z_{\text{eff}} = 500$).

2.2.5 Depletion Attraction

As we described in Sect. 2.2.2, a fundamental characteristic of HS colloids is that the positions of the two colloidal particles do not overlap. In other words, there is a region in the dispersion where no particles can exist. This effect caused by the size of the particles is called the excluded volume effect. In the HS system, there is no interaction other than the excluded volume effect of the spheres. The excluded volume concept is important for the depletion attraction described below.

Consider a system in which a non-adsorbing polymer is added to a dispersion of colloidal particles (Fig. 2.8). For simplicity, we consider the dissolved polymer as a hard sphere with a diameter of σ.

Then, the center of the polymer chain cannot approach less than the distance of $\sigma/2$ from the particle surface. In other words, for the particle of radius a, there is no polymer in the region surrounded by the particle's surface and concentric spheres with a radius of $a + \sigma/2$. This is the excluded region of the polymer chain by the sphere. If the two colloidal particles are close enough to each other and the excluded regions of the polymer overlap, the polymer cannot enter this overlapping region,

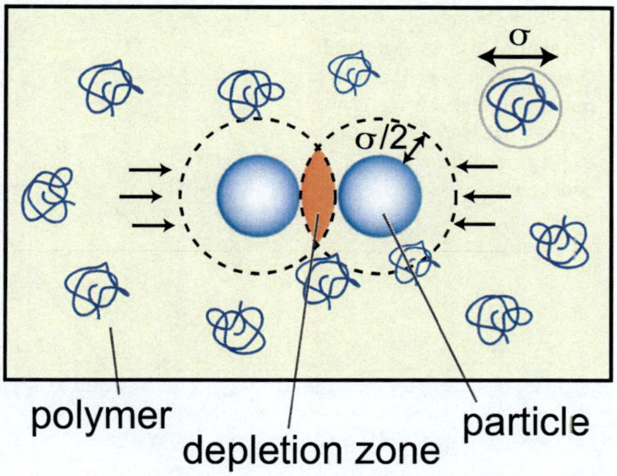

Fig. 2.8 Illustration of depletion-attraction system

which is referred to as the "depletion region." This difference in the polymer concentration generates an osmotic pressure difference between the inside and outside of the depletion region, resulting in an attractive force (depletion attraction) between the particles. Asakura and Osawa [29] theoretically predicted the existence of depletion attraction in the 1950s, and the research was developed by Vrij, Gast, Lekkerkerker, and Tuinier [30]. Assuming that the distance between the closest surfaces of the colloidal particles is $h(= r - 2a)$, the depletion potential $U_{\mathrm{AO}}\,(h)$ can be written as

$$U_{\mathrm{AO}}(h) = \begin{cases} -\Pi V_{OV}(h) & (0 \leq h \leq 2\sigma) \\ 0 & (h > 2\sigma) \end{cases}, \tag{2.16}$$

where Π is the osmotic pressure in bulk, which follows van't Hoff's law $\Pi = n_{\mathrm{b}} k_{\mathrm{B}} T$ in the dilute region, where n_{b} is the number density of polymers in bulk. V_{ov} is the volume of the depletion region and is given by (2.17).

$$V_{\mathrm{ov}}(h) = \frac{\pi}{6}(\sigma - h)^2(3a + \sigma + h/2). \tag{2.17}$$

When using the interparticle distance r and $R_{\mathrm{d}} = a + \sigma/2$, (2.16) and (2.17) can be described.

$$U_{\mathrm{AO}}(r) = \begin{cases} -\Pi V_{\mathrm{OV}}(r) & (2a \leq r \leq 2R_d) \\ 0 & (r > 2R_d) \end{cases}. \tag{2.18}$$

$$V_{\mathrm{ov}}(r) = \frac{4\pi}{3} R_d^3 \left[1 - \frac{3}{4}\frac{r}{R_d} + \frac{1}{16}\left(\frac{r}{R_d}\right)^3 \right] \tag{2.19}$$

The radius of gyration R_{g} of the polymer chain is often used as the radius of the depletant $\sigma/2$. Figure 2.9 shows an example of the potential curve ($2a = 100$ nm,

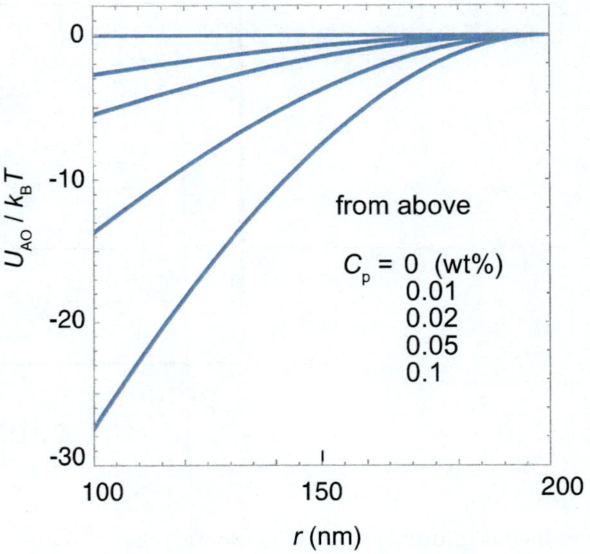

Fig. 2.9 Depletion (Asakura-Oosawa) potential curve (2.18). $2a = 100$ nm, polymer molecular weight = 29,000, $\sigma/2 = R_g = 50$ nm. C_p is the polymer concentration

polymer molecular weight = 29,000, $\sigma/2 = R_g = 50$ nm) at various values of polymer concentrations C_p. A detailed study considering the conformation of the dissolved polymer is discussed in the textbook by Lekkerkerker and Tuinier [30].

2.2.6 Total Potential and Stabilization of the Colloidal System

In Sects. 2.2.2, 2.2.3, 2.2.4 and 2.2.5, we described the pair potentials for the various interactions between colloidal particles. Since the potential energy is a scalar quantity, if several interactions are at work, the total potential U_{total} (r) can be calculated by simply adding up the potentials for each. In the following, we will discuss a few important cases. For simplicity, we consider a colloid in which only one kind of particle of radius a is dispersed.

HS repulsion + vdW attraction

Let us consider the case when vdW attraction (2.10) acts in addition to the HS repulsion, i.e., $U_{total}(r) = U_{HS}(r) + U_{vdW}(r)$. From (2.10), U_{vdW} (r) $= -\infty$ at $r = 2a$, but it is not achieved due to the roughness of the particle surfaces and the presence of solvation molecules at the particle surfaces. When the closest surface-to-surface distance of the two particles is put at $r = \delta$, the total potential curve has a minimum value at $r = 2a + \delta$. The potential curves for the HS repulsion and $U_{total}(r)$ are shown in Fig. 2.10a, b. If the minimum depth ΔU of U_{total} (r) is well below the thermal energy $k_B T$, the particles are dispersed without aggregation.

That is, the colloid is stable. On the other hand, when the minimum is deep enough, particles approach by Brownian motion and cannot escape from the minimum and

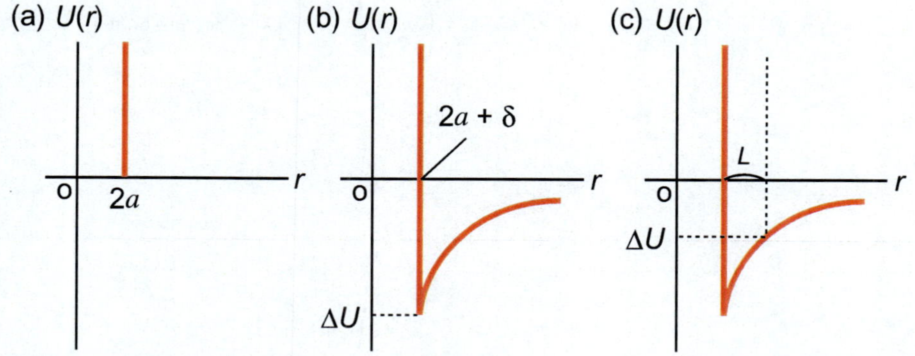

Fig. 2.10 **a** HS potential and **b** HS + vdW potential. **c** Potential curves for a particle with a steric hindrance layer of thickness L

aggregate. More precisely, the number density of particles $n(r)$ can be calculated based on the Boltzmann distribution $n(r) = n_0\exp(-U(r)/k_BT)$, as discussed in Sect. 2.2.1.

Steric repulsion + vdW attraction

In Sect. 2.1, we explain the steric stabilization of colloids. This technique can be explained by using potentials as follows. For simplicity, consider the case where molecules of length L are introduced on the surface of a particle with sufficiently low coverage. They can be surfactant and polymer. Assuming that the shape change of the molecule is negligible when the particles come into contact, the closest distance of the particles increases by the thickness L (Fig. 2.10c). Thus, ΔU is reduced by introducing the molecules and aggregations can be prevented if the minimum is shallow enough. When flexible molecules, including polymers, are used, the molecule layer's elastic deformation must be considered. Detailed studies have been reported on the behavior of polymers at interfaces [31].

Electrostatic repulsion + vdW attraction

Another stabilization method mentioned in Sect. 2.1 was electrostatic stabilization. In the DLVO theory described in Sect. 2.2.4, the stability of the colloidal system is discussed in terms of the vdW and electrostatic potentials. The DLVO theory is famous for explaining the phenomenon of aggregation of charged colloidal particles at high salt concentrations [1, 2, 4, 6]. In the following, we describe the concept of aggregation based on DLVO theory. When considering aggregation, the vdW force is approximated by the Eq. (2.10) at $r \to 2a$, since the interparticle distance is narrow. In addition, at high salt concentrations, the electrostatic interaction is not Yukawa type (2.15) but exponential type. In DLVO theory, Eq. (2.20) is used for the total potential as the sum of both.

$$U_{\mathrm{DLVO}} = -\frac{Aa}{12h} + \frac{64\pi a n_\infty k_BT}{\kappa^2}\tanh^2\left(\frac{ze_0\psi_\delta}{4k_BT}\right)\exp(-\kappa h). \tag{2.20}$$

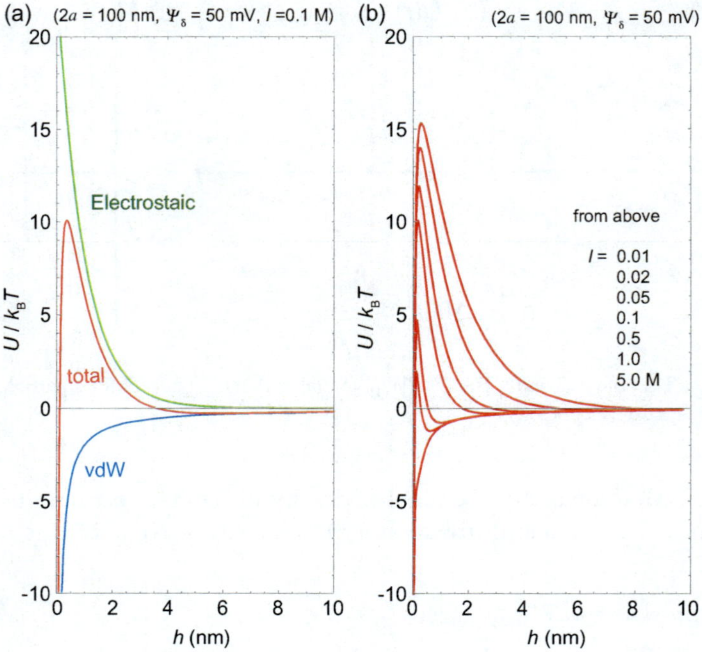

Fig. 2.11 **a** DLVO potential curve having two minima and one maximum. **b** Potential curves at various values of I

Here, n_∞ is the salt concentration in the bulk (i.e., the value at sufficiently large r), z is the ionic valence, and ψ_δ is the surface potential. In (2.20), the distance between the closest particle surfaces, $h = r - 2a$, is used as a variable.

When the vdW force is strong enough, as shown in Fig. 2.11, the potential curve has two minima and one maximum. If the vdW force is sufficiently stronger than the electrostatic repulsion, the total potential curve will have two minima and one maximum, as shown in the example in Fig. 2.11a.

When the particles are in contact ($h = 0$), they aggregate because there is a deep first minimum due to vdW attraction. However, if the potential maximum is sufficiently large, particles approaching from far away cannot pass through the energy barrier. Therefore, the particles cannot contact each other, and aggregation is prevented. This is the principle of electrostatic stabilization. However, when equilibrium is reached over a sufficiently long period of time, the distribution of particles follows the Boltzmann distribution law. If the potential curve has the shape shown in Fig. 2.11a, the particles will aggregate in the first minimum. Figure 2.12b shows the dependence of the potential curve on salt concentration. The higher the salt concentration, the smaller the maximum is and the more likely it is to aggregate.

Note that if the vdW force is sufficiently smaller than the electrostatic mutual phase, both the first and second minima may not exist. In this case, no aggregation occurs even in equilibrium and the colloid is stabilized.

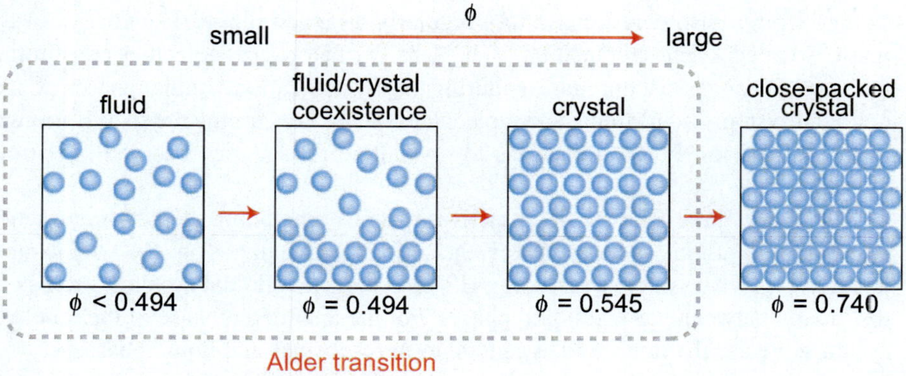

Fig. 2.12 Phase diagram of the HS colloid

Before the DLVO theory was proposed, it was known that protein molecules (which are charged) aggregate when the concentration of electrolyte in the solution is high. This phenomenon is called salting out. It is known that the effect of ions on salting out depends on the valence z, and that the ion concentration required for salting out is proportional to the sixth power of z, known as the Schulze-Hardy rule. Namely, divalent ions cause salting out 64 times more readily than monovalent ions. From the equation of $U_{\text{total}}(r)$, we can calculate the salt concentration at which the maximum of the potential equals 0 is proportional to z^6 [1, 2, 4, 6, 28]. DLVO theory is well known as a theoretical explanation of the Schulze-Hardy rule.

Column 2.2

The formation of the delta by salting out

We can see various examples of electrostatic stabilization in nature. One prominent example is the formation of a delta at an estuary. A delta is a terrain formed near the estuary by depositing materials carried by the river's flow. Fine grains of sand and rock are transported from the river to the estuary as muddy water. As mentioned in Chap. 1, muddy water is an example of colloids. Mud is a fine clastic material of rock mainly composed of metal oxide such as silica and has a surface charge in water like silica. Since the estuary is where the river joins the sea, the salt concentration is high near the estuary. The high salt concentration causes the precipitation of mineral particles and deposition of sediment, forming deltas.

2.3 Crystallization of Various Colloids

This section describes the self-assembly of colloidal particles into regular crystal lattice structures, that is, the crystallization of colloids. Crystallization is a phenomenon in which elements of the system (usually, they are atoms or molecules) change from a disordered phase to a regularly arranged phase. The change in the phase is called a "phase transition." The density of atoms or molecules constituting the system changes discontinuously during the crystallization. Similar phases exist in colloidal systems as in atomic and molecular systems concerning the spatial arrangement of particles. Phase transitions to crystalline phases are observed in various colloids.

There are three phases in attractive systems, including depletion-attraction systems, as in atomic systems, that is, the solid, liquid, and gas phases. On the other hand, in repulsive systems, including HS and charged colloidal systems, there is no distinction between gas and liquid phases. For the disordered phase of the repulsive systems, we use the term fluid, a generic term for the gas and liquid phases.

2.3.1 Hard-Sphere Colloids

Crystallization of the hard-sphere systems is similar to the phenomenon of spheres arranging regularly when spheres of uniform size are packed into a limited space. The packing fraction of a particle can be expressed in terms of the volume fraction ϕ defined by $\phi = Nv/V = (4/3)\pi a^3 N/V$. Here v is the volume of particles, N is the number of particles, and V is the system's volume.

As shown in Fig. 2.12, the arrangement of particles changes with increasing ϕ. At small ϕ is small, the colloid takes on a fluid phase with an irregular array of particles. On the other hand, as ϕ increases and the particles become crowded, they collide more frequently and gradually become unable to move. Thus, the average position of the particles becomes constant, and a periodic structure is formed. In other words, the colloidal system undergoes a phase transition from the fluid phase to the crystalline phase. The phase transition in the HS systems was discovered in computer simulations by Alder and Wainwright in1957 [32] and is called the Alder transition. The ϕ value at the freezing point (fluid to crystal phase transition) is $\phi_F = 0.494$, and at the melting point (crystal to fluid phase transition), $\phi_M = 0.545$. For $\phi_F < \phi < \phi_M$, both the fluid and crystal phases coexist.

According to thermodynamics, spontaneous changes in the system occur when volume and temperature are constant as the Helmholtz free energy $F = U - TS$ decreases. Here U is the system's internal energy (sum of the kinetic energy of particles and potential energy of particle–particle interaction), T is temperature, and S is entropy. In an isolated system with no energy or matter entering or leaving the system, spontaneous changes occur as entropy increases. This is known as the second law of thermodynamics. For constant pressure and temperature, Gibbs free energy $G = H - TS$ (H is the enthalpy) is the quantity that determines the spontaneous change. For constant volume and temperature, the change of the state is determined by Helmholtz free energy $F = U - TS$.

Now we assume the volume and temperature of the system are constant, and we consider the change in F before and after the crystallization, $\Delta F = \Delta U - T\Delta S$. There is no interparticle interaction in the HS systems unless the particles are in contact.

Furthermore, the kinetic energy of the system before and after the crystallization is the same. Thus, we have $\Delta U = 0$, and so ΔF is given by $\Delta F = -T\Delta S$. This implies that the equilibrium state, where F is minimized, is governed by S, as in the isolated system.

Thus, in the HS system, spontaneous changes occur as entropy increases ($\Delta S > 0$). The HS system in a fluid state will crystallize as entropy increases (because of $\Delta F = -T\Delta S < 0$). In other words, for $\phi_F \leq \phi$, the crystalline phase has a larger entropy than the fluid phase. This seems intuitively contrary to the fact that the more disordered system has the greater entropy, but it is explained as follows. According to statistical mechanics, S is given by $S = k_B \ln W$, where W is the number of states. There is only one arrangement of particles ($W = 1$) for a perfectly crystalline state. As ϕ increases, the number of possible configurations of particles gradually becomes limited, and the number of crystalline-like arrangements becomes greater than that of random structures. Thus, the S for crystalline-like particle arrangement becomes larger than the random arrangement.

The important points about the crystallization of the HS systems are summarized as follows. (1) First, the repulsion between the particles produces crystalline order; the attractive interaction is not always necessary for crystallization. (2) The phase behavior of the system is determined by S, and the only dominant parameter is ϕ. Particle size, for example, has nothing to do with phase behavior. (3) The magnitude of F varies with temperature, but the values of ϕ_F and ϕ_M are independent of temperature. (4) Colloidal crystallization occurs at $\phi_F = 0.494$, less than the maximum packing condition ($\phi = \phi_C = 0.740$) where particles are in contact with each other. In other words, a crystalline structure with gaps between particles is formed, and as $\phi \geq \phi_M$, the spacing between particles in the crystalline phase gradually narrows. At $\phi = \phi_C$, all particles are in contact.

As a model experimental system for the HS colloids, the polymethyl methacrylate/organic medium system is well known [33, 34]. Here, refractive index matching between the particles and the medium is used to keep the vdW forces sufficiently small.

In numerical simulations of the Alder transition, it has been reported that the colloidal crystal becomes FCC lattice when the interaction potential between particles is in the hard-sphere limit, and for softer potential, the lattice becomes BCC [35]. Experimentally, since the free energies of FCC and HCP structures are almost the same, HCP structures are mixed into FCC crystals as stacking faults.

2.3.2 *Charged Colloids*

As discussed in Sect. 2.2.4, dispersions of charged colloidal particles are stabilized due to the electrostatic repulsion. If the interparticle repulsion is weak, the arrangement of particles is irregular, but if the electrostatic force becomes sufficiently strong, the particles are regularly arranged into the crystalline structure [12–15, 17]. As in the hard-sphere repulsion systems, there is no distinction between gas and liquid in the electrostatic repulsion systems.

Because the electrostatic repulsion acts in a long range, the crystallization of charged colloids occurs at a much lower ϕ than in the hard-sphere colloids. Figures 2.13a and 2.14a show optical micrographs of the fluid and crystalline phases of aqueous dispersions of charged polystyrene particles ($2a = 430$ nm and $\phi = 0.03$ in both samples), respectively. In the crystalline phase, the medium is pure water, whereas in the fluid phase 100 μM NaCl is added to shield the electrostatic force between the particles.

The regularity of the particle arrangement is often evaluated using Fourier spectra and the radial distribution function $g(r)$. Fourier spectra obtained from image processing for the fluid and crystalline phases are shown in Figs. 2.13b and 2.14b. The $g(r)$-r plots are also shown in Figs. 2.13c and 2.14c, respectively. These analysis methods are discussed in detail in Chaps. 3 and 4.

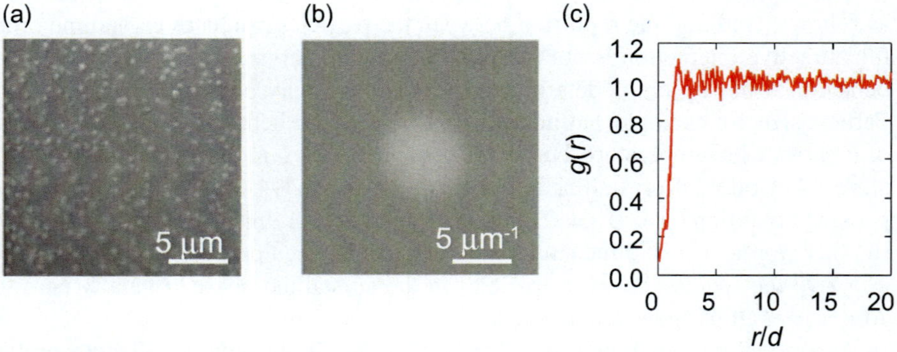

Fig. 2.13 **a** Optical micrograph, **b** the Fourier spectrum, and **c** the radial distribution function $g(r)$ of aqueous dispersions of charged polystyrene particles ($d = 430$ nm) in fluid phase. $\phi = 0.03$, [NaCl] = 100 μM

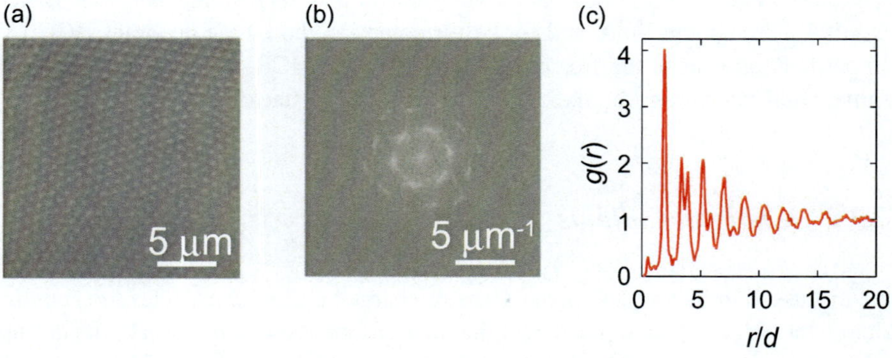

Fig. 2.14 **a** Optical micrograph, **b** the Fourier spectrum, and **c** $g(r)$-r plot of aqueous dispersions of charged polystyrene particles ($d = 430$ nm) in crystal phase. $\phi = 0.03$, salt-free condition

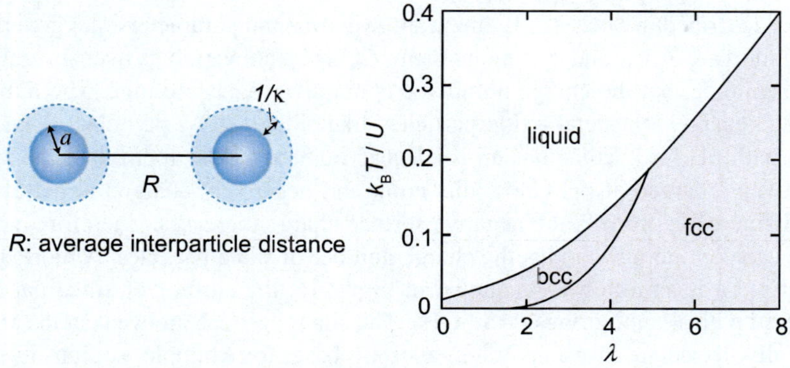

Fig. 2.15 Phase diagram of crystallization obtained by computer simulation (Robbins, Kremer, and Grest). The colored region presents the crystal phase

Many experimental and theoretical studies have been reported on the phase behavior of charged colloids [12, 13]. Robbins, Kremer, and Grest (RKG) [36] obtained the crystallization phase diagram in the two-parameter plane shown in Fig. 2.15 by computer simulation (molecular dynamics and lattice dynamics methods) using Yukawa-type potential. The vertical axis, $k_B T/U_Y(R)$, is the ratio of the electrostatic potential and thermal energy at the mean interparticle distance $r = R$ and is called the reduced temperature. The horizontal axis, $\lambda = \kappa R$, is the ratio of the interparticle distance to the Debye length and is referred to as the coupling parameter. Under weak interactions, the calculated crystal-fluid phase boundary agrees with the experiments (Chap. 5). At $T = 0$ and at the Coulomb limit ($\lambda = 0$, no electrostatic screening), the BCC lattice was stable. On the other hand, as λ increases, the range of the potential becomes narrower and the importance of the nearest neighbor interaction increases. At $\lambda = 1.72$, the FCC structure was more stable than the BCC lattice of the same density, because the nearest neighbor distance of the particles was larger and the free energy of FCC was smaller. Even when the FCC structure was taken at $T = 0$, an increase in T leads to a phase transition to BCC.

Experimentally, BCC, FCC, and hexagonally close-packed (HCP) structures have been identified in charged colloids. For polystyrene colloids, good agreement between experimental and RKG phase diagram crystal types has been reported [37]. However, especially for silica colloids, BCC crystals have sometimes been observed even under the conditions of FCC crystal formation in the RKG phase diagram [38]. This may be due to the larger size distribution of silica particles (~5% in standard deviation) than that of polystyrene particles. Note that BCC may initially form as a non-equilibrium phase during the crystal growth process and change to FCC over time [39].

Recently, a more precise phase diagram of the charged system is obtained by numerical simulation by Dykstra and coworkers [40]. We will also present a crystallization study using Brownian dynamics (see Fig. 4.6) in the simulation section of this book.

As described in Sect. 2.2.4, important experimental parameters that govern crystallization are Z, C_s, and ϕ. Among them, C_s and φ are variables over several orders of magnitude, but the charge number Z is not always easy to tune. The number of surface charges on metal oxide particles, like silica (SiO_2) particles, is known to vary with pH [41]. Silica has an isoelectric point at pH = approximately 2 and is positively charged at pH below this point and negatively charged at pH above it; as pH increases from 2, the negative surface charge increases. Therefore, adding a base can continuously adjust the charge number of silica particles. A more detailed description is as follows. As shown in Fig. 2.16, the surface of silica particles is covered with silanol groups (≡Si–OH). The silanol groups are weak acids and only partially dissociate in water. When a strong base, for example, sodium hydroxide NaOH is added to the silica colloids, silanol groups are neutralized to form Na-salt (≡Si–O^- Na^+). Since the Na-salt dissociates almost entirely in water, adding the base increases the number of charges on the silica particles. Thus, if the appropriate C_s and ϕ are chosen, the silica colloid should crystallize upon the addition of the base [17, 42, 43]. Methods for determining the number of charges on silica particles and other related methods are described in Chap. 3. It was discovered by chance that silica colloids crystallize when alkali is added. See Column 2.3.

Column 2.3

Crystallization of silica colloid on additions of base

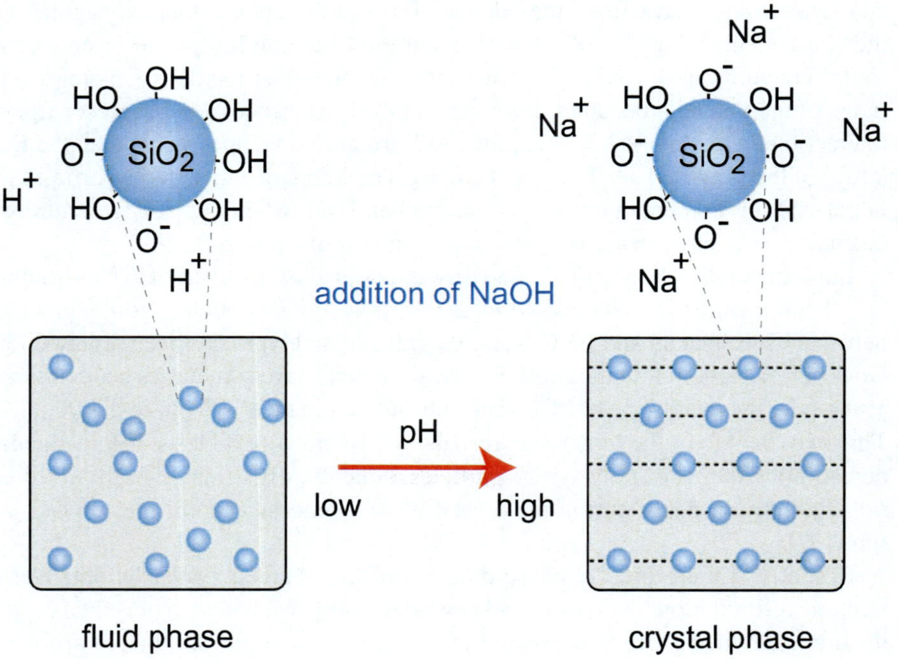

Fig. 2.16 Control the charge numbers of colloidal silica particles by adding a base

The crystallization of silica colloids by the addition of bases was found by chance. Although this is the authors' personal experience, we would like to share it with readers, in particular, young students, in the hope that it will be useful for their research. Several decades ago, the senior author (J.Y.) was conducting experiments on charged colloidal crystallization using commercial colloidal silica dispersions but was troubled because it was not crystallized at all. Since electrostatic repulsion between particles is shielded by trace amounts of impurity ions, he tried to be careful enough to deionize them for the experiment. In particular, plastic containers were used because basic impurities (NaOH, KOH, etc.) would leach out if glass containers were used. However, even after sufficient purification, no crystals were formed. It was found that the charge number of the particles was too low. He decided that the concentration should be higher. To determine the exact concentration by drying out, he heated and evaporated the water. Since the temperature in this experiment exceeded 100 °C, he used a glass bottle instead of a plastic container. Then the silica colloids showed Bragg diffraction, which originated from the crystals. Investigation of the cause revealed that the alkali from the glass container wall increased the charge number of the silica particles, causing them to crystallize. Subsequent experiments with NaOH reproduced the phenomenon.

The charge number of hydrophobic particles, e.g., polystyrene particles, can be adjusted by adding ionic surfactants that adsorb on the particle surface. Palberg's group experimentally examined the effect of charge number on colloidal crystallization using surfactants that adsorb tightly to hydrophobic particles, thereby testing the theory of effective charge [44]. If not all of the ionic surfactant added is adsorbed, the non-adsorbed portion increases the salt concentration. Since the amount of adsorption varies with temperature, the change in temperature will result in a change in Z and C_s. If these amounts are well adjusted, the electrostatic interaction can be controlled by temperature [45]. This will be discussed in detail in Chap. 5.

We obtained a crystallization phase diagram of silica colloids defined by the three variables Z, C_s, and ϕ, by observing the crystal structure under various conditions. The effects of these three parameters on electrostatic interactions are shown schematically in Fig. 2.17. Electrostatic interaction increases proportionally to the square of Z. Also, the interaction increases with ϕ because the interparticle distance r becomes smaller when ϕ is large. On the other hand, as C_s increases, κ increases and the electrostatic interaction is shielded. In determining the phase diagram, the Z value is first adjusted by adding NaOH to samples with different ϕ values. Then, we added NaCl and determined the C_s at the crystal-fluid phase boundary by spectroscopy and X-ray scattering. The results are shown in Fig. 2.18 [43]. As the charge number, we used the effective charge number Z_{eff}, evaluated from the electrical conductivity (see Chap. 3). The rectangular symbol in Fig. 2.18 is the phase boundary, and the region where the C_s value is lower than the phase boundary is the crystalline phase. Under the experimental conditions shown in Fig. 2.18, the colloidal crystal had BCC structures. The experimental details, including the estimations of Z_{eff} and the determination of crystal structures, will be described in Chap. 3.

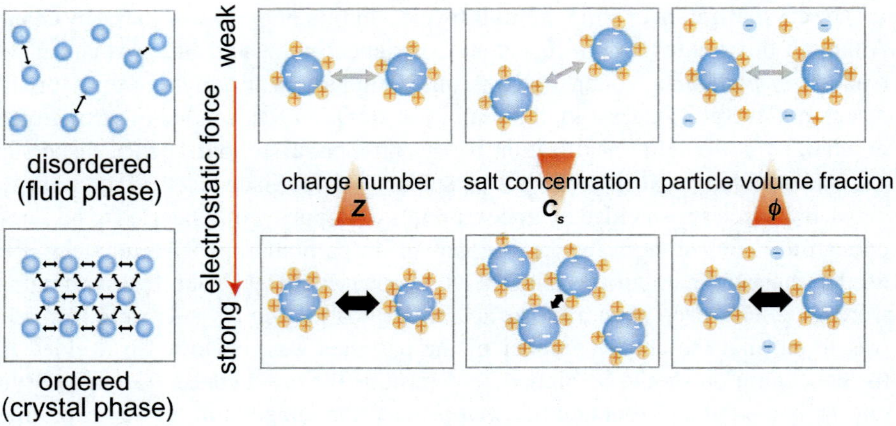

Fig. 2.17 Illustrations of the effects of Z, C_{S}, and ϕ on the electrostatic interactions

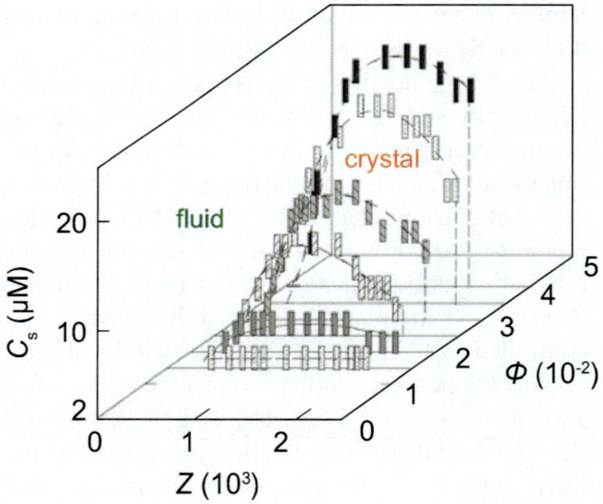

Fig. 2.18 Crystallization phase diagram of charged silica colloid ($d = 120$ nm) defined Z, C_{S}, and ϕ

2.3.3 Depletion Attraction

Depletion forces acting between colloidal particles result in crystallization under appropriate conditions. Hachisu et al. [46] reported that charged polystyrene particles formed colloidal crystals in their aqueous dispersion, when a small amount of sodium polyacrylate (NaPAA), a polyelectrolyte, was added. Figure 2.19 shows the colloidal crystals of the polystyrene particles (a diameter of 600 nm, $\phi = 0.0067$) in the presence of 0.08 wt% NaPAA, having a molecular weight of about 800,000. The photograph in Fig. 2.19 was taken with an inverted microscope and showed the crystal structure formed on the bottom surface of the sample cell.

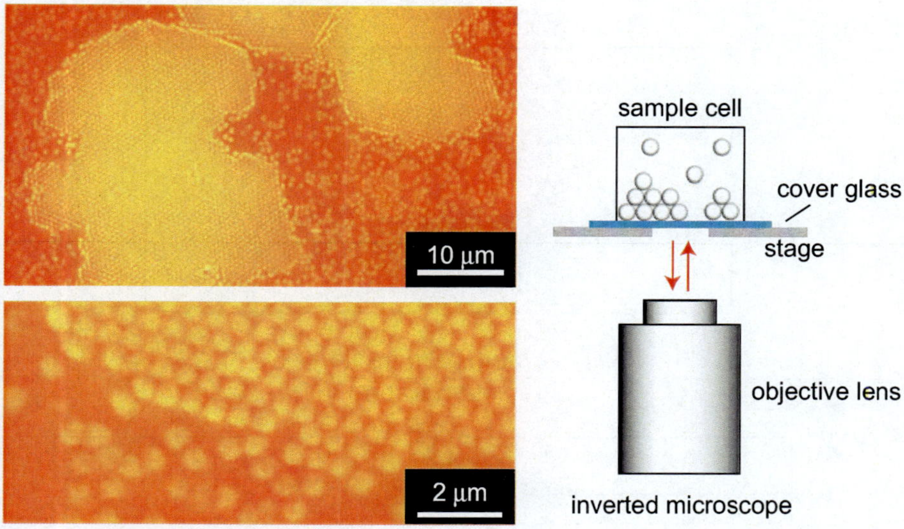

Fig. 2.19 Optical micrograph of the colloidal crystals of the polystyrene particles (d = 600 nm, 0.67 vol%), NaPAA (molecular weight of ~800,000, $C_p = 0.08$ wt%)

In colloidal crystals of the HS and charge repulsion systems, colloidal particles in crystals were not in contact with each other, and the crystal structure was uniform throughout the system. On the other hand, in the depletion-attraction system, island-shaped crystals are formed. The crystallization of colloids due to repulsion is similar to the melt crystallization of molecular systems. Examples include the solidification of molten metal upon cooling. On the other hand, the crystallization of attractive colloids corresponds to the solution crystallization—for example, the crystallization of NaCl from saturated solutions.

Lekkerkerker and Tuinier proposed the free volume theory (FVT) of the colloids as a theoretical framework for depletion attraction [30]. Their model is shown in Fig. 2.20a. The FVT theory assumes that an osmotic equilibrium is established between the depleting agent phase (reservoir) and the colloidal particles + depleting agent. The FVT clarifies the distribution of depletant between phases in each state and also allows for more than three particles. The overlap of the depletant layers is taken into account. Thus, results for relatively thick depletion layers are accurate.

When a depletant is a linear polymer, the radius of gyration R_g of the polymer is often used to measure the depletant size. The size ratio of the polymer to the colloidal particles is defined as $q = R_g/a_p$. Figure 2.20b, c shows an example of phase diagram ($q = 0.2$) obtained from FVT [30]. The horizontal axis is ϕ in Fig. 2.20b, and the vertical axis is the volume fraction of depletant in the reservoir (ϕ_{dR}). Figure 2.20c is the phase diagram defined as ϕ and depletant concentration in the sample, ϕ_d. When $\phi_d = 0$, it corresponds to the crystallization conditions of a rigid sphere system. $0.491 < \phi < 0.541$ is the coexistence region of the crystalline and liquid phases, and with increasing ϕ_d, the coexistence region becomes wider. The dotted lines are tie lines,

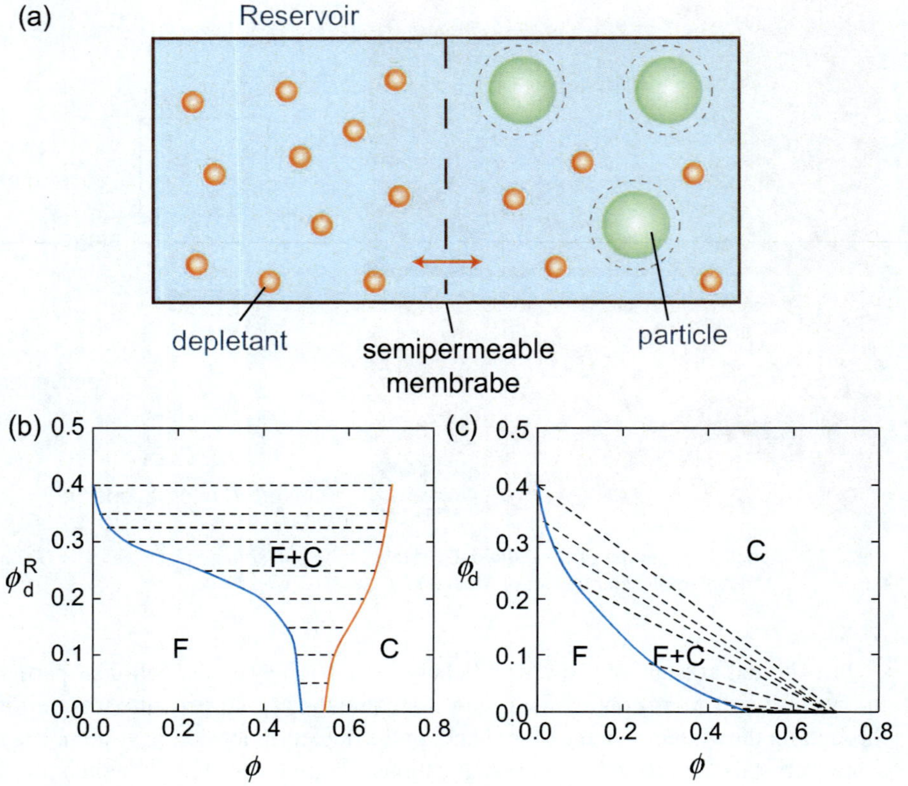

Fig. 2.20 **a** Model considered in the FVT. **b**, **c** are phase diagrams obtained from FVT ($q = 0.2$). $\phi_d{}^R$ and ϕ_d are depletant concentrations in the reservoir and the sample

indicating the two phases in the coexistence region. Examples of various structures formed by depletion attraction are shown in Chap. 6.

2.4 Opal-Type Colloidal Crystals

Here we would like to briefly introduce opal crystals. Closest-packed opal-type colloidal crystals (Fig. 2.21a) have attracted much attention as photonic materials. They are mainly fabricated by sedimentation [47], two-dimensional evaporation, or spin-coating methods. In opal structures, the particles are in contact with each other, and the particle size determines the lattice plane spacing. See Chap. 3 for details on how to make opals.

Note that inverse opals (Fig. 2.21b), which are crystalline arrays of spherical vacancies, can be obtained from opal structures [48, 49]. When opal crystals are dried, spaces are created between individual particles. These spaces are filled with

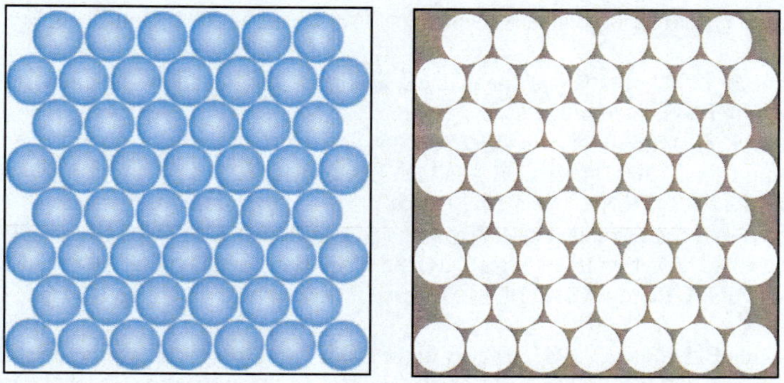

Fig. 2.21 **a** Closest-packed opal-type colloidal crystal and **b** an inverse opal structure

substances such as metals to construct inverse opal structures using the opal structure as a template. If organic polymer particles are used, they can be removed by heating and sintering. In the case of silica particles, they can be dissolved by fluoric acid treatment. The inverse opal structure can be applied to photonic crystals, catalysts, and templates with regularly arranged vacancies.

References

1. Everett DH (1988) Basic principles of colloid science. The Royal Society of Chemistry, London
2. Russel WB, Saville DA, Schowalter WR (1989) Colloidal dispersions. Cambridge University Press, Cambridge
3. Hunter RJ (2001) Foundations of colloid science, 2nd edn. Oxford University Press, Oxford
4. Hiemenz PC, Rajagopalan R (1997) Principles of colloid and surface chemistry, 3rd edn. CRC Press, Boca Raton
5. Masliyah JH, Bhattacharjee S (2006) Electrokinetic and colloid transport phenomena. Wiley, New Jersey
6. Berg JC (2010) An introduction to interfaces and colloids, the bridge to nanoscience. World Scientific, Singapore
7. Napper DH (1984) Polymeric stabilization of colloidal dispersions. Academic Press, London
8. Anderson VJ, Lekkerkerker HNW (2002) Nature 416:811–815
9. Yethiraj A, van Blaaderen A (2003) Nature 421:513–517
10. Israelachvili JN (2011) Intermolecular and surface forces, 3rd edn. Academic Press, Massachusetts
11. Parsegian VA (2006) Van der walls forces. Cambridge University Press, New York
12. Pieranski P (1983) Contemp Phys 24:25–73
13. Sood AK (1991) Solid state physics. In: Ehrenreich H, Turnbull D (eds). Academic Press, New York
14. Gast A, Russsel WB (1998) Phys Today 51:24–30
15. Ise N, Sogami I (2005) Structure formation in solution. Springer, Heidelberg
16. Li F, Josephson DP, Stein A (2011) Angew Chem Int Ed 50:360–388
17. Yamanaka J, Okuzono T, Toyotama A (2013) Colloidal crystals. In: Kinoshita S (ed) Pattern formations and oscillatory phenomena. Elsevier, Amsterdam

18. Vogel N et al (2015) Chem Rev 115:6265–6311
19. Aoyama Y, Sato N, Toyotama A, Okuzono T, Yamanaka J (2022) Bull Chem Soc Jpn 95:314–324
20. Widom B (2002) Statistical mechanics, a concise introduction for chemists. Cambridge University Press, Cambridge
21. Autumn K et al (2000) Nature 405:681–685
22. Autumn K et al (2002) Proc Natl Acad Sci USA 99:12252–12256
23. Loskill P (2013) J R Soc Interf 10:20120587
24. Sato N, Aoyama Y, Yamanaka J, Toyotama A, Okuzono T (2017) Sci Rep 7:6099
25. Debye PW, Hückel E (1923) Phys Z 24:185. [Translated and Typeset by Braus MJ (2019)]
26. Bockris, JOM, Reddy AKN (1973) Modern electrochemistry, volume 1: ionics. Plenum, New York
27. Derjaguin B, Landau L (1941) Theory of the stability of strongly charged lyophobic sols and of the adhesion of strongly charged particles in solutions of electrolytes. Acta Phys Chem URSS 14:633–662. See also Prog Surf Sci 43:30–59
28. Verwey EJW, Overbeek JThG (1948) Theory of the stability of lyophobic colloids. Elsevier, Amsterdam
29. Asakura S, Osawa F (1954) On interaction between two bodies immersed in a solution of macromolecules. J Chem Phys 22:1255–1256
30. Lekkerkerker HNW, Tuinier R (2011) Colloids and the depletion interaction. Springer, Amsterdam
31. Fleer GJ, Cohen Stuart MA, Scheutjens JMHM, Cosgrove T, Vincent B (1993) Polymers at interfaces. Chapman and Hall, London
32. Alder BJ, Wainwright TE (1957) J Chem Phys 27:1208–1209
33. Pusey PN, van Megen W (1986) Nature 320:340–342
34. Bartlett P, van Megen W (1994) In: Metha A (ed) Granular matter. Springer, New York, pp 195–257
35. Löwen H (1994) Phys Rep 237:249–324
36. Robbins MO, Kremer K, Grest GS (1988) J Chem Phys 88:3286–3312
37. Monovoukas Y, Gast AP (1989) J Colloid Interface Sci 128:533
38. Murai M, Okuzono T, Yamamoto M, Toyotama A, Yamanaka J (2012) J Colloid Interf. Sci. 370:39–45
39. Kakihara C, Toyotama A, Okuzono T, Yamanaka J, Ito K, Shinohara T, Tanigawa M, Sogami I (2015) Int J Microgravity Sci Appl 32:320205-1~320205-4
40. Hynninen A-P, Dijkstra M (2003) Phys Rev E 68:021407-1~021407-8
41. Iler RK (1979) The chemistry of silica, 1st edn. Wiley-Interscience, New York
42. Yamanaka J, Koga T, Ise N, Hashimoto T (1996) Phys Rev E 53:R4314-4317
43. Yamanaka J, Yoshida H, Koga T, Ise N, Hashimoto T (1998) Phys Rev Lett 80:5806–5809
44. Palberg T, Mönch W, Bitzer F, Piazza R, Bellini T (1995) Phys Rev Lett 74:4555–4558
45. Toyotama A et al (2014) Chem Mater 26:4057−4059
46. Kose A, Hachisu S (1976) J Colloid Interf Sci 55:487–498
47. Valsov YA, Bo X-Z, Sturm JC, Norris DJ (2001) Nature 414:289–293
48. Velev OD, Kaler EW (2000) Adv Mater 12:531–534
49. Stein A, Li F, Denny NR (2008) Chem Mater 20:649–666

Chapter 3
Experimental Methods

Abstract In this chapter, we describe important experimental methods for studying colloidal self-assembly. They include preparation of colloidal samples, characterization of colloidal particles, techniques for colloidal structure formation, and methods of structural analysis. First, the synthesis of uniformly sized polystyrene particles is introduced. Next, we describe purification methods and characterization methods for particle volume fraction, diameter, and surface charge number. Then we describe the preparation of various colloidal crystals. Furthermore, we explain methods for observing crystal structures using optical microscopy, spectroscopy, and scattering techniques. Useful tips for experiments on colloidal systems are also presented.

Keywords Particle synthesis · Purification of colloids · Characterization of particles · Colloidal crystal · Spectroscopy · Microscopy · Scattering experiment

3.1 Preparation of Colloidal Samples

Experiments on self-assembly of colloidal systems often require particles with a narrow particle size distribution. For example, the preparation of colloidal crystals usually needs samples with a standard deviation of less than 10%. For polystyrene (PS) [1], poly(methyl methacrylate) [2], and silica (SiO_2) particles [3, 4], methods for synthesizing particles of uniform size have been established. Recently, the synthesis of microgel spheres has also been developed. Here, we introduce the synthesis of monodisperse PS particles and microgel spheres.

3.1.1 Synthesis of Polystyrene Particle

Various methods have been reported for synthesis of PS particles [5, 6]. Soap-free emulsion polymerization is a relatively simple method to obtain uniformly sized (monodisperse) PS particles [7, 8]. This method synthesizes spherical particles by dispersing styrene monomers as oil droplets in a polar solvent and radical

© The Author(s), under exclusive license to Springer Nature Singapore Pte Ltd. 2023
J. Yamanaka et al., *Colloidal Self-Assembly*, Lecture Notes in Chemistry 108,
https://doi.org/10.1007/978-981-99-5052-2_3

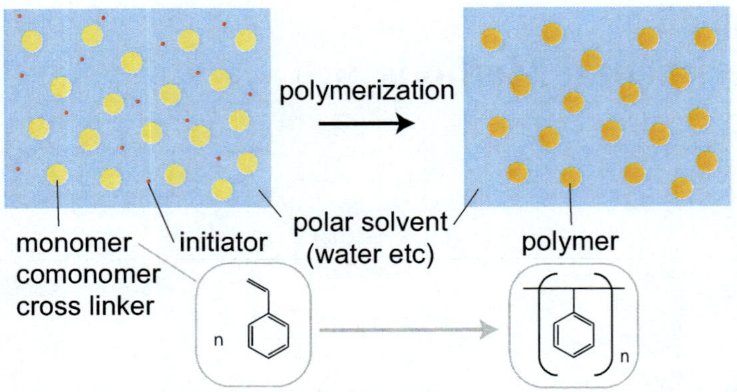

Fig. 3.1 Illustration of the synthesis of polystyrene particle by soap-free emulsion polymerization

copolymerization without using an emulsifier (surfactant), which is often used in particle synthesis as shown in Fig. 3.1. A small amount of divinyl benzene, DVB, is added as a cross-linking agent. As no surfactant is added, no emulsifier is adsorbed on the particle surface, facilitating purification. Ionic comonomers can be added and copolymerized with styrene. This provides surface charges which maintain the colloid stable without aggregation. The polymerization is initiated by adding a radical polymerization initiator to the reaction solution and heating the sample.

A typical example of a synthesis of negatively charged PS particles is described below. Figure 3.2 shows a typical experimental setup. Four-necked separable flasks are used as reaction vessels for the sequential addition of reagents. A cooling tube is attached to the flask to prevent evaporation of the reaction solution. Nitrogen gas is passed through the solution because the radicals produced are lost through the reaction with oxygen.

The scheme in Fig. 3.2 shows the reagents for the synthesis, i.e., (a) styrene, (b) sodium styrene sulfonate (NaSS), (c) potassium persulfate (KPS), and (d) DVB. Before synthesis, styrene and DVB are washed with 1 M aqueous sodium hydroxide (NaOH) solution to remove the polymerization inhibitor. NaSS is employed as the ionic comonomer. A mixture of 210 mL water + 5 mL methanol, 20 mL styrene, 0.1 g NaSS, and 1 mL DVB was added. The mixture was kept in a nitrogen atmosphere at 80 °C in a thermostatic bath and stirred with a rotation speed of 600 rpm for about 30 min. Then the anionic initiator, KPS, is added. After about one hour, the reaction solution begins to turn opaque because of the formation of particles. The reaction is allowed to continue for about 8 h. After synthesis was completed, the lid of the flask was removed, covered with filter paper, and maintained at room temperature to remove unreacted styrene and other volatile reagents. Large agglomerates formed during the reactions are removed using filter paper.

Particle size changed by varying concentrations of styrene and NaSS, methanol, and reaction time. Under the conditions mentioned above, the diameter of PS particles is about 110 nm.

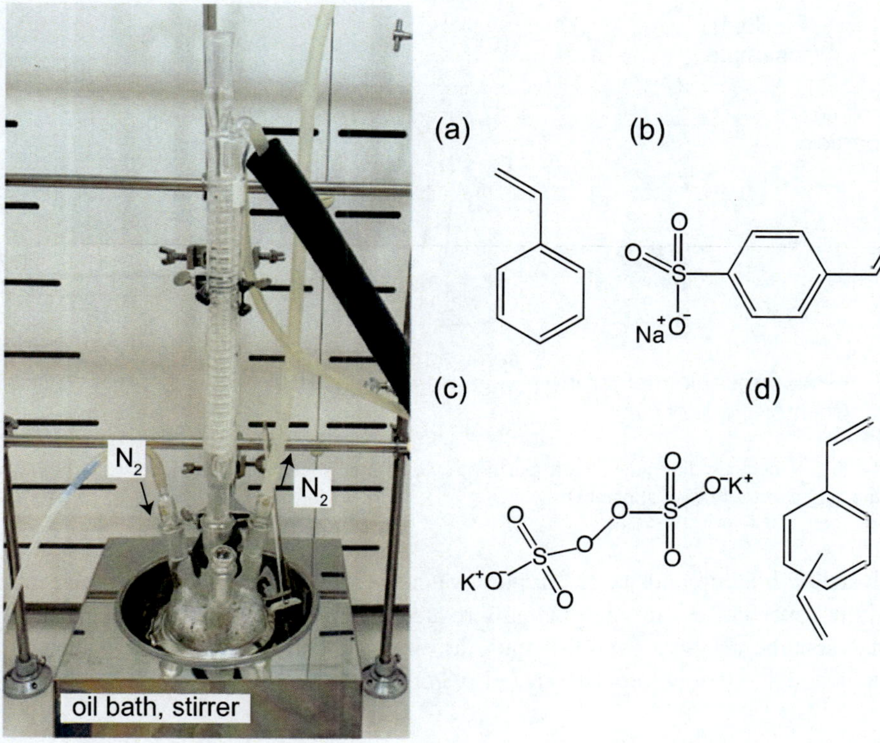

Fig. 3.2 Appearance of the apparatus (left) for the polystyrene particle synthesis and reagents used **a** styrene (monomer), **b** sodium styrene sulphonate (comonomer), **c** potassium persulphate (initiator), and **d** divinylbenzene (crosslinker)

3.1.2 Sample Purification

The colloidal samples thus synthesized contain impurities, such as unreacted monomers. Commercial colloidal dispersions also contain various impurities; for example, salts and surfactants added as stabilizers. Since these impurities may affect particle characterization and structure formation, purification should be carried out.

As mentioned in Chap. 2, the electrostatic interactions in charged colloids are shielded by the presence of ions in the sample. Therefore, the deionization of the sample is quite important. In the following, we describe the purification of aqueous colloidal dispersions.

3.1.2.1 Dialysis

Dialysis using semipermeable membranes allows passing through only small molecules or ions below a specific size (Fig. 3.3a). Low-molecular-weight impurities are removed by placing the sample in a cellulose dialysis tube. The tube is

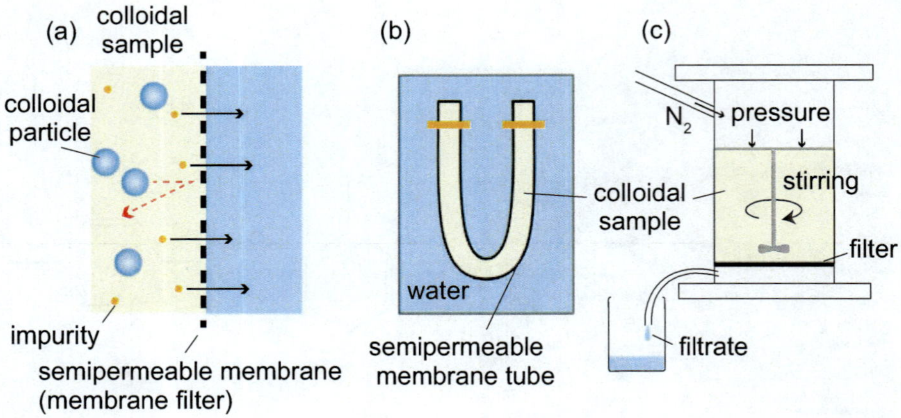

Fig. 3.3 Schematic diagrams of **a** impurities removal by a semipermeable membrane, **b** dialysis tube, and **c** ultrafiltration apparatus

closed with a clip and soaked in pure water vat (Fig. 3.3b). Repeated replacement of the water in the vats can gradually reduce the concentration of impurities in the dialysis tube. Dialysis continues until the electrical conductivity of the water in the vat is equal to that before dialysis (~1 μS/cm).

3.1.2.2 Ultrafiltration

Ultrafiltration is another purification method that uses a semipermeable membrane. Figure 3.3c shows a schematic diagram of the device. A semipermeable membrane is equipped at the bottom of a cylindrical container, and the sample solution is pressurized with nitrogen gas to remove the impurities. Because the pressure is applied, purification can be accomplished in a shorter time than dialysis. The impurities concentration in the sample can be reduced by adding pure water and repeating filtration.

The colloidal particles may concentrate and agglomerate near the membrane; therefore, the sample must be stirred during the purification. The purification is completed when the electrical conductivity of the filtrate is close to that of pure water. Ultrafiltration can also be used to concentrate samples and solvent substitution.

3.1.2.3 Ion Exchange

The ion exchange method is used to remove ionic impurities in the sample. Usually, a mixed-bed of cation and anion exchange resin beads is added to the sample. They exchange cationic and anionic impurities into H^+ and OH^- ions, respectively. The resulting H^+ and OH^- ions form water molecules; thus, we can remove impurity ions from the medium (Fig. 3.4). Care should be taken not to stir the sample vigorously,

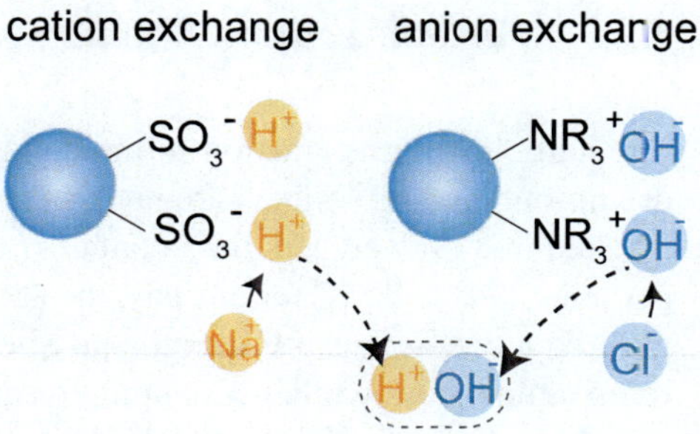

Fig. 3.4 Illustration of the exchange of Na^+ and Cl^- ions with H^+ and OH^- ions by cation and anion exchange resins

as this will crush the ion exchange resin beads. The purified sample is stored in a refrigerator with coexisting ion exchange resins. Since commercially available ion exchange resins swell with water, add the resin beads to the colloidal sample and wait several days to allow the resin to reach swelling equilibrium before characterization, such as the determination of particle volume fraction ϕ.

3.1.2.4 Experimental Tips

Because alkali ions (Na^+ ions, K^+ ions, etc.) may leach from the glass wall, glass containers should not be used when ionic impurities should be avoided, especially in experiments on charged colloidal systems. As described in Chap. 2, the addition of an alkaline solution to the silica colloid increases the dissociation of silanol groups on the particle surfaces, which increases the particle charge number Z. Therefore, silica colloids may form charged colloidal crystals when introduced in glass containers. It is necessary to use quartz or plastic containers instead of glass wares. Similarly, containers made of Teflon are recommended for storing water and other materials to avoid contamination. Carbon dioxide in the air produces carbonic acid when dissolved in water, which acts as an ionic impurity. The experiment should be conducted under a nitrogen or argon atmosphere if the ionic concentration needs to be strictly controlled. Ultrapure water prepared by Milli-Q system (Millipore Co., Ltd.), or other purification apparatus is commonly used for the same reason.

3.2 Characterization of Colloid

Volume fraction, diameter, and surface charge number of the colloidal particle are fundamental quantities to characterize colloids. We describe the methods to determine these quantities in the following.

3.2.1 *Particle Volume Fraction*

The particle volume fraction in the colloidal dispersion can be determined by the drying-out method. A fixed amount of the purified sample (usually about 0.5–1 mL) is taken in a preweighed glass container and dried in an oven. In the case of silica particles, heat at 80 °C for one day, then at 130 °C for one day to dry thoroughly. The glass container is then transferred into a box containing drying agents and returned to room temperature while maintaining its dry state. The weight of the particles in the sample is determined from the difference in the weight of the sample bottle before and after drying. The ϕ value of the particles is determined by the dried weight and density of the particle.

3.2.2 *Particle Size*

3.2.2.1 Dynamic Light Scattering

Dynamic light scattering (DLS) method determines the diffusion coefficient of particles by detecting the temporal fluctuation of the scattered light intensity from the particles [9–11]. See Appendix E for a definition and explanation of the diffusion coefficient. Because the diffusion coefficient of the particle is inversely proportional to the particle size, one can determine the particle size from the intensity fluctuations of the scattered light at time t, $I(t)$. The autocorrelation function $g^{(2)}(\tau)$ for the time interval τ is obtained from the intensity $I(t)$ and $I(t + \tau)$.

$$g^{(2)}(\tau) = \frac{\langle I(t) \cdot I(t+\tau) \rangle}{\langle I(t) \rangle^2}. \tag{3.1}$$

At smaller τ, the correlation is larger because particles do not move significantly. The correlation becomes weaker over time. Therefore, $g^{(2)}(\tau)$ decays with τ.

$g^{(2)}(\tau)$ is also expressed as follows.

$$g^{(2)}(\tau) = 1 + \beta \left| g^{(1)}(\tau) \right|^2. \tag{3.2}$$

Here, $g^{(1)}(\tau)$ is the first-order autocorrelation function, and β is an instrument-dependent constant. When the particles are monodispersed sphere, $g^{(1)}(\tau)$ is an exponential function and is expressed using the decay constant Γ as follows:

$$g^{(1)}(\tau) = \exp(-\Gamma \tau). \tag{3.3}$$

Γ is related to the translational diffusion coefficient D and the scattering vector q given as $q = \frac{4\pi n_0}{\lambda_0} \sin\left(\frac{\theta}{2}\right)$. Here n_0 is the refractive index of the solvent, λ_0 is the wavelength of the incident laser light, and θ is the angle between the line connecting

Fig. 3.5 Schematic diagram of dynamic light scattering (DLS) measurement

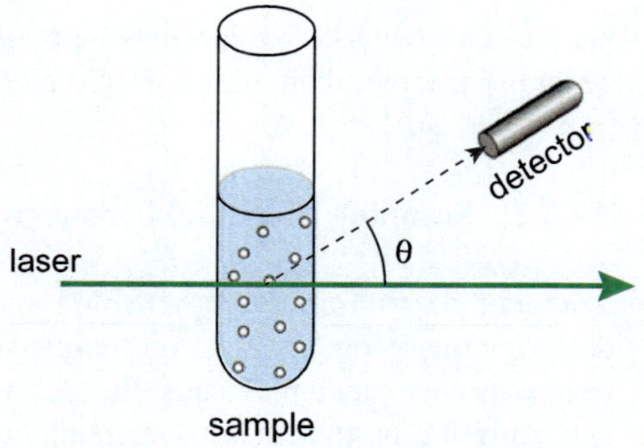

the sample and the detector and the laser light (Fig. 3.5). Γ is given by

$$\Gamma = q^2 D. \tag{3.4}$$

Thus, the diffusion coefficient of the particles can be determined from the scattering experiment.

From the Stokes–Einstein relation [12], the particle radius a_{p} is determined from the D value as

$$D = \frac{RT}{N_A}\frac{1}{6\pi\eta a_{\mathrm{p}}}, \tag{3.5}$$

where R is the gas constant, N_{A} is the Avogadro's number, and η is the viscosity of the medium. Note that this equation holds for monodispersed spherical particles. As with monodisperse samples, the particle size is determined from D using Eq. (3.5). There are several methods for determining the decay constant, including the cumulant method and the histogram method [9–11]. In the case of polydisperse samples with distribution in particle size, $g^{(1)}(\tau)$ is obtained by combining exponential decay curves.

$$g^{(1)}(\tau) = \int G(\Gamma)\exp(-\Gamma\tau)\mathrm{d}\Gamma. \tag{3.6}$$

The distribution $G(\Gamma)$ is the relative intensity of light scattered and depends on the volume fraction and size of the scatterer. Note that the Einstein–Stokes relation holds only when the interaction between particles is negligible [12]. Therefore, to determine the particle size DLS measurements must be performed under dilute particle concentrations. The electrostatic interactions between the particles are long-ranged and influence the Brownian motion of the particle. To screen the interactions sufficiently, one must add salts of approximately ~ 100 μM. Additions of salt at too high concentrations might cause aggregations of the particles. Since light scattering by dust in the sample prevents the precise determination of the diffusion coefficient, dust must be cut off using a membrane filter with a pore size sufficiently larger than the particle size. The particle size determined by DLS is a hydrodynamic size and generally is larger than that measured by SEM, which will be described.

3.2.2.2 Scanning Electron Microscopy

Scanning electron microscopy (SEM) is an observation method of tiny objects by detecting the secondary electrons emitted from the sample upon irradiation of electron beams in a vacuum. Because the measurement details are described in Sect. 3.4.1, here only the particle size measurement is described. The sample is prepared by slowly drying dilute colloidal particle dispersions on carbon tape so that particles do not overlap to avoid agglomeration. The surface of the dried sample is observed by depositing a vapor of metal, such as gold. Figure 3.6 shows an SEM image of a silica particle (particle diameter $d = 1.5$ μm, ×3500). Images are obtained at several locations on the sample to measure d of about 50–100 particles and determine the mean value and standard deviation.

3.2.3 Particle Surface Charge

Electrical conductivity [13–15] and zeta potential [16, 17] measurements are often used to determine the number of charges on the particle surface. Other methods using viscoelastic measurements have also been reported [18].

Fig. 3.6 Scanning electron microscope image of silica particles ($d = 1.5$ μm)

3.2.3.1 Determination of Charge Number by Electrical Conductivity

3.2.3.2 Electrical Conductivity

When the electric current under an applied voltage V to an object is I, the electrical resistance R is obtained by using Ohm's law $V = IR$. The value of R varies depending on the size and shape of the sample. For example, for rectangular parallelepiped material, R is proportional to the length l and inversely proportional to the cross-sectional area A

$$R = \rho \frac{l}{A}. \tag{3.7}$$

The constant ρ in Eq. (3.7) is a substance-specific quantity independent of size and shape called resistivity.

The reciprocals of R and ρ are called conductance (G) and conductivity (K), respectively, which represent ease of the flow of the electric current.

$$G \equiv \frac{1}{R}, \tag{3.8}$$

$$K \equiv \frac{1}{\rho}. \tag{3.9}$$

The unit of K is $1/\Omega\text{m} = \text{S/m}$. Here $\text{S} = 1/\Omega$ is called Siemens, after German scientist W. von Siemens. When a parallel plate electrode of area A is immersed in the solution with an interval l (Fig. 3.7), Eq. (3.7) is applicable by assuming that the electric field is perpendicular to the plate. Usually, we can apply commercial electrical conductivity meters to measure electrical conductivity by an alternating voltage between two electrodes.

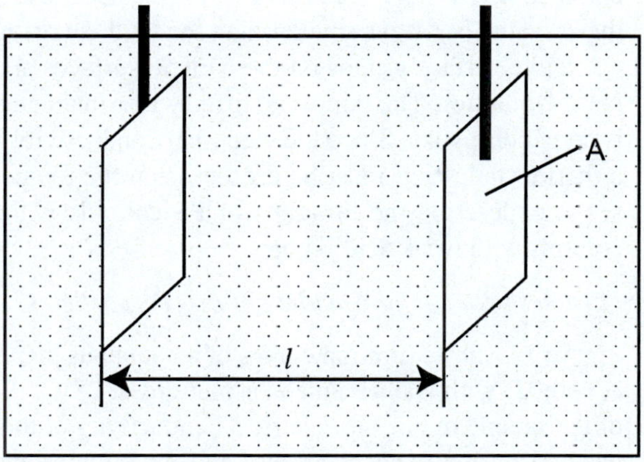

Fig. 3.7 Schematic of electrical conductivity measurement with electrodes. Electrodes of area A are placed a distance l apart from each other

The value K is expressed using the molar conductivity Λ (Scm^2/mol), a substance-specific value, and the molar concentration of the solute C (mol/l).

$$K = \frac{\Lambda C}{1000}, \tag{3.10}$$

where Λ depends on the solvent's viscosity, temperature, and ion concentration. A factor 1000 in the denominator of Eq. (3.10) is needed to express K in S/cm when values of Λ and C are given in Scm^2/mol and mol/l. The electrical conductivity of an aqueous solution is determined as the sum of conductivities of all ions in the samples, that is,

$$K = \sum_i \frac{\Lambda_i C_i}{1000}. \tag{3.11}$$

Estimation of charge number from electrical conductivity

The charge number on the particles can be determined by electrical conductivity titration using acid–base reactions (neutralization reactions) of the surface charges of colloidal particles in the dispersion. We explain the conductivity titration and the data analysis of conductivity titration for particles with acidic groups, often used in experiments. The titrations for low-molecular-weight strong and weak acid solutions are described in Appendix C.

PS particles have sulfate and sulfonate groups, which are strong acids, on their surfaces. Silica particles have weakly acidic silanol groups (Fig. 3.8). The titration curves of these particles show the characteristics of polyacids. Meanwhile, the counterions of a linear polyelectrolyte are attracted to the polyions, reducing their effective charge number Z_e. In polymer sciences, this is called counterion condensation [19]. The same phenomenon occurs at charged interfaces, solid–liquid interfaces, and spherical colloidal particles. In colloid and interface science, the layer of these enriched ions is called the Stern layer [20]. Because of the counterion condensation, the Z_e value is always smaller than the total charge number Z_a [21].

Acidic surface groups on the particle surfaces are titrated with an alkaline (e.g., NaOH) solution. The sample is stirred with a magnetic stirrer, and an alkaline solution is added dropwise. Since the electrical conductivity is larger at higher temperature, a thermostatic bath must be used to keep the temperature constant. The following sections describe the titrations of PS colloids with strong acid groups and silica colloids with weak acid groups.

Particles with strongly acidic groups (PS particles)

Figure 3.9 shows titration curves of an aqueous dispersion of PS particles ($d = 380$, $\phi = 0.02$) with NaOH aqueous solutions at 25°C. ΔK is the electrical conductivity of the sample minus the electrical conductivity of the medium. The sample has been purified by ion exchange resin, and the counterions of the colloidal particles are exchanged to H^+ (Fig. 3.10i). They are referred to as H-type particles. The particle surface is covered with sulfuric or sulfonic acid groups (Fig. 3.8a); they are strong acids and dissociate almost completely $-OSO_3H \rightarrow -OSO_3^- + H^+$ or $-SO_3H \rightarrow$

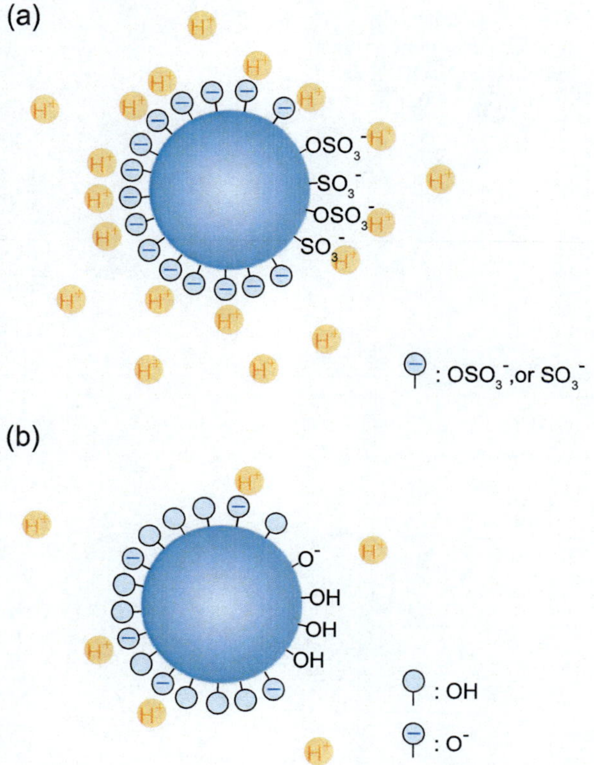

Fig. 3.8 Illustrations of the surface charges of **a** polystyrene particle and **b** silica particle

$-SO_3^- + H^+$. Due to the counterion condensation mentioned above, only a part of the dissociating group provides effective charges. Z_e is calculated from the ΔK value in the absence of NaOH as follows. Since ΔK is the sum of the counterion and the particle contribution, it is expressed as

$$\Delta K = 10^{-3}(\Lambda_{H^+}C_c + \Lambda_p C_p), \tag{3.12}$$

where Λ is the molar conductivity (Scm^2/mol) and C_c and C_p are molar concentration of counterions and surface charges on the particle (mol/l), respectively. We assume that counterions, which are not trapped in the Stern layer, have the same molar conductivity (i.e., the same mobility) as free ions. For Na-type colloidal particles with Na ions as counterions, the contribution of the particle and counterion to electrical conductivity is nearly equal (the transport number of the particle is ~0.5) [14]. That is, we can assume $\Lambda_{Na^+}C_c = \Lambda_p C_p$ Thus, Eq. (3.12) can be written as $\Delta K = 10^{-3}(\Lambda_{H^+} + \Lambda_{Na^+})C_c$, assuming that the effective charge numbers for H^- and Na^+ type particles are the same. At sufficiently low ion concentrations, we can safely assume that the molar conductivities of ions are those at infinite dilution ($\Lambda_{H^+} = 349.8$ Scm^2/mol and $\Lambda_{Na^+} = 50.1$ Scm^2/mol at 25 °C). Thus, we can estimate the counterion concentration C_c (mol/L) by the value of ΔK. When the number

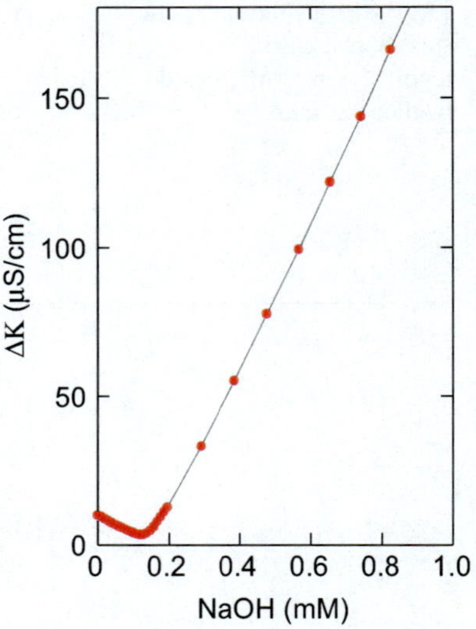

Fig. 3.9 Conductivity titration curve of polystyrene colloid with NaOH ($d = 380$ nm, $\phi = 0.02$)

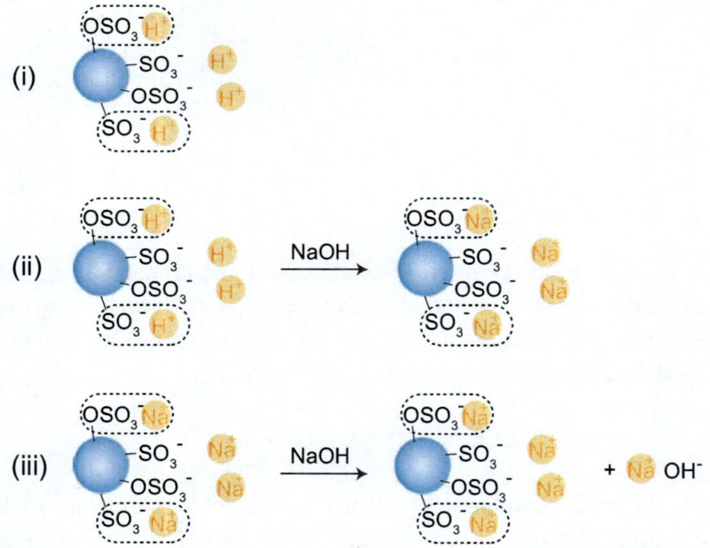

Fig. 3.10 Illustration of polystyrene particles on additions of NaOH

density of particles is N (1/cm^3), the total number of charges in 1 cm^3 of colloid is $N{\cdot}Z_e$, which equals the number of all counterions in 1 cm^3. Therefore, we obtain

$$Z_e = \frac{C_c N_A}{1000N}, \tag{3.13}$$

where N_A is the Avogadro number. $N = \frac{1\times\phi}{\frac{4}{3}\pi a_p^3}$.

In Fig. 3.9, the ΔK of a PS particle dispersion of 380 nm diameter ($\phi = 0.02$) is ~ 9 8 μS/cm when the NaOH concentration is 0, so $\Lambda_{H^+} + \Lambda_{Na^+} \sim 400$ Scm^2/mol, $\Delta K = 10^{-3}(\Lambda_{H^+} + \Lambda_{Na^+})C_c$, resulting in C_c ~ 2.45 × 10^{-5} M. Also, since N ~ 6.96 × 10^{11}, we obtain Z_e ~ 2.11 × 10^4 from Eq. (3.13).

By additions of NaOH, the dissociable groups on the particle surfaces are gradually neutralized. The counterion H^+ is exchanged for Na^+ (Fig. 3.10ii). Note that because the dissociation groups are strong acids, the charge number is constant regardless of the degree of neutralization. Because of $\Lambda_{H^+} > \Lambda_{Na^+}$, the electrical conductivity of the sample decreases by the ion exchange from H^+ to Na^+. In the case of a low-molecular-weight strong acid, such as sulfuric acid, the electrical conductivity decreases almost linearly according to $\Delta K = \frac{(-349.8+50.1)c_{NaOH}}{1000}$, where C_{NaOH} is the NaOH concentration. The slope of the titration curve of PS colloid is smaller than the case of a low-molecular-weight acid because of the counterion condensation. When all H^+ in the particle is exchanged for Na^+, the particle is neutralized, and ΔK takes a minimum value. Z_a is obtained from the C_{NaOH} at the neutralization point. The number of Na^+ ions added in 1 cm^3 of sample is $\frac{C_{NaOH} N_A}{1000}$. This equals the number of counterionic H^+ ions initially in the sample. Because of the electrical neutrality, Z_a equals the number of counterions per particle. Thus, we have

$$Z_a = \frac{C_{NaOH} N_A}{1000 N}. \tag{3.14}$$

At C_{NaOH} higher than the neutralization point, NaOH added is present in excess in the sample as Na^+ and OH^- (Fig. 3.10iii). Therefore, ΔK increases linearly with [NaOH].

Particles with weakly acidic groups (silica particles)

The surface of silica particles is covered with weakly acidic silanol groups Si–OH. In general, dissociable groups of polyacids have different dissociation constants and become less dissociable as neutralization proceeds. For example, phosphoric acid H_3PO_4 has three OH groups as dissociable groups. However, it takes more electrostatic energy to dissociate two of them than to dissociate one of them. This is because another negative charge must be generated on the negatively charged molecule. Thus, the dissociation constants of the three OH groups gradually increase with the degree of neutralization. In the case of silica particles, the particle surface is covered with so many silanol groups that pK_a can be regarded as a function that varies continuously with the degree of neutralization [4]. This property as a weak poly acid appears in the conductivity titration curve of silica particles. An example of a titration curve is shown in Fig. 3.11 with red symbols. The titration was performed at 25°C with 0.01 M NaOH solution against 15 mL of silica colloid (d = 80 nm, volume fraction $\phi = 0.02$). The results of the pH titration are also shown with blue symbols.

The conductivity titration curve of silica colloids can be analyzed in the following three regions. Schematic drawings of the dissociable groups of silica particles in each region are shown in Fig. 3.12.

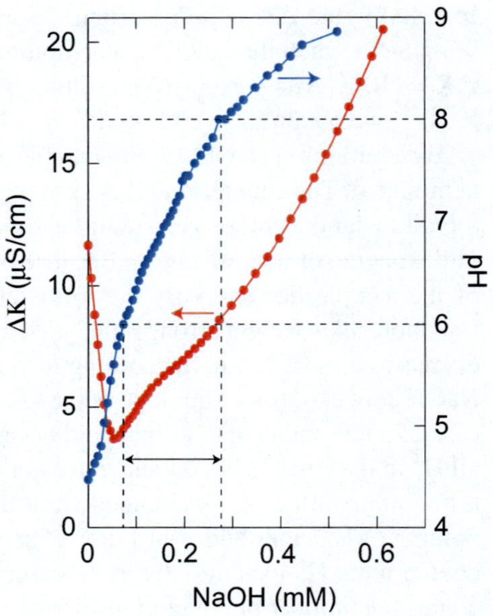

Fig. 3.11 Conductivity (red) and pH (blue symbols) titration curves of silica colloids with NaOH (d = 80 nm, $\phi = 0.02$)

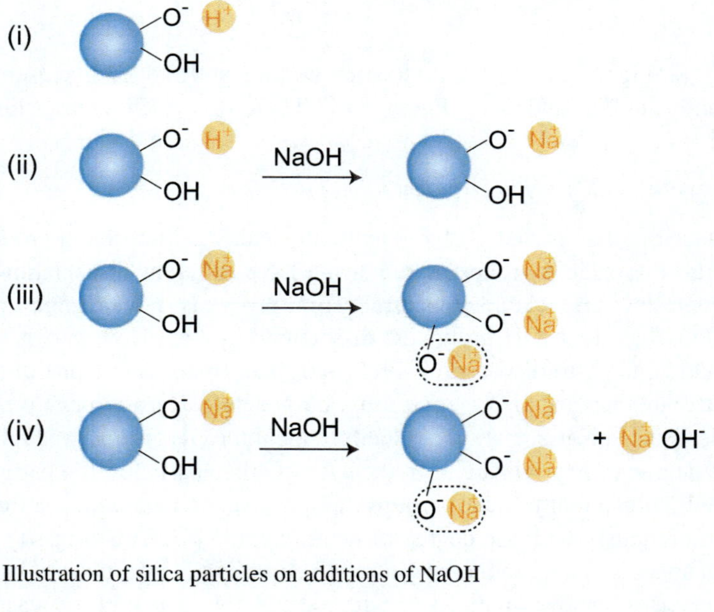

Fig. 3.12 Illustration of silica particles on additions of NaOH

(i) Without addition of NaOH (Fig. 3.12i).

Silanol groups are only partially dissociated. A part of counterions are strongly trapped on the particle surface to form Stern layer and do not contribute to electrical conduction. As with the PS particles, the Z_e of the particle is obtained from ΔK. In this case, ΔK is 11.63 and Z_e ~ 234 is obtained.

(ii) With addition of NaOH.

Additions of NaOH cause the exchange of counterions H^+ and Na^+ (Fig. 3.12ii), that is, $-OH + NaOH \rightarrow -O^-Na^+ + H_2O$. ΔK is reduced by counterion exchange as in the case of strong acids. At the same time, however, the added NaOH reacts to neutralize the undissociated silanol groups, gradually increasing the charge number of the silica (Fig. 3.12iii). When a sufficiently large amount of NaOH is added, the reaction of NaOH with all silanol groups on the particle surface is substantially completed, and the added NaOH is present in the sample in the form of Na^+ and OH^- (Fig. 3.12iv). Then ΔK increases almost in proportion to the amount of added NaOH.

The titration curve indicates the value of the charge number of silica particles as a function of NaOH concentration. Here, only the region of $6 \leq$ pH is targeted because of ease of analysis; in this pH region, $[H^+] \leq 10^{-6}$ M, and so $[H^+] < < [Na^+]$. Therefore, we can safely assume that the counterions of the silanol groups are Na^+. If NaOH is present in the medium in excess, it would also complicate the analysis. Therefore, the analysis is performed for pH ≤ 8, where the concentration of free NaOH is also not larger than 10^{-6} M. Thus, in $6 \leq \text{pH} \leq 8$, we can assume that all added NaOH reacts with silanol groups, and all counterions of silanol groups can be regarded as Na^+. The values of Z_e and Z_a are obtainable by (3.13) and (3.14) as in the case of particles with strongly acidic dissociative groups.

From the Z_e and Z_a values obtained for silica and PS particles of d ~ 100 nm, we have obtained an empirical relationship $\log Z_e = C_1 \log Z_a - C_2$. [14, 22, 23] C_1 and C_2 are coefficients determined from experiments. Figure 3.13a-1, a-2 shows the Z_a-Z_e relationship for silica particles 1 and 2 ($d = 120$ nm and 110 nm) on linear and logarithmic scales [22]. C_1 and C_2 are 0.51 and 1.23 for silica particles 1 and 0.52 and 1.16 for silica 2, respectively. For these two particles with similar particle sizes, the values do not differ significantly; however, the relationship varies depending on particle size.

Counterion condensation is more significant for larger particles but does not occur with low-molecular-weight acids. In the case of strong acids, Z_e should be equal to Z_a in the limit of particle size $\rightarrow 0$. Figure 3.13b-1, b-2 shows results for silica particles of various particle sizes on linear and logarithmic scales. When dealing with different particle sizes, using the particle charge density (=Ze/particle surface area, e is an elementary charge) σ instead of Z. We obtained an empirical relation $\ln\sigma_e = 0.49\ln\sigma_a - 1.0$ for silica and polystyrene particles of about 100 nm diameter [23]. The relationship between σ_a and σ_e for silica particles of various sizes (16–110 nm) is shown in Fig. 3.13b-1, b-2 [24]. The data for particles ($d = 110$ nm) are the same as for silica 2 in Fig. 3.13a-1, 2. A straight line with $\sigma_a = \sigma_e$ is also shown as a comparison. As the particle size decreases, σ_e approaches σ_a.

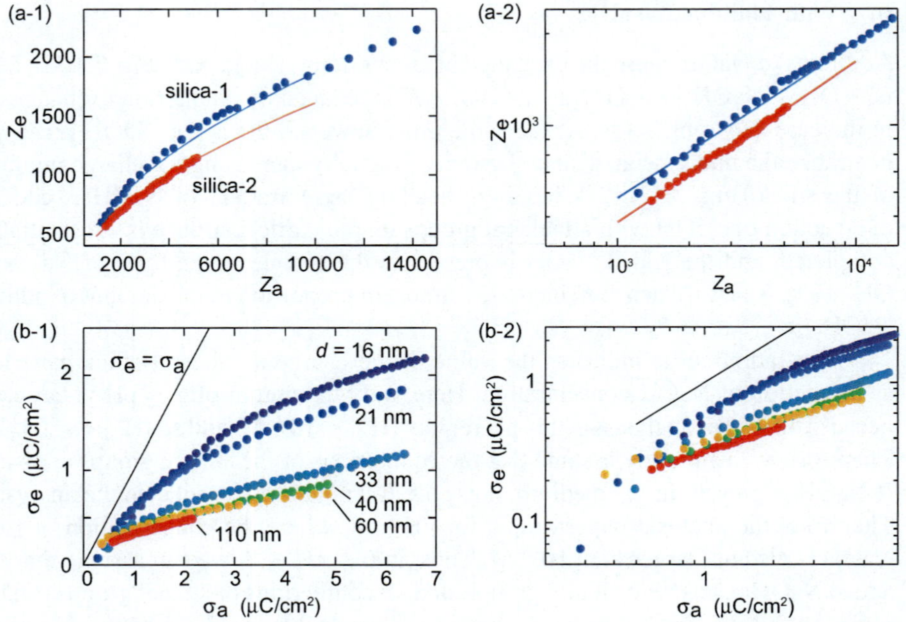

Fig. 3.13 Linear and logarithmic plots of the relationships **a**-1,2 between total charge number Z_a and effective charge number Z_e for silica particles of $d = 120$ nm (silica1) and 110 nm (silica 2), and **b**-1,2 between σ_a and σ_e for silica particles of different particle sizes

3.2.3.3 Determination of Charge Number by Electrophoresis

The electrostatic potential is a fundamental quantity in electrostatics (see Appendix A). The electrostatic potential at the particle surface is referred to as the surface potential ψ_0. ψ_0 can be related to the electrophoretic velocity of the particles under an electric field. Because the particles move with the surrounding solvent molecules in the dispersion, the electrostatic potential determined in the electrophoresis experiment is not ψ_0, but the potential at the plane of the solvent molecules moving the particles. This plane is called the slipping plane, and the potential at the slipping plane is called the zeta potential ζ [16, 25, 26]. Figure 3.14a illustrates the potential ψ around the particle. Note that the magnitude of ψ_0 is larger than ζ. In the following, we describe the microscopic electrophoresis method, in which the motion of colloidal particles under the electric field is observed directly by an optical microscope. Methods other than electrophoresis for measuring zeta potential include ultrasonic and electrophoretic light scattering.

Figure 3.14b shows an electrophoretic measurement cell seen from the top. An electric field E is applied to the sample using two electrodes at both cell ends. The negatively charged particles, seen as white dots, move toward the cathode (Fig. 3.14c). Since glass cell walls are usually negatively charged, like silica particles, positively charged counterions are present near the cell walls. Under the electric field, these counterions of glass move toward the anode (an enlargement image of the area near

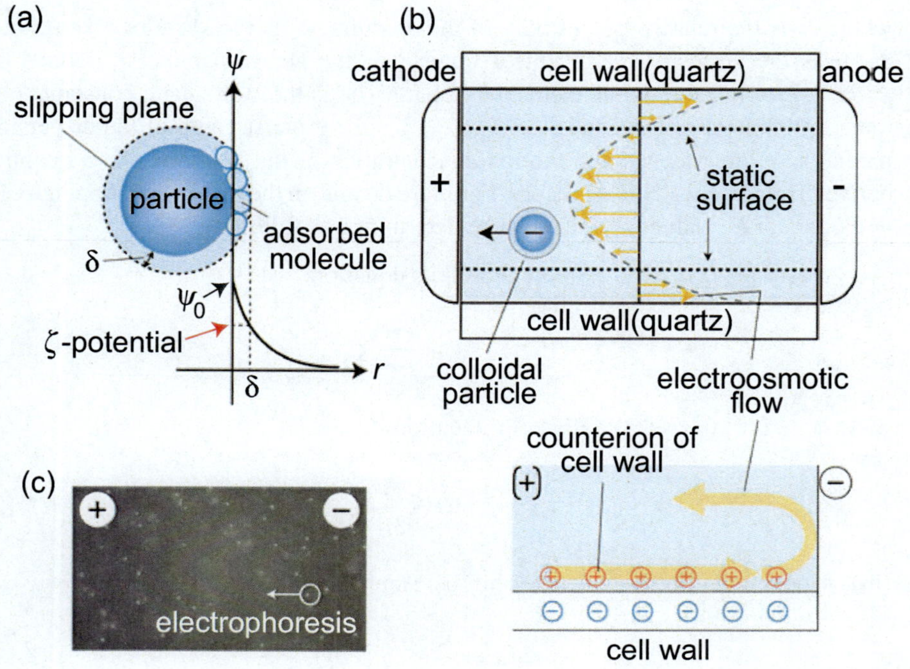

Fig. 3.14 Schematic diagrams of **a** potential around a colloidal particle, **b** a cell for electrophoretic ζ-potential measurement and electroosmotic flow, and **c** a photograph of electrophoresis of colloidal particles

the wall is shown in Fig. 3.14b, bottom). Because of the frictional force between the ion and solvent molecules, the ions cause the motion of solvent molecules, resulting in a flow near the cell wall. Furthermore, since the cell is a closed system, this flow causes a counterflow near the center of the cell. Such flows in the sample cell are referred to as electroosmotic flow [27] (orange arrows). The position where the velocity of the electroosmotic flow is zero is called the stationary layer (two locations are indicated by dashed lines in Fig. 3.14b). The velocity of the particles is measured at the stationary layers to avoid electroosmosis flows. The particle velocity v at which the particles move can be observed under a microscope. The electrophoretic mobility u is the particle velocity per unit electric field and is defined by the following equation,

$$u = v/E. \tag{3.15}$$

The relationship between u and ζ depends on the ratio $a_p/(1/\kappa) = \kappa a_p$ of the particle radius a_p to the thickness $1/\kappa$ of the diffuse electric double layer and the magnitude of scaled electrophoretic mobility E_m

$$E_m = \frac{3\eta zeu}{2\varepsilon_r\varepsilon_0 k_B T}, \tag{3.16}$$

where ε_r is the relative permittivity of the medium, ε_0 is the vacuum permittivity, and η is the viscosity coefficient of water. κ is the Debye parameter, defined by $\kappa^2 = e^2 I/\varepsilon_r \varepsilon_0 k_B T$. e is the elementary charge, k_B is the Boltzmann constant, T is the absolute temperature, and $I = \sum z_i^2 c_i/2$ is the ionic strength of the dispersion media. z_i, c_i are valence and the ion concentration of the i-th ion. κ is a quantity derived from Debye–Hückel theory. For more details on the theory, see Appendix B.

For small E_m values, ζ value can be determined as follows.

(i) $a \gg 1/\kappa$, $(\kappa a \gg 1)$: Smoluchowski's equation

$$u = \frac{\varepsilon_0 \varepsilon_r}{\eta} \zeta, \tag{3.17}$$

(ii) $a \ll 1/\kappa$, $(\kappa a \ll 1)$: Hückel's equation

$$u = \frac{2\varepsilon_0 \varepsilon_r}{3\eta} \zeta, \tag{3.18}$$

(iii) Under condition between (i) and (ii): Henry's equation [28].

$$u = \frac{\varepsilon_0 \varepsilon_r}{\eta} \zeta f(\kappa a_p). \tag{3.19}$$

Here, $f(\kappa a_p)$ is called the Henry coefficient, and Ohshima et al. reported that it could be approximated by the following equation [29]

$$f(\kappa a_p) = \frac{2}{3}\left[1 + \frac{1}{2\left\{1 + \frac{2.5}{\kappa a_p (1 + 2e^{-\kappa a_p})}\right\}^3}\right]. \tag{3.20}$$

The Henry equation corrects the other two equations to account for the distortion of the external electric field, both of which are proportional to ζ, where u is proportional to ζ. When the E_m values are large, relaxation effects caused by the deviation of the diffuse electric double layer from spherical symmetry due to the particle's motion must be considered. The computer program for the numerical analysis is given by O'Brien and White [30]. For any value of ζ and $\kappa a_p \geq 10$, the Ohshima-Healy-White equation can be used as an approximate analytical expression [31]. Figure 3.15 shows the range of parameters to which these theories are applicable [25].

We usually use dilute colloids of charged particles at a salt concentration of 10 μM and apply a voltage of 20 mV for the measurement. For particles of $d = 100$ nm, κa_p is around 1.7 and ζ is calculated using the Henry equation or numerical analysis by O'Brien and White.

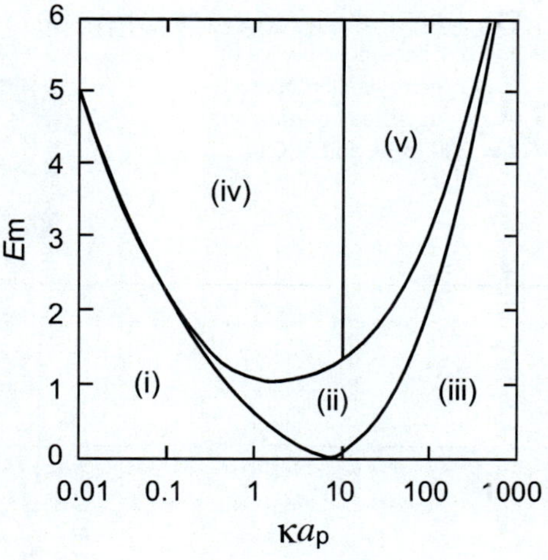

Fig. 3.15 Applicable regions of (i) Hückel equation, (ii) Henry equation, (iii) Smoluchowski equation, (iv) O'Brien-White's model, and (v) Ohshima-Healy-White equation

3.3 Formations of Colloidal Crystals

This section describes the preparation of the colloidal crystals. It was already reported in the 1950s that particles with a narrow size distribution form colloidal crystals [32]. Luck, Klier, and Wesslau observed that dispersions of monodisperse polystyrene particles exhibit bright iridescent colors, which they attributed to the Bragg diffraction from the colloidal crystals [33]. For the mechanism of colloidal crystallization in various systems, please refer to Chap. 2. This section presents simple crystallization experiments using silica and PS particles, often used in our laboratory.

3.3.1 *Opal-Type Crystal*

Opal-type crystals have a regular arrangement of particles in contact with each other [34]. The Bragg condition for opals is given by $2nd_{\text{hkl}}\sin\theta = \text{m}\lambda$ $2nd_{\text{hkl}}\sin\theta = \text{m}\lambda$. Here n is the refractive index of the sample, d_{hkl} is the (hkl) lattice spacing of the crystal, θ is the angle of the incident light, m is an integer, and λ is the diffraction wavelength. (For detail, see Sect. 3.4.2.1). Generally, a colloidal particle size of 200–300 nm is appropriate for making opal-type crystals that exhibit visible Bragg diffraction. About 0.1 ml of colloidal particle dispersion with $\varphi \sim 0.05$ is dropped onto a flat substrate such as a glass plate and allowed to dry slowly at room temperature. In order to produce opal-type crystals with a well-defined structure, it is desired that the evaporation rate of water is slow enough and that the conditions are chosen so that the particles are placed in a thermodynamically stable arrangement (crystal lattice points). The evaporation rate can be controlled by adjusting the temperature and humidity of the environment. When opal crystals form, bright diffractive colors appear. By changing the particle size and refractive index, samples with different

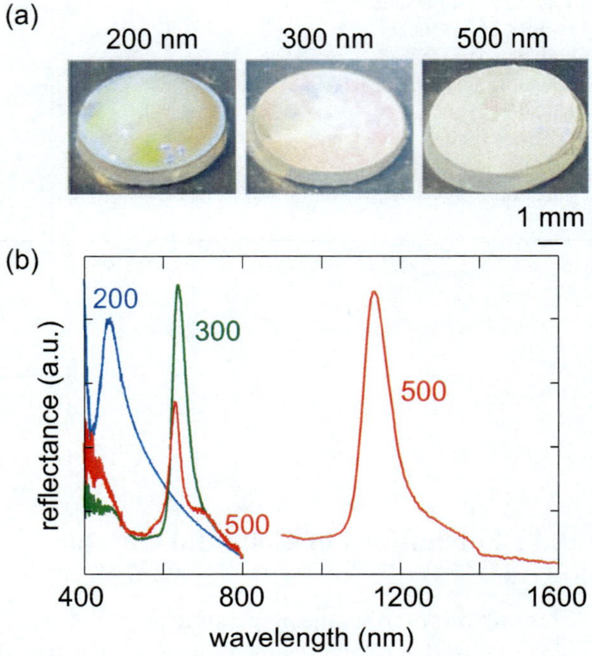

Fig. 3.16 **a** Appearances and **b** reflectance spectra of opal-type crystals prepared by drying silica colloids with $d = 200$, 300, and 500 nm

diffraction colors can be formed. Figure 3.16 shows photos of the appearances of opal-type crystals produced by slowly drying dispersions of three types of PS particles with different particle sizes and their respective reflectance spectra. Crystal structures exhibiting different colors are formed. The Bragg diffraction peaks are observed at different wavelengths in the spectra. Secondary peaks are also observed in the crystal structure of the 500 nm particles. The relationship between reflectance spectrum measurements and crystal structure is described in Sect. 3.4.2. Opal-type colloidal crystals are obtainable by precipitating the particles by centrifugation.

3.3.2 Charged Colloidal Crystals

As discussed in Chap. 2, the driving force of crystallization of charged colloid is an increase in electrostatic interaction between the particles. The major experimental parameters that determine the magnitude of the interaction are the ϕ value, Z, and the salt concentration of the system C_s [13, 35]. Here, we explain how to control crystallization according to these parameters. In addition, methods for creating a crystallization phase diagram are also described.

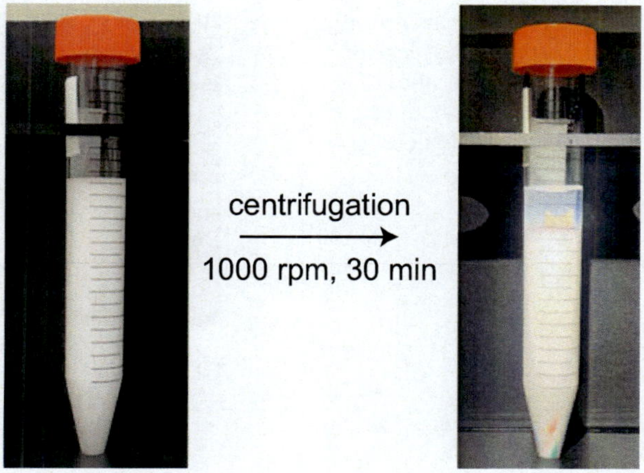

Fig. 3.17 Photographs of silica colloids ($d = 300$ nm, $\phi = 0.04$) before and after centrifugation (1000 rpm, 30 min)

3.3.2.1 Crystallization by Controlling Particle Volume Fraction

Increasing the ϕ value of the colloidal sample favors crystallization because the interparticle distance is shorter and interparticle interactions are greater. Ultrafiltration, described in Sect. 3.1.2.2, is sometimes used to concentrate colloidal samples, but centrifugation, which concentrates by sedimentation of particles, is a simple and commonly used method. Choosing a rotational speed that is not too fast is necessary to prevent particle agglomeration. Figure 3.17 shows the appearance of colloidal silica dispersion ($d = 300$ nm, $\phi = 0.04$, [NaOH] = 1 mM). After centrifugation at 1000 rpm for 30 min crystallization, resulting structural colors were observed at the bottom of the cell. This was due to the increase in concentration associated with particle sedimentation.

Figure 3.18 shows the appearances of charged colloidal crystals of silica dispersion ($d = 100$ nm) at three different concentrations and their reflection spectra. The higher the particle concentration, the smaller the lattice spacing of the crystal and, therefore, the shorter the Bragg diffraction wavelength.

3.3.2.2 Crystallization by Controlling Charge Number

On increasing the Z value, the electrostatic repulsion between particles becomes stronger, and the colloid crystallizes under appropriate conditions. For particles with a weakly acidic surface charge, such as silica particles, Z can be tunable by changing pH. Well-purified silica particle dispersions are weakly acidic (Fig. 3.8b). As mentioned in Sect. 3.2.3, the silanol groups on the particle surface partly dissociate to $-O^-H^+$. When a base such as NaOH is added to the silica colloid, the weakly acidic silanol groups are neutralized to form sodium salts. The sodium salt of silanol groups dissociates completely to $-O^-\ Na^+$, increasing the Z of the silica particles. Therefore, at appropriate ϕ and C_s, silica colloids crystallize [14, 36, 37].

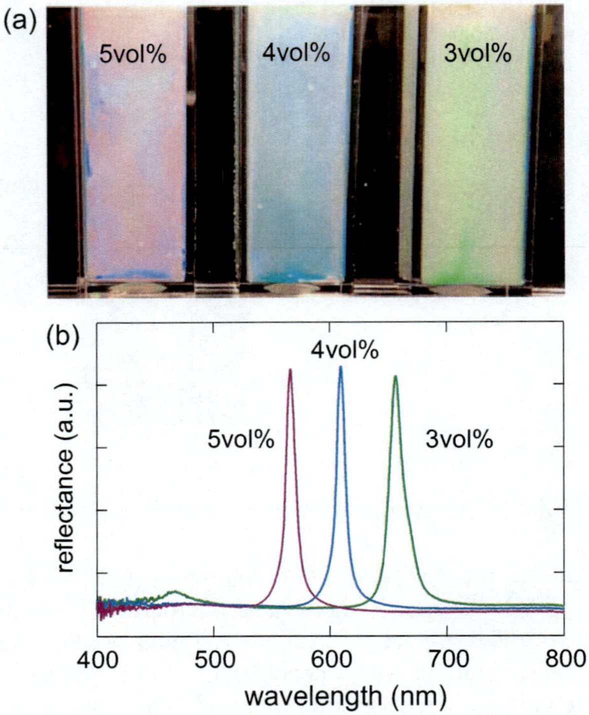

Fig. 3.18 **a** Appearances and **b** reflectance spectra of charged colloidal silica (d = 100 nm) crystals of three different concentrations (ϕ = 0.03, 0.04, 0.05)

For example, silica colloid having $d = 100$ nm and $\phi = 0.03$ crystallizes at NaOH = a few 10 μM.

Although particles with strong acidic surface charges do not increase in charge number with the addition of base, the Z of various types of particles can be tuned by the addition of ionic surfactants [38, 39]. Z of the cationic particle with a weak basic surface can be tuned by pH as well. The colloids have been reported to crystallize with the addition of HCl [40].

3.3.2.3 Crystallization by Decreasing Salt Concentration

Ions present in the dispersion medium shield electrostatic interactions between charged particles. Thus, as C_s decrease, electrostatic interactions become stronger. In dilute charged colloids of a few volume%, crystallization often requires less than C_s 10 μM. Generally, C_s is reduced by deionization using ion exchange resins. Precleaned ion exchange resins with a mixture of cation and anion exchange resins are commercially available.

3.3.2.4 Crystallization Phase Diagram

The phase diagram can be determined as follows. First, samples of various ϕ values are crystallized by adding a fixed amount of NaOH. Then, salt (NaCl) is added gradually to determine the salt concentration at which the crystals melt. By plotting them, a two-parameter plane phase diagram is obtained, as shown in Fig. 3.19a. In addition, crystallization with increased Z is possible by adding various concentrations of NaOH with a constant ϕ value. The salt concentration at the solid–liquid phase boundary is then determined when NaCl is added gradually, and the crystals are melted, yielding the phase diagram shown in Fig. 3.19b. The phase diagram provides the basic information for the controlled crystallization experiments described in Chap. 5. Yamanaka et al. reported a phase diagram of three parameters: C_s, Z, and ϕ [41].

3.3.3 *Crystallization by Depletion Attraction*

As described in Chap. 2, in colloid and polymer mixtures, there exist regions between the particles where the added polymer cannot be present (depletion zone). The osmotic pressure difference between the depletion zone and the bulk causes an attractive force (depletion attraction) between the particles [42, 43]. This attraction causes the particles aggregate to crystallize [43, 44]. Polyelectrolytes are particularly useful as depletants because they spread widely in water and do not adsorb with particles of the same sign due to electrostatic repulsion.

Figure 3.20a, b shows examples of two-dimensional (2D) and three-dimensional (3D) crystals obtained by depletion attraction. PS particle dispersion (d = 600 nm, $\phi = 0.02$) crystallizes when 0.1 wt% polymeric polyacrylic acids (molecular weight ~ 10^6) are added (Fig. 3.20a). Crystals are often observed at the cell bottom and walls because the depletion attraction between particle–wall is twice as large as the depletion attraction acting between particle and particle. When the depletion attraction is sufficiently strong, a bulk, 3D crystal structure is formed.

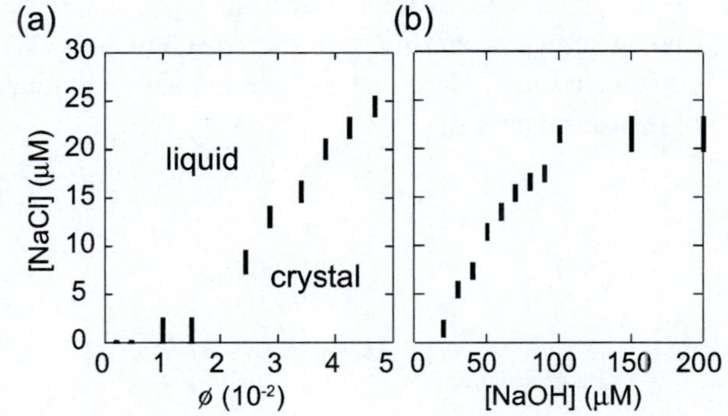

Fig. 3.19 Crystallization phase diagrams of silica colloid (d = 110 nm). **a** [NaOH] = 75 μM and **b** ϕ = 0.034

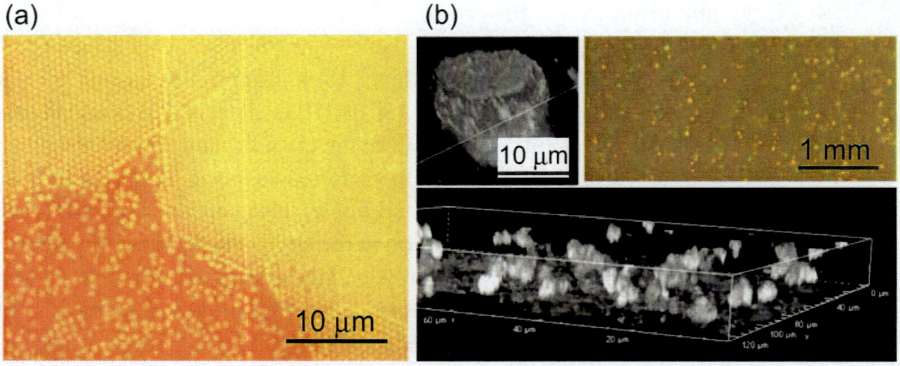

Fig. 3.20 Micrographs of colloidal crystals of polystyrene particles formed by depletion attraction. Depletant: polyacrylic acid, M_w ~ 800,000, polymer concentration 0.1wt% **a** optical microscope image of crystals on the bottom of the cell ($d = 600$ nm) and **b** 3D crystal formed in the bulk ($d = 200$ nm), CLSM, and stereomicroscope images

3.4 Characterization of Crystal Structure

Colloidal particles with several hundred nm or larger can be observed in situ and in real time as single-particle images by optical microscopy easily. An inverted reflective optical microscope that allows observation of the bottom of the sample cell is helpful for observation. Confocal laser scanning microscopy, which can observe the 3D structures, can also be used to observe the inner structure of the sample without the influence of cell walls.

The structure of colloidal crystals can also be determined by scattering experiments (e.g., X-rays, neutron, and synchrotron radiation) similar to atomic and molecular crystalline materials. However, the information on the structure of colloidal systems of large size appears in the region of small scattering vectors. Hence, scattering experiments need to be carried out in the ultra-small-angle region. When the diffraction wavelength lies in the visible to near-infrared region, it can be detected by spectroscopic measurements (reflection and transmission) corresponding to diffracted light from the crystal. Optical microscopy provides local information, whereas scattering and spectroscopy provide statistical information on the crystal structure. Here, these experimental techniques are presented with examples of measurements.

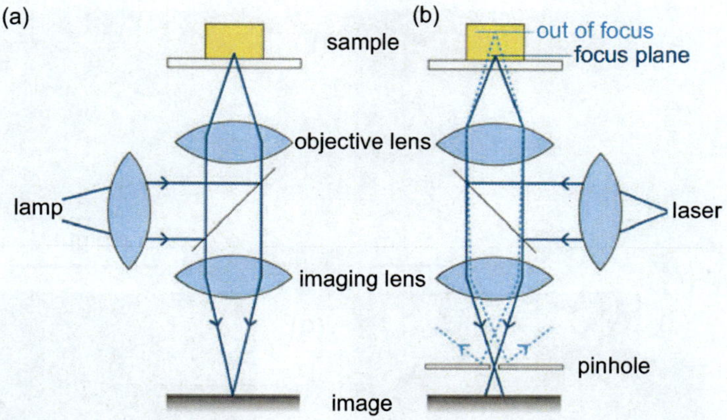

Fig. 3.21 Schematic diagrams of the optical system of **a** an inverted microscope and **b** a confocal microscope

3.4.1 Microscopy

3.4.1.1 Optical Microscope

An inverted microscope (Fig. 3.21a), which enables observation of the sample near the container bottom, is useful for studying colloidal systems. Photographs and videos can be recorded and analyzed by using a digital camera. For particles not labeled with fluorescent dyes, particles with a diameter of 300 nm or more are observable if the refractive index (n) difference between the particle and the medium is large enough. The n values of water, silica, and PS particles are 1.33, 1.46, and 1.59, respectively. Thus, PS particles are easier to observe than silica particles in aqueous dispersions. Metal particles, including gold colloidal particles, appear very bright because of the surface plasmon resonance (See Chap. 6). The position of a single gold particle can be determined even at particle sizes as small as 80 nm. Figure 3.22a–c shows optical microscopy images of PS particles (d = 260 nm), silica particles (d = 300 nm), and gold particles (d = 80 nm). Here (a) and (b) are those crystallized samples. By introducing fluorescent dyes, colloidal particles are distinguishable, even in a multicomponent system (Fig. 3.22d) (d = 1250 nm for each).

3.4.1.2 Confocal Laser Scanning Microscope

For conventional optical microscopes, the information obtained is limited to that near the cell wall. For observing colloidal particles of several hundred nm in diameter, it is necessary to use a high magnification objective lens but the higher the lens magnification, the shallower the focal depth. Sometimes cell walls strongly influence the structure of colloidal systems, as colloidal crystals tend to orientate on the cell wall, and crystal growth is likely to start from the sample cell wall. For this reason,

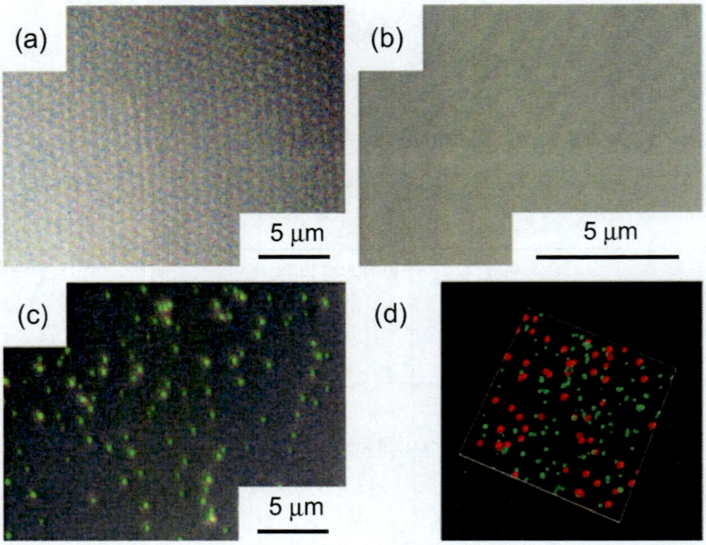

Fig. 3.22 Microscopic images of various colloidal particles. **a–c** are inverted microscope images; **d** is CLSM image, **a** and **b** are in crystalline state. **a** polystyrene particles, $d = 260$ nm **b** silica particles, $d = 300$ nm, **c** gold particles, $d = 80$ nm, **d** fluorescent (red, green) polystyrene particles $d = 1250$ nm

confocal laser scanning microscopy (CLSM) is helpful, because it allows observation at depths away from the wall surface [45, 46]. The confocal microscopes have a pinhole in front of the photodetector positioned conjugate to the illumination point on the sample (Fig. 3.21b). When the sample is in focus in a single plane, it is also in focus in the conjugate plane. Only this focused light passes through the pinhole and is detected. Light coming from depths other than the focal point is not in focus at the pinhole position; therefore, most of light is cut off. This allows only information at the focal point to be acquired. The 2D scanning at the different focal points can obtain 2D sliced images at each position. By reconstructing the images thus obtained, a 3D image is obtained. CLSM allows individual particles and their higher-order organization in a sample to be observed down to hundreds of micrometers inside.

Occasionally, the colloidal particles used may be smaller than the resolution limit of an optical microscope (several hundred nm). In this case, the particles captured by the CLSM are projections of images magnified on the focal plane, which correctly show the position of individual particles but not their size. Examples of CLSM images of charged colloidal crystals (PS particles, $d = 480$ nm, $\phi = 0.025$) are shown in Fig. 3.23a, b. It can be observed that the particles are regularly arranged in 3D. Figure 3.23b shows an image of a glass bead of about 60 μm diameter submerged in a colloidal crystal. Colloidal crystals are observed oriented on the cell wall and the glass beads.

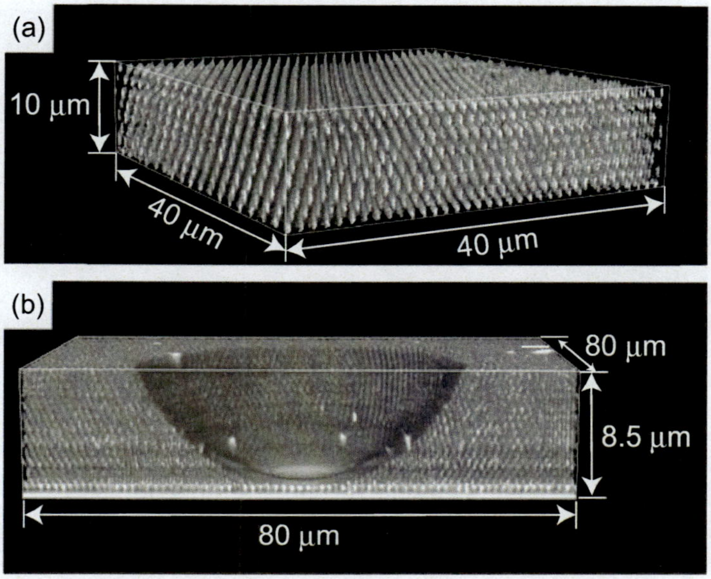

Fig. 3.23 CLSM images of **a** charged colloidal crystals of polystyrene particles (d = 480 nm, ϕ = 0.025) and **b** a glass bead (d ~ 80 µm) in the colloidal crystal

3.4.1.3 Scanning Electron Microscope

Transmission electron microscopy (TEM) and scanning electron microscopy (SEM) enable observation at much higher magnifications than optical microscopy. This section describes SEM observations. SEM observes the surface structure of a sample by detecting the secondary electrons produced on the sample when an electron beam is irradiated onto it. As the intensity of the secondary electrons generated depends on the angle at which the electron beam is incident on the sample surface, the slight roughness of the sample surface can be detected as secondary electron intensity. By gradually shifting (scanning) the electron beam across the surface of the sample, SEM provides a 3D image of the surface with a deep depth of focus and significant magnification (Fig. 3.6). The resolution of an SEM is determined by the diameter of the electron beam spot on the sample (probe size), with the resolution of an SEM being typically 0.5–4 nm.

The sample chamber of SEM is vacuumed for observation so that the electrons produced by the electron source reach the sample without colliding with gas molecules. For this reason, it is necessary to carry out pretreatment, such as removing water to prevent the sample from deforming under vacuum conditions. In addition, to prevent the sample from being charged by electron irradiation and to increase the amounts of secondary electrons generated, the sample surface is usually coated by depositing a metal such as gold or platinum–palladium.

Furthermore, using high-resolution SEM allows finer structures, e.g., the shape of the particle surface, to be observed. Figure 3.24a shows a PS particle with a

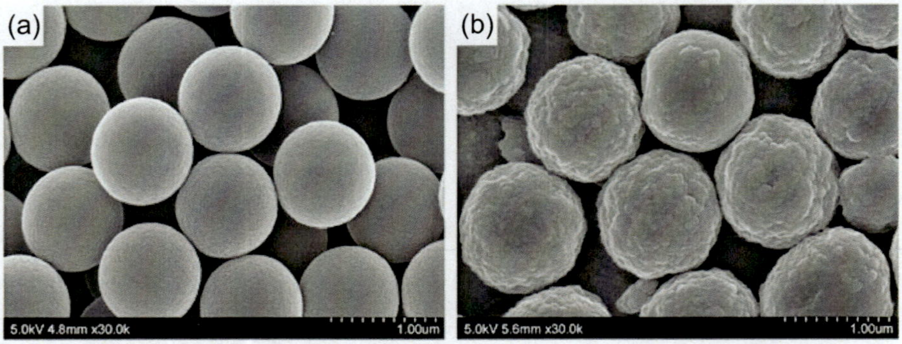

Fig. 3.24 High-resolution SEM images of **a** polystyrene particles (d ~ 1.46 μm) and **b** titania particles synthesized by the sol–gel method (d ~ 1.18 μm)

diameter of approximately 1.46 μm observed at 30,000×. It shows irregularities on the particle surface that could not be observed at 3500× (Fig. 3.6). For TiO_2 particles synthesized by the sol–gel method, the surface convolutions are more pronounced than in PS particles (Fig. 3.24b).

3.4.1.4 Image Processing Method

The arrangements of the colloidal particles can be analyzed from microscopic images. Figure 3.25a-1 and a-2 shows images acquired with an inverted microscope. Both are dispersions of PS particles ($\phi = 0.03$) with a particle size of 430 nm, images of the fluid state with randomly arranged particles, and the crystalline state with a regular structure, at [NaCl] = 100 μM and 0 μM, respectively. The Fourier spectra of these images, transformed by image processing software, are shown in Fig. 3.25b-1 and b-2. The Fourier spectrum of the fluid state shows a halo, while the crystalline state shows a six-fold symmetric spot. Figure 3.25c shows the radial distribution function $g(r)$. The red and blue curves indicate the fluid and crystalline states, respectively. The radial distribution function $g(r)$ indicates the density probability that other particles exist at a distance r from the center of a particle. $g(r)$ is expressed by (3.21),

$$g(r) = \frac{n(r)}{2\pi r \rho_{\text{area}} \Delta r}, \tag{3.21}$$

where $n(r)$ is the number of particles present in a circle from the center of the particle to a distance r ~ $(r + \Delta r)$ and ρ_{area} is the average particle density of the system (number of particles per unit area)(Fig. 3.26). For fluid, $g(r)$ does not show a clear peak, but for the crystalline state, sharp peaks are seen because the particles are aligned at regular intervals.

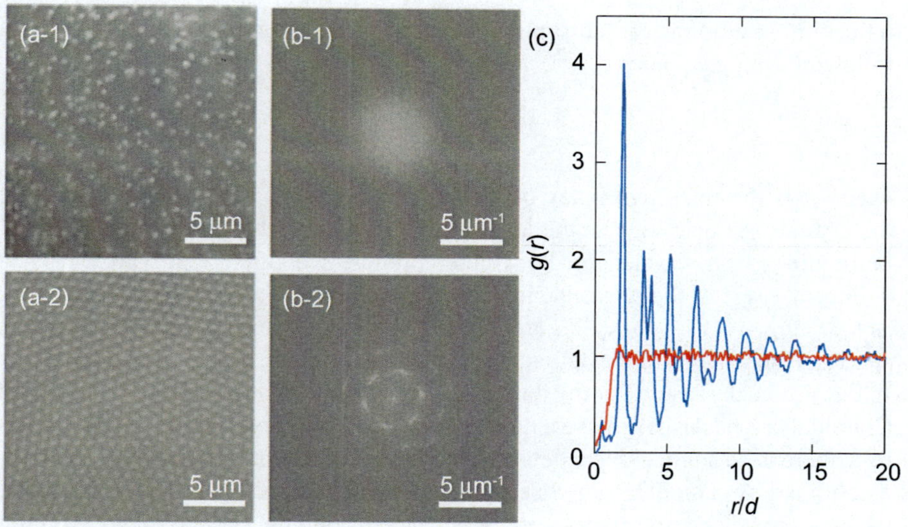

Fig. 3.25 Microscopic images of colloidal samples (polystyrene particles, $d = 430$ nm, $\phi = 0.03$). Microscopic images of **a**-1 the liquid state at [NaCl] = 100 μM, **a**-2 the colloidal crystalline state at [NaCl] = 2 μM. **b**-1, **b**-2 Fourier spectra of (**a**-1) and (**a**-2), respectively. **c** Radial distribution functions $g(r)$ for (**a**-1) and (**a**-2), shown in red and blue, respectively

Fig. 3.26 Illustration of $n(r)$

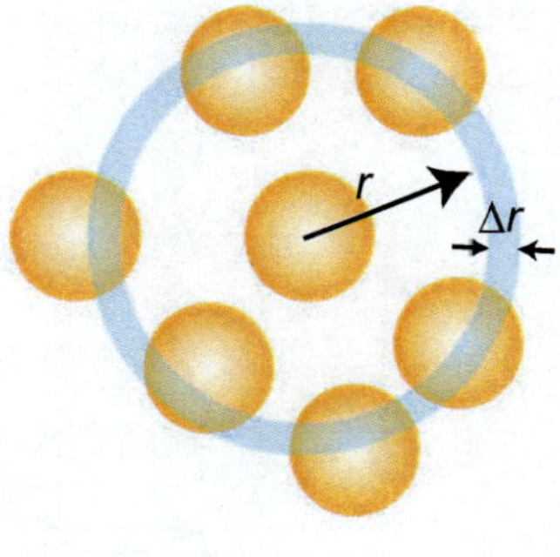

3.4.2 Spectroscopy

3.4.2.1 Reflectance and Transmission Spectroscopy

Colloidal crystals show structural colors when their Bragg diffraction wavelengths are in the visible light region. The diffraction wavelengths can then be determined by reflectance or transmission spectroscopy. White light, composed of a wide range of wavelengths, is irradiated onto the sample, and the intensity of the reflected or transmitted light at each wavelength is measured. These profiles plotted against wavelength are the reflectance and transmission spectra, respectively. Because light is reflected at diffraction wavelengths, it is observed as a peak in reflectance spectrum measurements and a dip in transmission spectra (transmittance is reduced by reflection, scattering, and absorption). The diffraction conditions are given by the following Bragg equation

$$2nd_{hkl}\sin\theta = \mathrm{m}\lambda, \tag{3.22}$$

where n is the refractive index of the sample, d_{hkl} is the (hkl) lattice spacing of the crystal (see below), θ is the angle of the incident light, m is an integer, and λ is the diffraction wavelength. For dilute colloids, n is often approximated by the volume average of the refractive indices of particles, $n_{\mathrm{r,p}}$, and medium (water), $n_{\mathrm{r,w}}$, as $n = \phi n_{\mathrm{r,p}} + (1 - \phi)n_{\mathrm{r,w}}$. Figure 3.27a, b shows the transmission and reflectance spectra, respectively, of the colloidal crystal (silica colloid, $d = 110$ nm, $\phi = 0.035$). In Fig. 3.27a, the results for the fluid state are shown as dashed lines for comparison. Colloidal crystals usually have a randomly oriented polycrystalline structure. Therefore, there is no angular dependence concerning the diffraction wavelength, which is determined by ϕ and the sample's refractive index. Figure 3.18 shows examples of the appearance and reflectance spectra of charged colloidal crystals at three different concentrations. Thus, the diffraction wavelength varies depending on the ϕ value.

Charged colloidal crystals are known to have a face-centered-cubic (FCC) or body-centered-cubic (BCC) lattice structure (Fig. 3.28a, b). The crystal structure and lattice parameters can be determined from the spectra. Crystals have various lattice planes, represented by a set of three numbers (hkl) called Miller indices. Figure 3.28c shows examples. The spacing d_{hkl} between each lattice planes is calculated using the lattice constant a, as follows.

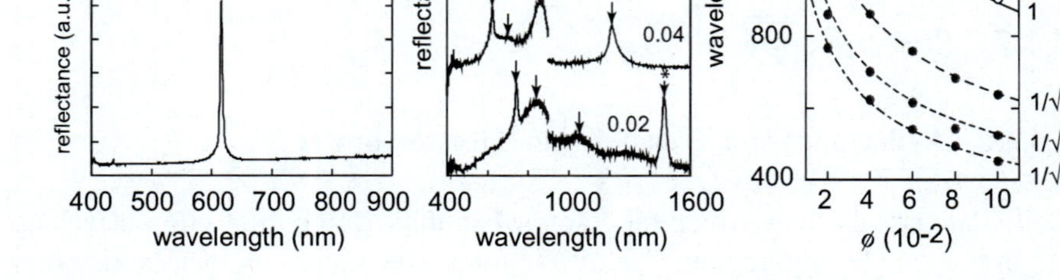

Fig. 3.27 **a** Transmission spectrum (dotted line indicates the liquid state) and **b** reflection spectrum of a silica colloidal crystal ($d = 110$ nm, $\phi = 0.035$). **c** Reflection spectra of colloidal crystals of various values of ϕ (silica particle, $d = 216$ nm, $Z = 733$, salt-free). Arrows indicate peak positions, and asterisks indicate the first peaks. **d** ϕ dependence on the peak wavelengths of the spectra in (**c**) shown as symbols. White symbols are the first peaks. Curves are calculated assuming a BCC structure. Reprinted with permission from Ref. [48]. Copyright (2012) Elsevier

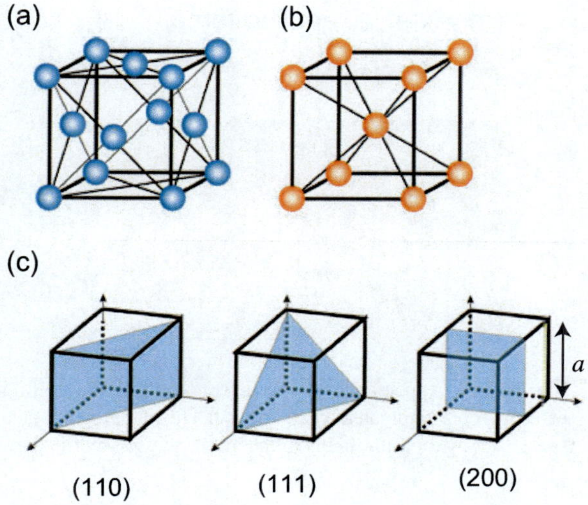

Fig. 3.28 Schematic diagrams of **a** face-centered-cubic lattice, **b** body-centered-cubic lattice, and **c** Miller indices of crystal planes

$$d_{hkl} = \frac{a}{\sqrt{h^2 + k^2 + l^2}}. \tag{3.23}$$

In BCC and FCC crystals, it is known that light is reflected in the following specific lattice planes.

BCC: (110), (200), (211) ……

FCC: (111), (200), (220) ……

The diffraction wavelengths λ corresponding to these lattice plane spacings are in Eq. (3.22). When these λs are normalized by the first peak wavelengths, the ratio of the λ corresponding to the above specific lattice spacing is as follows.

$$\text{BCC} : \frac{d_{hkl}}{d_{110}} = 1, \frac{1}{\sqrt{2}}, \frac{1}{\sqrt{3}} \cdots$$

$$\text{FCC} : \frac{d_{hkl}}{d_{111}} = 1, \frac{\sqrt{3}}{2}, \frac{\sqrt{6}}{4}.$$

Figure 3.27c shows an example of the reflectance spectra of colloidal crystals at various values ϕ, with their λ indicated by arrows [47]. The relations of λ and ϕ in the spectra are shown in Fig. 3.27d. In this case, the ratio of λ indicates that this spectrum is due to the BCC structure. For isotropic diffraction up to the fourth order, the experimental λ values indicated by the symbols are in good agreement with the theoretical values indicated by the solid and dashed lines

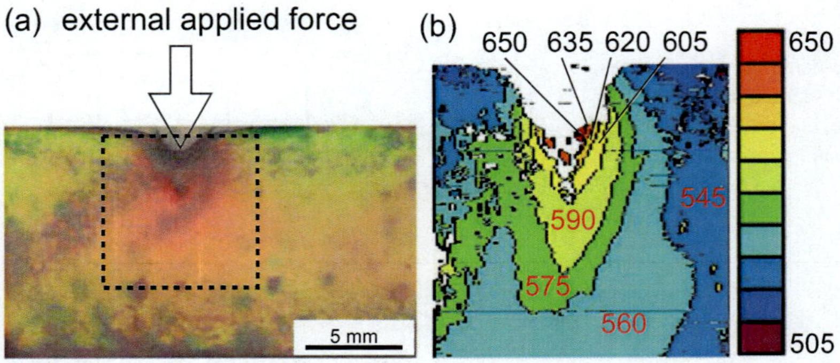

Fig. 3.29 **a** Appearance of the deformation of the colloidal crystal immobilized in gel due to external forces indicated by an arrow. **b** The mapping image of the peak wavelengths by 2D imaging spectroscopy measurements in the area enclosed by the dotted line in (**a**)

3.4.2.2 2D Imaging Spectroscopy

2D imaging spectroscopy is a device that uses a special spectrometer that allows spectra on a line to be evaluated at once. By scanning the sample, a two-dimensional distribution of transmitted or reflected light intensity is acquired. By processing the data, mapping at specific wavelengths can be carried out, by which the uniformity of the sample can be visualized [48]. For example, under mechanically compressed, the deformation of a colloidal crystal immobilized in a flexible polymer gel can be visualized as a shift of the Bragg peak/dip wavelengths. Figure 3.29a shows a gel-fixed colloidal crystal with mechanical deformation applied to the arrowed area. 2D mapping of the Bragg diffraction wavelengths in the region framed in a black dotted line is shown in Fig. 3.29b. The figures in the mapping image represent Bragg diffraction wavelength (in nm). It shows how the wavelength gradually changes from the point where the deformation is applied. The changes in diffraction wavelength correspond to the gel deformation, and the mapping image represents the stress distribution.

3.4.3 Kikuchi–Kossel Diffraction

When the light from various directions is incident on the sample simultaneously, the diffraction of the crystals can be observed as a 2D pattern, a Kikuchi–Kossel pattern. Kikuchi, working on electron diffraction, observed the diffraction patterns in thin single crystal mica [35, 49, 50]. Meanwhile, Kossel et al. observed the excitation of electron beams through the copper crystal lattice and the pattern obtained by diffraction of the resulting X-rays by a single crystal of copper [51]. Both studies utilize diffraction by atomic crystal lattice planes, and the principles can be applied to the structural analysis of colloidal crystals as follows.

A laser beam is irradiated onto the crystal sample from a direction perpendicular to the horizontal plane of the sample cell (Fig. 3.30a). A light diffuser on the cell causes the incident light to become a conical divergent light and diffuse into the colloidal sample having a lattice plane of the Miller index (*hkl*). The light incident at angles satisfying the Bragg condition is reflected, while light components incident at other angles pass through the lattice plane (Fig. 3.30a). A screen is placed behind the crystal sample. Since the laser beam does not reach the area where the cone with the half apex angle $\alpha_{hkl} = \pi/2 - \theta$ at the Bragg angle θ intersects the screen, this area can be observed as a dark curve. The wavelength of the laser used is often in the visible region. For example, He–Ne laser (wavelength 543.5 nm, 632.8 nm) or Ar laser (wavelength 488.0 nm) is often used due to the lattice plane spacing of the colloidal crystal [21].

Diffraction from various lattice planes of the crystals produces a pattern of dark curves called the Kikuchi–Kossel diffraction image. The pattern depends on the

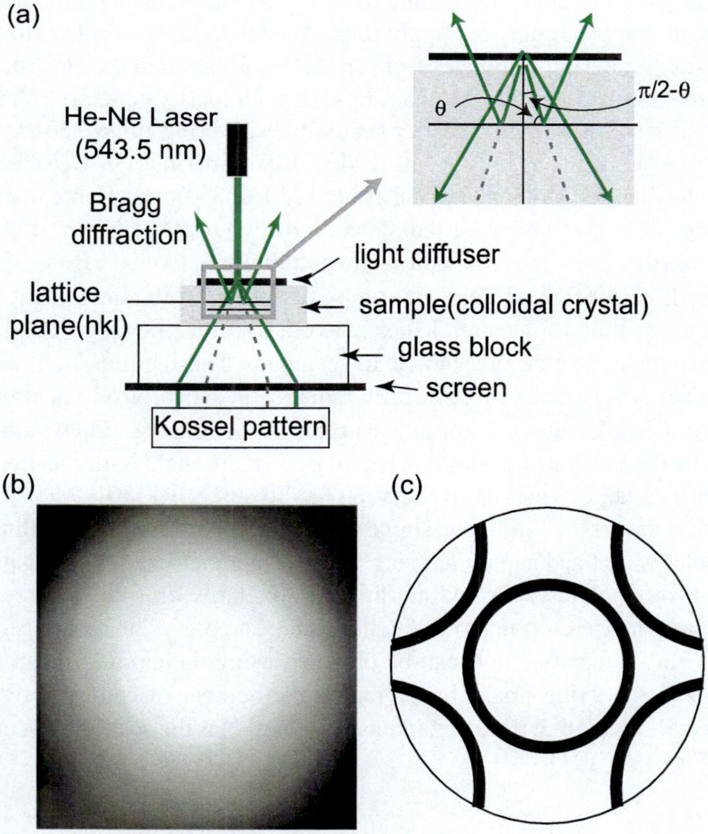

Fig. 3.30 **a** Schematic diagram of Kossel line measurement; **b** example of Kossel line of colloidal silica crystal ($d = 110$ nm, $\phi = 0.035$, BCC structure), He–Ne laser (543.5 nm); **c** Kossel pattern calculated for the crystal shown in (**b**)

lattice spacing d_{hkl} and the wavelength of the irradiating laser, and it also reflects the symmetry of the crystal. In other words, it provides information on the 3D structure within the colloidal crystal sample. When the colloidal sample is polycrystal, a concentric pattern called Debye rings [52, 53] is obtained. In this case, the d can be determined from the Debye ring radius.

Figure 3.30b, c shows an example of the Kossel pattern of a wall-oriented silica colloid single crystal ($d = 110$ nm, ϕ ~ 0.035, BCC) irradiated by a He–Ne laser at 543.5 nm and the pattern obtained from a theoretical calculation, respectively. The inner circle is the (110) plane, and the four outer curves show the Kossel pattern from the {110} family.

3.4.4 Scattering Experiments (USAXS)

Light scattering is a standard method for studying structures with micron to submicron length scales but may not apply to samples with high turbidity (low transmission). X-ray scattering originates from spatial modulation in the electron density of the sample and is, therefore, relevant to such colloidal systems. In relatively large micron-sized colloids, X-rays scatter at minimal scattering angles (2θ–0.01°), necessitating measurements by ultra-small-angle X-ray scattering (USAXS). For measurements at low angles, a unique optical system called the Bonse-Hart camera [54] was developed [55]. The optical system consists of two triple-reflecting channel-cut Si crystals and functions as a monochromator and analyzer. Even if a Bonse-Hart camera is not used, USAXS measurements can be performed if the instrument has a sufficiently long sample-to-detector distance. In other words, the longer the distance, the more accurate the scattering pattern at lower angles. This requires intense beams, and measurements are made using powerful radiation at synchrotron radiation facilities.

Figure 3.31a–c shows scattering patterns of colloids (samples: silica colloid, particle size = 105 nm, $\phi = 0.03$) obtained by a small-angle X-ray scattering system at SPring-8, a large synchrotron radiation facility at JASRI [56].

The data are for the fluid state, single crystal, and (poly)microcrystalline structure. All samples were fixed in polymer gels. Six-fold symmetrical scattering images were observed for single crystals and multiple ring patterns for (poly)microcrystals. No special patterns were observed in the fluid state samples. The relationship between the scattering intensity and θ can be obtained using the circular mean of the one-dimensional scattering image by averaging the number of counts observed on the detector and calculating the mean intensity. q (nm^{-1}) is the scattering vector, defined below (Fig. 3.32) [21].

$$q = \frac{4\pi}{\lambda}\sin\theta, \tag{3.24}$$

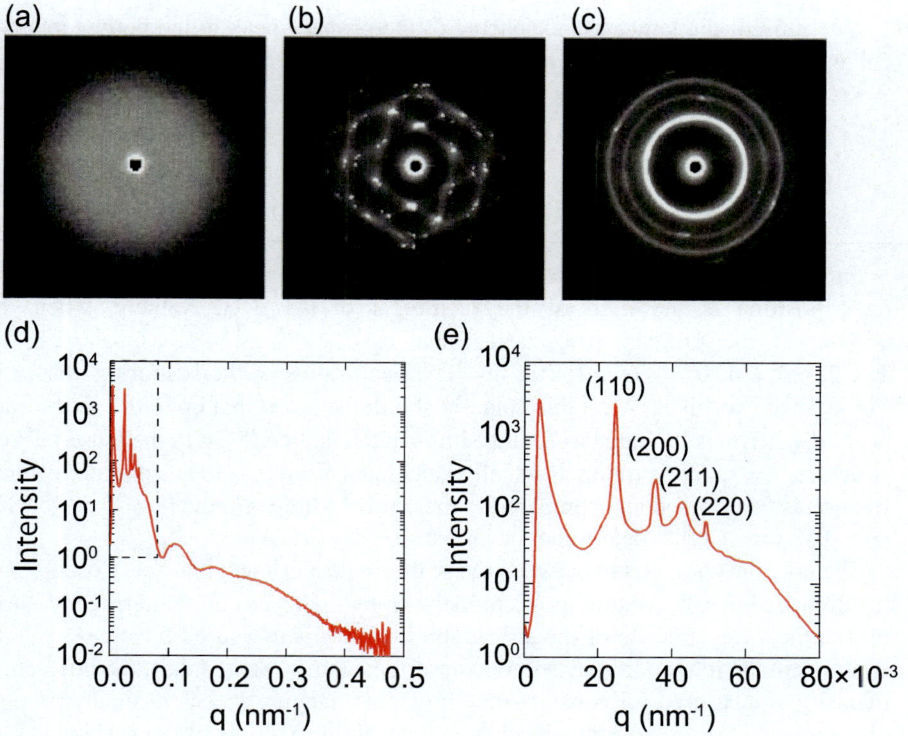

Fig. 3.31 Example SAXS patterns of silica colloids ($d = 120$ nm, $\phi = 0.035$, [Py] = 45 μM) **a** in liquid state, **b** single crystal sample, **c** polycrystalline sample, **d** SAXS profile of the sample in (**b**), and **e** enlargement within the dotted line in (**d**)

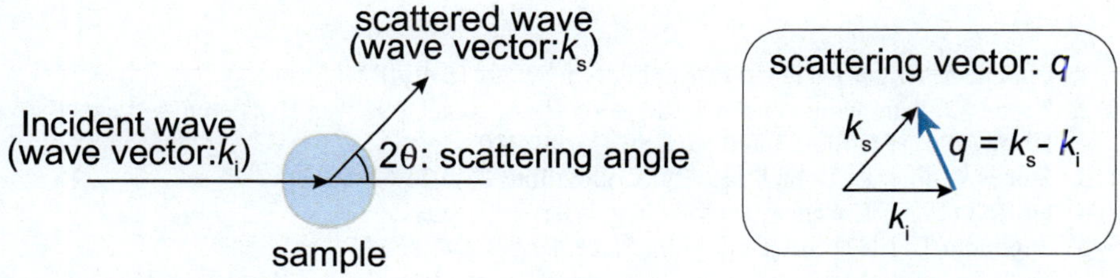

Fig. 3.32 Schematic diagram of small-angle X-ray scattering (SAXS) measurements and scattering vectors

where λ is the wavelength of the incident light in a vacuum. The scattering curve obtained from Fig. 3.31b is shown in Fig. 3.31d, and a magnified view of the area enclosed by the dotted line is shown in Fig. 3.31e. Generally, the more complete the periodicity of the crystal, the sharper the observed diffraction peaks. As the long-range periodic structure disturbance increases, it disappears from the higher-order (wide-angle) peaks. Thus, the more perfect the crystal is, the more ordered structure is kept up for long distances, and diffraction peaks are observed up to higher orders.

The ratio of the lattice plane spacing to the primary peak in the body-centered-cubic lattice (BCC) (d_{hkl}/d_{110}) is expressed as [57].

$$(hkl) = (110), (200), (211), (220), (310), (222), (321), (400), \cdots$$

$$d_{hkl}/d_{110} = 1, 1/\sqrt{2}, 1/\sqrt{3}, 1/\sqrt{4}, 1/\sqrt{5}, 1/\sqrt{6}, 1/\sqrt{7}, 1/\sqrt{8}, \cdots$$

In the scattering curve, the abscissa is of the order of nm^{-1}, so the peak position is the ratio of the reciprocal of d_{hkl}/d_{110}. Namely, when the q-value of the primary peak is 1, then $1, \sqrt{2}, \sqrt{3}, \sqrt{4}, \sqrt{5}, \sqrt{6}, \sqrt{7}, \sqrt{8}, \cdots$ $1, \sqrt{2}, \sqrt{3}, \sqrt{4}, \sqrt{5}, \sqrt{6}, \sqrt{7}, \sqrt{8}, \cdots$. The peak ratios of the scattering curves in Fig. 3.31 are consistent with this value. It should be noted that up to the sixth-order peak, d_{hkl} / d_{110} is the same as for the simple cubic lattice (SC). In the close-packed structure, the sphere's radius is usually calculated from d_{110}. The sphere's volume fraction is determined and compared to the critical filling volume ($\phi = 0.52$ for SC, $\phi = 0.68$ for BCC) to determine the structure [21].

We note the comparison between the scattering experiment and spectroscopy. As mentioned above, the scattering vector is given by Eq. (3.24). In scattering measurements, the θ-dependence of the diffraction intensity is measured for constant λ. In contrast, in the reflection spectroscopy described above (Sect. 3.4.2), the diffraction intensity is measured for constant $\theta(= \pi/2)$ on changing λ. λ is inversely proportional to q, so the ratio of the diffraction wavelengths in the spectrum is shown as an inverse number.

References

1. Chonde Y, Kriger IM (1981) J Appl Polym Sci 26:1819–1827
2. Bosma G, Pathmamanoharan C, de Hoog EHA, Kegel WK, van Blaaderen A, Lekkerkerker HNW (2002) J Colloid Interface Sci 245:292–300
3. Stöber W, Fink A, Bohn E (1968) J Colloid Interf Sci 26:62–69
4. ler RKI (1979) The chemistry of silica. Wiley-Interscience
5. Sugimoto T (1987) Adv Colloid Interface Sci 28:65–108
6. Fitch RM (1997) Polymer colloids: a comprehensive introduction. Academic Press
7. Juang MS-D, Krieger IM (1976) J Polym Sci Polym Chem Ed 14:2089–2107
8. Kim JH, Chainey M, El-Aasser MS, Vanderhoff JW (1992) J Polym Sci, Part A Polym Chem 30:171–83
9. Johnson CS, Gabriel DA (1981) Laser light scattering. Dover Publications Inc., New York
10. Pusey PN, Tough RJA (1985) Dynamic light scattering. In: Pecora R (ed). Plenum Press, New York
11. Hassan PA, Rana S, Verma G (2015) Langmuir 31:3–12
12. Miller CC (1924) Proc R Soc Lond A 106:724–749
13. Sood AK (1991) Solid State Phys 45:1–73
14. Yamanaka J, Hayashi Y, Ise N, Yamaguchi T (1997) Phys Rev E 55:3028
15. Hessinger D, Evers M, Palberg T (2000) Phys Rev E 61:5493
16. Hunter RJ (1981) Zeta potential in colloid science, principles and applications. Academic Press

17. Russel WB, Saville DA, Schowalter WR (1989) Colloidal dispersions. Cambridge University Press
18. Palberg T, Würth M (1996) J. Phys. I France 6:237–244
19. Manning GS (1969) J Chem Phys 51:924–933
20. Israelachvili JN (2011) Intermolecular and surface forces. Academic Press
21. Ise N, Sogami IS (2005) Structure formation in solution, ionic polymers and colloidal particles. Springer-Verlag, Berlin Heidelberg
22. Yoshida H, Yamanaka J, Koga T, Koga T, Ise N, Hashimoto T (1999) Langmuir 15:2684–2702
23. Yamanaka J, Yoshida H, Koga T, Ise N, Hashimoto T (1999) Langmuir 15:4198–4202
24. Yamanaka J, Hibi S, Ikeda S, Yonese M (2004) Mol Simul 30:149–152
25. Ohshima H (2006) Theory of colloid and interfacial electric phenomena. Elsevier
26. Ohshima H (2010) Biophysical chemistry of biointerfaces. Wiley
27. Mori S, Okamoto H (1980) Fusen 27:117–126 (in Japanese)
28. Henry DC (1931) Proc. R. Soc. A 133:106–129
29. Ohshima H (2001) J Colloid Interface Sci 239:587–590
30. O'Brien RW, White LR (1978) J Chem Soc Faraday Trans 2 74:1607–1626
31. Ohshima H, Healy TW, White LR (1980) J Colloid Interface Sci 90:17–26
32. Alfrey T, Bradford EB, Vanderhoff JW, Oster G (1954) J Opt Soc Am 44:603–609
33. Hiltner PA, Krieger IM (1969) J Phys Chem 73:2386–2389
34. Colvin VL (2001) MRS Bull 26:637–641
35. Pieranski P (1983) Contemp Phys 24:25–73
36. Yamanaka J, Murai M, Iwayama Y, Yonese M, Ito K, Sawada T (2004) J Am Chem Soc 126:7156–7157
37. Wette P, Klassen I, Holland-Moritz D, Herlach DM, Schöpe HJ, Lorenz N, Reiber H, Palberg T, Roth SV (2010) J Chem Phys 132:131102
38. Toyotama A, Yamamoto M, Nakamura Y, Yamazaki C, Tobinaga A, Ohashi Y, Okuzono T, Ozaki H, Uchida F, Yamanaka J (2014) Chem Mater 26:4057–4059
39. Herlach DM, Palberg T, Klassen I, Klein S, Kobold R (2016) J Chem Phys 145:211703
40. Nagano R, Toyotama A, Yamanaka J (2011) Chem Lett 40:1366–1367
41. Yamanaka J, Yoshida H, Koga T, Ise N, Hashimoto T (1998) Phys Rev Lett 80:5806
42. Asakura S, Osawa F (1954) J Chem Phys 22:1255–1256
43. Lekkerkerker HN, Tuinier R (2011) Colloids and the depletion interaction. Springer, Netherland
44. Kose A, Hachisu S (1976) J Colloid Interface Sci 55:487–498
45. Dinsmore AD, Weeks ER, Prasad V, Levitt AC, Weitz DA (2001) Appl Opt 40:4152–4159
46. Wilson T (2003) Confocal microscopy. In: Biomedical photonics handbook. CRC Press
47. Murai M, Okuzono T, Yamamoto M, Toyotama A, Yamanaka J (2012) J Colloid Interface Sci 370:39–45
48. Kanai T, Sawada T, Toyotama A, Kitamura K (2005) Adv Funct Mater 15:25–29
49. Kikuchi S (1928) Jpn J Phys 5:83–96
50. Peng L-M, Dudarev SL, Whelan MJ (2004) High-energy electron diffraction and microscopy. Oxford University Press
51. Tixier R, Waché C (1970) J Appl Cryst 3:466–485
52. Zhang J-T, Chao X, Liu X, Asher SA (2013) Chem Commun 49:6337–6339
53. Smith NL, Coukouma A, Dubnik S, Asher SA (2017) Phys Chem Chem Phys 19:31813
54. Snyder RL (1999) X-Ray diffraction in "X-ray characterization of materials". Wiley-VCH
55. Koga T, Hart M, Hashimoto T (1996) J Appl Cryst 29:318–324
56. Sugao Y, Onda S, Toyotama A, Takiguchi Y, Sawada T, Hara S, Nishikawa S, Yamanaka J (2016) Jpn J Appl Phys 55:087301
57. Winey KI, Thomas EL, Fetters LJ (1991) J Chem Phys 95:9367–9375

Chapter 4
Numerical Simulation Methods

Abstract In this chapter, we explain a basic concept of numerical modeling and describe useful algorithms and techniques to analyze numerical data obtained by computer simulations of colloidal systems. Before describing practical simulation methods for colloidal systems in detail, we briefly introduce the method of molecular dynamics including some techniques which are also useful for simulations of colloidal systems. Because the size of colloidal particles is so large compared to the solvent molecules surrounding them, the degrees of freedom of rapid molecular motions should be eliminated. Instead, a stochastic description of the colloidal particles' motions is needed. As such method, we illustrate the Brownian dynamics and Monte Carlo methods in a comprehensible manner. In particular, a few examples of numerical simulations for colloidal systems are presented with pseudocodes of the critical algorithms and some pictures/movies obtained by the simulations.

Keywords Numerical simulation · Numerical data analysis · Brownian dynamics · Monte Carlo method

Note: Throughout this chapter, we use a lot of mathematical symbols to represent *physical quantities* that include units of the quantities. For example, when the volume of a material is measured and found to be 1.23 L, the volume of this material is $V = 1.23\text{L}$. Using another unit m^3, for example, we can express the volume as $V = 1.23 \times 10^{-3}\text{m}^3$, so we have many expressions: $V = 1.23\,L = 1230\,\text{mL} = 1.23 \times 10^{-3}\text{m}^3 = \ldots$. These expressions indicate the same physical quantity, although the factors in front of the units are different depending on those units. Therefore, we can perform mathematical operations without worrying about the units of physical quantities. However, we need to be aware of the *physical dimensions* of physical quantities. The basic physical dimensions (in mechanics) are [length], [mass], and [time]. The dimensions of physical quantities are composed of these basic physical dimensions. For example, $[\text{volume}] = \left[\text{length}\right]^3$, $[\text{force}] = [\text{mass}]\left[\text{length}\right]/[\text{time}]^2$, $\left[\text{pressure}\right] = [\text{force}]/\left[\text{length}\right]^2 = [\text{mass}]/\left(\left[\text{length}\right][\text{time}]^2\right)$, $\left[\text{energy}\right] = [\text{mass}]\left(\left[\text{length}\right]/[\text{time}]\right)^2\left(= \left[\text{pressure}\right][\text{volume}]\right)$. It should be noted that addition

© The Author(s), under exclusive license to Springer Nature Singapore Pte Ltd. 2023

J. Yamanaka et al., *Colloidal Self-Assembly*, Lecture Notes in Chemistry 108,
https://doi.org/10.1007/978-981-99-5052-2_4

is only possible between physical quantities that have the same physical dimension. Pressure and volume cannot be added together, for example.

4.1 Molecular Simulation: An Example

Nowadays, computer simulation is an indispensable tool to connect theories and experiments in the research of colloidal sciences. There are many software packages for computer simulations, and we can easily find and get them from the Internet. Usually, we have our own experimental data to analyze or interesting phenomena that we want to understand. However, the computers themselves could not immediately solve these desired matters. We need a model that provides the method of data analysis or describes the phenomenon. The model should be composed of mathematical expressions or equations derived from the physical principles.

For example, considering a simple molecular system of volume V which contains N molecules, position $\boldsymbol{r}_i$, and velocity $\boldsymbol{v}_i$ of each molecule $i = 1, 2, \ldots, N$ change in time t according to the classical mechanics, that is the Newton's law, as

$$\frac{\mathrm{d}\boldsymbol{r}_i}{\mathrm{d}t} = \boldsymbol{v}_i, \quad m_i \frac{\mathrm{d}\boldsymbol{v}_i}{\mathrm{d}t} = \boldsymbol{F}_i, \tag{4.1}$$

where m_i is the mass of molecule i and $\boldsymbol{F}_i$ is the force acting on molecule$i$. We do not consider the rotational and internal degrees of freedom of each molecule, so the molecule can be regarded as a spherical particle, and we call it a "particle" in this chapter. Note that $\boldsymbol{r}_i, \boldsymbol{v}_i, \boldsymbol{F}_i$ are vector quantities and have their components $\boldsymbol{r}_i = (x_i, y_i, z_i)$, $\boldsymbol{v}_i = \left(v_i^x, v_i^y, v_i^z\right)$, $\boldsymbol{F}_i = \left(F_i^x, F_i^y, F_i^z\right)$ in a Cartesian coordinate system in three dimensions, for example. In (4.1), the force $\boldsymbol{F}_i$ is, in many cases, assumed to be given by the sum of pair interaction force $\boldsymbol{F}_{ij}$,

$$\boldsymbol{F}_i = \sum_{j \neq i} \boldsymbol{F}_{ij} = -\frac{\partial}{\partial \boldsymbol{r}_i} \sum_{j \neq i} u\left(\left|\boldsymbol{r}_i - \boldsymbol{r}_j\right|\right), \tag{4.2}$$

where $u\left(\left|\boldsymbol{r}_i - \boldsymbol{r}_j\right|\right)$ is the potential energy between i and j particles and Newton's third law $\boldsymbol{F}_{ij} = -\boldsymbol{F}_{ji}$ is satisfied. The symbol $\frac{\partial}{\partial \boldsymbol{r}_i}$ in (4.2) is the vector operator of partial differentiation that means $\frac{\partial u}{\partial \boldsymbol{r}_i} = \left(\frac{\partial u}{\partial x_i}, \frac{\partial u}{\partial y_i}, \frac{\partial u}{\partial z_i}\right)$. When an external field, such as gravity, is applied, $\boldsymbol{F}_i$ should be replaced by $\boldsymbol{F}_i + \boldsymbol{g}_i$, where $\boldsymbol{g}_i = m_i \boldsymbol{g}$ in the case of gravity ($\boldsymbol{g}$ is the gravitational acceleration).

For the above model to be closed, the pair potential $u(r)$ in (4.2) as a function of interparticle distance r should be given. For charge-neutral molecules (such as argon), the Lennard–Jones potential

$$u_{\mathrm{LJ}}(r) = 4\epsilon\left[\left(\frac{\sigma}{r}\right)^{12} - \left(\frac{\sigma}{r}\right)^{6}\right] \tag{4.3}$$

is often used. Here, σ is the diameter of the particle, and ε is the depth of the potential well. The first and second terms on the right-hand side of (4.3) correspond to the core repulsion and the van der Waals attraction, respectively. The set of (4.1)–(4.3) forms a model for the dynamics of the molecular system. In this model, the number of particles N, the total energy (kinetic and potential energies of molecules), and the volume of the system V are constant. The ensemble of such kinds of systems is called a microcanonical ensemble in statistical mechanics. There are other ensembles for more practical situations. However, here we deal with the microcanonical system since it is the most primitive one. See [1–3] for the details.

Now we show how to carry out numerical simulations based on the above model. Before describing numerical simulation methods for the above model system, we rewrite (4.1)–(4.3) by using dimensionless variables, making mathematical and numerical treatment easier and our physical insight clearer. Hereafter, we consider the case that $m_i = m$ for $i = 1, 2, \ldots, N$, for simplicity. The physical quantities $\boldsymbol{r}_i$, $\boldsymbol{v}_i$, m, and u have the physical dimensions: length, velocity, mass, and energy, respectively. If we choose σ, m, and ε as characteristic quantities (or basic units) of length, mass, and energy, respectively, we can define the dimensionless variables

$$\hat{\boldsymbol{r}}_i \equiv \boldsymbol{r}_i/\sigma, \quad \hat{\boldsymbol{v}}_i \equiv \boldsymbol{v}_i/(\sigma/\tau), \quad \hat{u} \equiv u/\epsilon, \tag{4.4}$$

for $\boldsymbol{r}_i$, $\boldsymbol{v}_i$, u, respectively. Here, we have introduced $\tau \equiv \sigma\sqrt{m/\epsilon}$ as a time unit, because $\varepsilon = mv^2$ and $v = \sigma/\tau$. Note that the choice of the basic units is not unique: We may choose the mean interparticle distance $(V/N)^{\frac{1}{3}}$ as a unit of length, for example. Using the dimensionless variables given by (4.4), the equations of motion (4.1) are expressed, in the case of $u = u_{\mathrm{LJ}}$, as

$$\frac{\mathrm{d}\hat{\boldsymbol{r}}_i}{\mathrm{d}\hat{t}} = \hat{\boldsymbol{v}}_i, \quad \frac{\mathrm{d}\hat{\boldsymbol{v}}_i}{\mathrm{d}\hat{t}} = -\frac{\partial}{\partial \hat{\boldsymbol{r}}_i}\sum_{j\neq i}\hat{u}_{ij}, \tag{4.5}$$

with

$$\hat{u}_{ij} = 4\left[\left(\frac{1}{\hat{r}_{ij}}\right)^{12} - \left(\frac{1}{\hat{r}_{ij}}\right)^{6}\right], \tag{4.6}$$

where $\hat{t} \equiv t/\tau$ and $\hat{r}_{ij} \equiv \left|\hat{\boldsymbol{r}}_i - \hat{\boldsymbol{r}}_j\right|$. In (4.5) and (4.6), no parameters appear. There are a lot of algorithms for numerical integration of (4.5). We briefly describe a basic algorithm by Verlet (see [1–3]).

To solve (4.5) and (4.6) numerically, we use discretized time t_n instead of continuous time t, that is,

$$t \to t_n = n\Delta t, \quad n = 0, 1, 2, \ldots \tag{4.7}$$

where Δt is a time step; hereafter, we omit "hat" attached to the dimensionless variables. Expanding $\boldsymbol{r}_i(t_n \pm \Delta t)$ in a Taylor series as

$$\boldsymbol{r}_i(t_n \pm \Delta t) = \boldsymbol{r}_i(t_n) \pm \tfrac{\mathrm{d}\boldsymbol{r}_i}{\mathrm{d}t}(t_n)\Delta t + \tfrac{1}{2}\tfrac{\mathrm{d}^2\boldsymbol{r}_i}{\mathrm{d}t^2}(t_n)\Delta t^2 \pm \ldots, \tag{4.8}$$

or, using (4.5) we have

$$\boldsymbol{r}_i(t_n + \Delta t) = \boldsymbol{r}_i(t_n) + \boldsymbol{v}_i(t_n)\Delta t + \tfrac{1}{2}\boldsymbol{F}_i(t_n)\Delta t^2 + \ldots, \tag{4.9}$$

$$\boldsymbol{r}_i(t_n - \Delta t) = \boldsymbol{r}_i(t_n) - \boldsymbol{v}_i(t_n)\Delta t + \tfrac{1}{2}\boldsymbol{F}_i(t_n)\Delta t^2 - \ldots \tag{4.10}$$

From (4.9) and (4.10), we obtain

$$\boldsymbol{r}_i(t_n + \Delta t) + \boldsymbol{r}_i(t_n - \Delta t) = 2\boldsymbol{r}_i(t_n) + \boldsymbol{F}_i(t_n)\Delta t^2 + O(\Delta t^4), \tag{4.11}$$

$$\boldsymbol{r}_i(t_n + \Delta t) - \boldsymbol{r}_i(t_n - \Delta t) = 2\boldsymbol{v}_i(t_n)\Delta t + O(\Delta t^3), \tag{4.12}$$

where $O(\Delta t^4)$ represents the higher-order terms that means $|O(\Delta t^4)/\Delta t^4| < \infty$ as $\Delta t \to 0$, or, roughly speaking, $O(\Delta t^4) = \boldsymbol{C}_4(t_n)\Delta t^4 + \boldsymbol{C}_6(t_n)\Delta t^6 + L$ with finite coefficients $\boldsymbol{C}_4(t_n)$, $\boldsymbol{C}_6(t_n)$, L, and we can neglect these terms. Therefore, we can update the particle positions from (4.11) as

$$\boldsymbol{r}_i(t_n + \Delta t) = 2\boldsymbol{r}_i(t_n) - \boldsymbol{r}_i(t_n - \Delta t) + \boldsymbol{F}_i(t_n)\Delta t^2, \tag{4.13}$$

without knowledge of the velocity $\boldsymbol{v}_i$. The velocity can be calculated from (4.12) as

$$\boldsymbol{v}_i(t_n) = \tfrac{1}{2\Delta t}[\boldsymbol{r}_i(t_n + \Delta t) - \boldsymbol{r}_i(t_n - \Delta t)]. \tag{4.14}$$

For more advanced algorithms, see the standard textbooks [1–3].

Since one of the purposes of molecular simulations is to calculate and predict physical quantities of thermodynamic systems, we need to carry out simulations of very large systems that are expected to show correct behavior in thermodynamic limit, $N \to \infty$ and $V \to \infty$. However, we cannot treat infinitely large systems in computer simulations, and we should introduce boundary conditions for the finite system. To avoid the boundary effects, periodic boundary conditions are often used. Consider N particles whose positions are $\boldsymbol{r}_1, \boldsymbol{r}_2, \ldots, \boldsymbol{r}_N$, and they are confined in a rectangular box with dimensions L_x, L_y, and L_z in x, y, and z directions, respectively. Under the periodic boundary conditions, the box at the center that contains molecules $\boldsymbol{r}_i$ ($i = 1, 2, \ldots, N$) is surrounded by "copies" of the central box, which are periodically arranged (Fig. 4.1). The positions of the particles $\boldsymbol{r}_i^{(a,b,c)}$ in the box indexed by three integers a, b, c are given as

$$\boldsymbol{r}_i^{(a,b,c)} = \boldsymbol{r}_i + a\boldsymbol{L}_x + b\boldsymbol{L}_y + c\boldsymbol{L}_z, \tag{4.15}$$

where $\boldsymbol{L}_x = (L_x, 0, 0)$, $\boldsymbol{L}_y = (0, L_y, 0)$, $\boldsymbol{L}_z = (0, 0, L_z)$, and the positions in the central box are $\boldsymbol{r}_i^{(0,0,0)} = \boldsymbol{r}_i$. Using the periodic boundary conditions, we can carry

Fig. 4.1 Schematic picture of periodic boundary conditions in two dimensions. Actual calculations are carried out in the centered box. A circle of radius r_c shows the interaction range in which there are particles that interact with the particle at the center of the circle

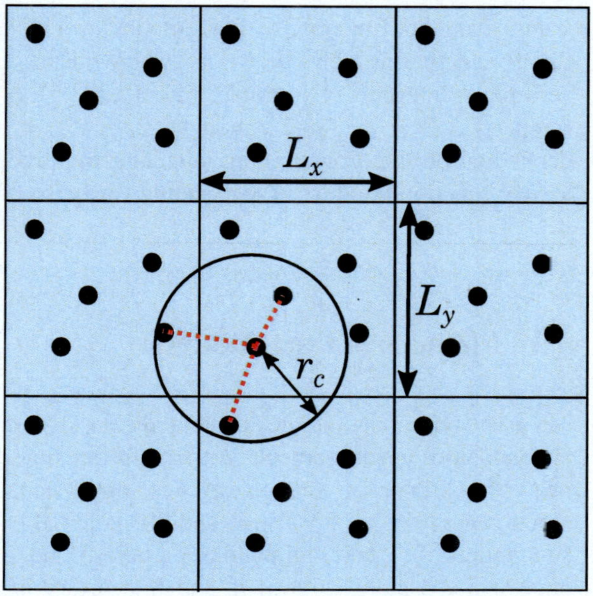

out the simulations of infinite systems, approximately, with information of molecules in the central box, that is, $(a, b, c) = (0, 0, 0)$, as shown below.

Under the periodic boundary conditions, we must calculate the force $\boldsymbol{F}_i$ acting on a particle i which is the sum of pair interaction forces, $\sum_j \boldsymbol{F}_{ij}$, where j is all particles (other than i) including those in image boxes (a, b, c). It is impossible to calculate the sum of an infinite number of interaction forces. However, it is possible if the interaction is short-ranged. In this case, we set a cutoff length r_c so that the particle i interacts with all particles j satisfying $r_{ij} < r_c$, and the total interaction energy $\sum_j u_{ij}$ does not increase anymore as r_c increases. In a typical molecular simulation, the cutoff length is determined as $r_c \approx 3\rho^{-1/3}$, where $\rho \equiv N/V$ is the number density. When we calculate $r_{\rm ij}$ for fixed $\boldsymbol{r}_i$ in the central box, there are at least 27 candidates (in a 3D system) for the position of particle j, that is, $\boldsymbol{r}_j^{(a,b,c)}$ for a, b, c each taking -1, 0, and $+1$. The distance r_{ij} between particles i and j is given by the minimum of $\left|\boldsymbol{r}_i - \boldsymbol{r}_j^{(a,b,c)}\right|$. It should be noted that the above simple method to calculate interactions between particles is inappropriate for systems of long-range interactions, such as Coulomb interactions. For those systems, we should use a particular method such as the Ewald method (see [1, 2]).

When a particle located at $(x_i(t), y_i(t), z(t))$ at time t moved to a new position $(x_i(t+\Delta t), y_i(t+\Delta t), z_i(t+\Delta t))$ after Δt, it must be checked whether the new position is inside or outside the centered box. If $x_i(t+\Delta t) > L_x$, for example, the new position should be $x_i(t+\Delta t) - L_x$, that is, $x_i(t+\Delta t)$ is replaced by $\mathrm{mod}(x_i(t+\Delta t), L_x)$. The components in other directions should be updated in the same manner. Note that the function $\mathrm{mod}(x, y)$ is usually defined for integers x and y and returns the remainder of x divided by y, whereas for floating-point numbers

x and y, another function, such as fmod(x, y) in C language, is used. The function fmod(x, y) returns the value $x - ny$, where n is the quotient of x/y, rounded toward zero to an integer. For example, fmod$(7.2, 3.5) = 0.2$, fmod$(7.2, -3.5) = 0.2$, fmod$(-7.2, 3.5) = -0.2$, fmod$(-7.2, -3.5) = -0.2$. These values may vary depending on the programming language and processing system used. When you use the function fmod(x, y), you should verify the definition and actual operation of it.

4.2 Methods of Data Analysis

The basic output data of the above molecular dynamics simulations are positions $\boldsymbol{r}_i$ and velocities $\boldsymbol{v}_i$ (or momenta $m\boldsymbol{v}_i$) at discrete times. The equations of motion (4.1) or (4.5) conserve the total energy E of the system. The energy E is a sum of the kinetic energy $K = \frac{1}{2}\sum_i^N m\boldsymbol{v}_i^2$ and the potential energy $U = \sum_{\langle ij\rangle} u_{ij}$, where the summation $\sum_{\langle ij\rangle}$ takes all pairs of particles i and j. The total energy $E = K + U$ is always constant, but each K and U can.vary in time and fluctuates around the respective average value in equilibrium. The statistical mechanics indicates that the ensemble average of a dynamical quantity is equivalent to the long-time average of that in equilibrium. For example, $\langle K\rangle = \langle K\rangle_t$, where $\langle\cdot\rangle$ and $\langle\cdot\rangle_t$ mean the ensemble average and the long-time average, respectively. The long-time average of $K(t)$ in equilibrium is easily calculated as

$$\langle K\rangle_t \equiv \lim_{\tau\to\infty} \frac{1}{\tau}\int_0^\tau K(t)dt \approx \frac{1}{M}\sum_{n=0}^{M-1} K(t_n), \tag{4.16}$$

where the time region $0 \le t \le \tau$ is sufficiently large so that the system can be regarded as in equilibrium and M is the sample number which should be large enough to consider the system to be in equilibrium.

The average of kinetic energy K is related to the temperature T of the system. According to the law of equipartition of energy, the average of kinetic energy per degree of freedom is equal to half of the thermal energy, that is,

$$\left\langle \tfrac{1}{2}m\left(v_i^\alpha\right)^2\right\rangle = \tfrac{1}{2}k_\mathrm{B}T, \tag{4.17}$$

where v_i^α is α-component of the velocity of molecule i and k_B is the Boltzmann constant. Since the present molecular system is invariant under any translational motion of the whole system with a constant velocity (Galilean invariance), we can always choose the velocities $\boldsymbol{v}_i$ such that $\sum_i \boldsymbol{v}_i = 0$. This constraint reduces the degrees of freedom of the velocities from $3N$ to $3(N - 1)$ in three dimensions. From (4.17), we find

$$\frac{1}{2}m\left\langle\sum_{i=1}^{N}|\boldsymbol{v}_i|^2\right\rangle = \frac{3(N-1)}{2}k_BT \approx \frac{3N}{2}k_BT. \tag{4.18}$$

This equation gives an expression of temperature that can be calculated from the data obtained. Note that using the dimensionless variables, the dimensionless temperature is given as $\hat{T} \equiv T/(\epsilon/k_{\mathrm{B}}) = \left\langle\sum_i \hat{\boldsymbol{v}}_i^2\right\rangle/[3(N-1)]$.

In order to obtain a microscopic expression of the pressure P, here we consider a "wall" around the system and use the virial theorem of classical mechanics [4]. Define

$$G(t) = \sum_i m_i\boldsymbol{v}_i \cdot \boldsymbol{r}_i, \tag{4.19}$$

and its time derivative is

$$\frac{\mathrm{d}G}{\mathrm{d}t} = \sum_i\left(m_i\frac{\mathrm{d}\boldsymbol{v}_i}{\mathrm{d}t}\cdot\boldsymbol{r}_i + m_i\boldsymbol{v}_i\cdot\boldsymbol{v}_i\right) = \sum_i(\boldsymbol{F}_i\cdot\boldsymbol{r}_i + m_i\boldsymbol{v}_i\cdot\boldsymbol{v}_i). \tag{4.20}$$

Here, we have used the equations of motion (4.1). Taking a long-time average of (4.20), we have

$$\lim_{\tau\to\infty}\frac{1}{\tau}[G(\tau) - G(0)] = \left\langle\sum_i \boldsymbol{F}_i\cdot\boldsymbol{r}_i\right\rangle_t + 2K_t. \tag{4.21}$$

The left-hand side of (4.21) vanishes, if $G(t)$ has an upper bound, which is a plausible hypothesis. Then we have the virial theorem,

$$\langle K\rangle_t = -\frac{1}{2}\left\langle\sum_i \boldsymbol{F}_i\cdot\boldsymbol{r}_i\right\rangle_t. \tag{4.22}$$

Now, we divide the force $\boldsymbol{F}_i$ exerted on a particle i into two contributions from interactions with other particles $\sum_j \boldsymbol{F}_{ij}$ and the external interaction with the walls $\boldsymbol{F}_i^{\mathrm{ext}}$. The latter interaction exerts only at the wall surface and yields a pressure $-P\boldsymbol{n}\mathrm{d}S$ on a surface element $\mathrm{d}S$ located at $\boldsymbol{r}$ toward the outward unit vector $\boldsymbol{n}$ after the long-time average. Therefore, we find

$$\left\langle\sum_i^N \boldsymbol{F}_i^{\mathrm{ext}}\cdot\boldsymbol{r}_i\right\rangle_t = -P\int \boldsymbol{r}\cdot\boldsymbol{n}\mathrm{d}S = -P\int \nabla\cdot\boldsymbol{r}\mathrm{d}V = -3PV. \tag{4.23}$$

Here, we have used the Gauss theorem and the identity $\nabla\cdot\boldsymbol{r} = 3$, in the second and last identity, respectively. Equation (4.22) now reads

$$PV = Nk_BT - \frac{1}{3}\left\langle\sum_i \boldsymbol{r}_i\cdot\frac{\partial}{\partial\boldsymbol{r}_i}\sum_j u(r_{ij})\right\rangle_t. \tag{4.24}$$

The last term in (4.24) shows the effect of interactions between particles. In the absence of this term, (4.24) becomes the equation of state of an ideal gas.

Next, we focus on the structures of liquids and crystals. Positions of particles in a crystal are regularly arrayed. They have rotational and translational symmetries: The positions of particles after applying particular operations of rotations and translations can be superimposed on those before. There are many crystal structures, such as body-centered-cubic (BCC) and face-centered-cubic (FCC) structures corresponding to the different symmetry operations. On the other hand, the particles in a liquid are moving and their positions are disordered as long as we observe data obtained with computer simulations. How can we characterize a liquid structure? Since a particle in a liquid has a finite diameter σ, and the distance between two centers of particles cannot be less than σ, particles around a particle will form a shell-like structure on average. This intuitive observation leads to a concept of radial distribution function (also called pair correlation function) $g(r)$. The radial distribution function is proportional to the conditional probability that a particle is found at $\boldsymbol{r}$ ($|\boldsymbol{r}| = r$) when another particle is found at the origin. Equivalently, considering a spherical shell of radius r centered at $\boldsymbol{r}_i$ and counting the number of particles j that satisfy $r \le \left|\boldsymbol{r}_j - \boldsymbol{r}_i\right| < r + \Delta r$, we define the number density of particles in the shell $n_i(r) = N_i(r)/\left(4\pi r^2 \Delta r\right)$, where $N_i(r)$ is the number of particles in the shell. Averaging $n_i(r)$ over i we have $\overline{n}(r)$. Using the ensemble or time average $\langle \overline{n}(r) \rangle$ of $\overline{n}(r)$, we define the radial distribution function $g(r)$ as

$$g(r) = \langle \overline{n}(r) \rangle / \rho, \tag{4.25}$$

where $\rho \equiv N/V$ is the number density of particles in a uniform system. When r is large, particles are expected to be uniformly distributed so that $g(r) \to 1$ as $r \to \infty$. Since the interparticle distance cannot be less than the particle diameter σ due to the strong core repulsion, $g(r) \to 0$ as $r \to 0$. A typical shape of $g(r)$ for a liquid is shown in Fig. 4.2. Several peaks are found in $g(r)$ for a liquid, whereas only a single peak is found in gas. For a solid system, we will find many sharp peaks at particular positions corresponding to the crystal structure (see Fig. 4.5).

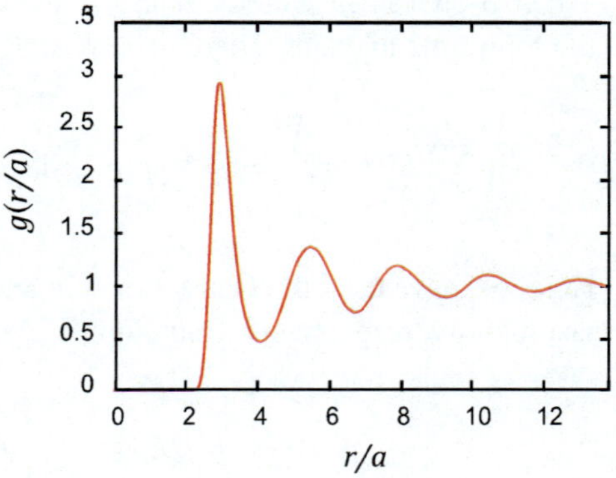

Fig. 4.2 Example of radial distribution for a colloidal system in liquid state as a function of the dimensionless center-to-center distance r/a between particles, where a is the particle radius

The radial distribution or pair correlation function is often used to measure the translational order of particles in a liquid. However, to evaluate an orientational order in a crystal or clusters, another kind of quantity should be introduced. In the two-dimensional melting problem, the bond-orientational order parameter has been introduced to measure the so-called Hexatic order as

$$\psi_6 = \left\langle e^{6i\theta} \right\rangle, \tag{4.26}$$

where θ is the orientation relative to some fixed reference axis, say x-axis, of the "bond" between two neighboring particles, and the average $\langle \cdot \rangle$ is taking over the neighboring particles and ensemble average in equilibrium. If the system has the perfect six-fold symmetry, or all particles reside on a triangular lattice, then $\psi_6 = 1$, and if the system is disordered, then $\psi_6 = 0$. Note that ψ_6 is a complex number.

A natural generalization of the bond-orientational order in three-dimensional systems has been proposed by Steinhardt et al. [5]; they have defined

$$Q_l \equiv \left[\frac{4\pi}{2l+1} \sum_{m=-l}^{l} |\langle Y_{lm}(\theta, \phi) \rangle|^2 \right]^{\frac{1}{2}}, \tag{4.27}$$

where $Y_{lm}(\theta, \phi)$ is the spherical harmonics (which are complex functions) of the polar angles θ, ϕ of the bond vector relative to some fixed coordinate system. Here, the bond vector is a vector $\boldsymbol{r}_{ji} = \boldsymbol{r}_j - \boldsymbol{r}_i$ from particle i to a neighbor particle j, and the average $\langle \cdot \rangle$ is taking over j for the local orientational order parameter Q_l. Typically, $(Q_4, Q_6) = \left(1.91 \times 10^{-2}, 5.75 \times 10^{-1}\right)$ for the FCC single cluster, $(Q_4, Q_6) = \left(3.64 \times 10^{-2}, 5.11 \times 10^{-1}\right)$ for the BCC single cluster. To evaluate crystal structures, we should average over (i, j). See [6] for details.

It should be valuable to mention how to relate the above numerical methods to experimental observations. In order to identify molecular array structures of materials, scattering experiments are often performed. In those experiments, a sample of material is irradiated with a light (X-ray, neutron, etc.), and the scattered light is detected. The incident light is a monochromatic plane wave $\Psi(\boldsymbol{r}, t)$ with wave vector $\boldsymbol{k}$ and angular velocity ω, with which we have $\Psi \propto \exp(i\boldsymbol{k} \cdot \boldsymbol{r} - i\omega t)$. This light is scattered by a particle located at $\boldsymbol{r}_i$ in the sample. The scattered light, which is a spherical wave with wave vector $\boldsymbol{k}'$, is detected at a position (R, θ), where R is the radial coordinate and θ is the polar angle. We assume the scatterings are elastic, i.e., $\left|\boldsymbol{k}'\right| = |\boldsymbol{k}|$. Introducing the scattering vector $\boldsymbol{q} \equiv \boldsymbol{k}' - \boldsymbol{k}$, we have $|\boldsymbol{q}| = 2|\boldsymbol{k}| \sin(\theta/2)$. Hereafter, we take the origin of the coordinate system at a particle position $\boldsymbol{r}_1 = 0$ in the sample and z-axis into the $\boldsymbol{k}$-direction, as shown in Fig. 4.3. R is assumed to be much larger than the sample size, so that the scattered light can be regarded as a spherical wave with a form $[f(\theta)/R] \exp(ik'R)$, where $k' \equiv \left|\boldsymbol{k}'\right|$ and the prefactor $f(\theta)$ contains precise information of the light source and the scatterer at $\boldsymbol{r}_i$. However, these are irrelevant when we are interested in a structure that consists of homogeneous particles. Only the phase factor $\exp(i\boldsymbol{k} \cdot \boldsymbol{r})$ is relevant in the present case. The scattering waves of two particles located at $\boldsymbol{r}_i$ and $\boldsymbol{r}_1$ $(i \neq 1)$ cause interference.

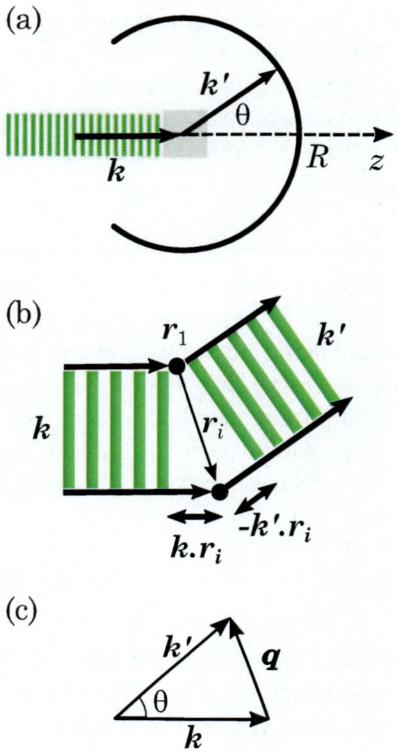

Fig. 4.3 **a** Schematic picture of scattering experiment. A sample placed in the center is irradiated with light of wave vector $\boldsymbol{k}$ and the scattered light with wave vector $\boldsymbol{k}^{'}$ is detected at (R, θ). **b** Scattering processes by the two particles located at $\boldsymbol{r}_1(=0)$ and $\boldsymbol{r}_i$ with phase shift $\boldsymbol{k}^{'} \cdot \boldsymbol{r}_i - \boldsymbol{k} \cdot \boldsymbol{r}_i$. **c** Relation among an incident wave vector $\boldsymbol{k}$, a wave vector scattered by a particle $\boldsymbol{k}^{'}$, and the scattering vector $\boldsymbol{q}$

Since phase lag between these two waves is $\boldsymbol{k}' \cdot \boldsymbol{r}_i - \boldsymbol{k} \cdot \boldsymbol{r}_i = \boldsymbol{q} \cdot \boldsymbol{r}_i$ (Fig. 4.3), so that the amplitude of the scattering wave is $[f(\theta)/R] \exp(i\boldsymbol{q} \cdot \boldsymbol{r}_i)$.

Since the observed wave at (R, θ) is given as a sum of all waves scattered at $\boldsymbol{r}_i$ $(i = 1, 2, \ldots, N)$, the scattering intensity is given as

$$I(\theta) = \left\langle \left| \frac{f(\theta)}{R} \sum_{i=1}^{N} \exp(i\boldsymbol{q} \cdot \boldsymbol{r}_i) \right|^2 \right\rangle, \tag{4.28}$$

where the angular bracket means to take a time average which can be replaced by an ensemble average. Hereafter, we consider a system in which N, V, and T are constant. In that system, an average over momentum space can be integrated out yielding only the configurational average as the problem. Therefore, for a quantity $A(\{\boldsymbol{r}_i\})$, $\langle A(\{\boldsymbol{r}_i\}) \rangle = \int A(\{\boldsymbol{r}_i\}) dP(\{\boldsymbol{r}_i\})$, where $\mathrm{d}P(\{\boldsymbol{r}_i\}) = p(\{\boldsymbol{r}_i\}) \mathrm{d}\boldsymbol{r}^N$ and $p(\{\boldsymbol{r}_i\})$ is the probability density or Boltzmann weight and $\mathrm{d}\boldsymbol{r}^N = \mathrm{d}\boldsymbol{r}_1 \mathrm{d}\boldsymbol{r}_2 \cdots \mathrm{d}\boldsymbol{r}_N$. The above equation can be expressed as

$$I(\theta) = \frac{|f(\theta)|^2}{R^2} \left\langle \sum_i \sum_j \exp(i\boldsymbol{q} \cdot (\boldsymbol{r}_i - \boldsymbol{r}_j)) \right\rangle = \frac{|f(\theta)|^2}{R^2} \left\langle N + \sum_{i \neq j} \exp(i\boldsymbol{q} \cdot (\boldsymbol{r}_i - \boldsymbol{r}_j)) \right\rangle. \tag{4.29}$$

The last term in (4.29) can be related to the radial distribution function $g(r)$. To this end, we introduce distribution or correlation functions of particles using the Dirac delta function $\delta(x)$, which is a continuous version of the Kronecker delta, that is, if $a < 0 < b$ then $\int_a^b \delta(x)\,\mathrm{d}x = 1$, otherwise $\int_a^b \delta(x)\,\mathrm{d}x = 0$. Therefore, $\int_{-\infty}^{\infty} f(x)\delta(x - x_0)\mathrm{d}x = f(x_0)$ for a continuous function $f(x)$. In three-dimensional space, $\delta(\boldsymbol{r}) = \delta(x)\delta(y)\delta(z)$ for $\boldsymbol{r} = (x, y, z)$. The local number density $\rho(\boldsymbol{r})$ of particles can be written as

$$\rho(\boldsymbol{r}) = \sum_{i=1}^{N} \delta(\boldsymbol{r} - \boldsymbol{r}_i). \tag{4.30}$$

Integrating (4.30) over the whole system with volume V, we can see

$$\int_V \rho(\boldsymbol{r})\mathrm{d}\boldsymbol{r} = \sum_{i=1}^{N} \int_V \delta(\boldsymbol{r} - \boldsymbol{r}_i)\mathrm{d}\boldsymbol{r} = N, \tag{4.31}$$

where $\mathrm{d}\boldsymbol{r} = \mathrm{d}x\mathrm{d}y\mathrm{d}z$ is an infinitesimal volume element and find the number density of a uniform system $\rho = N/V$. Taking the ensemble average of (4.30) for $\boldsymbol{r}_i$, we define the single-particle distribution as

$$\rho^{(1)}(\boldsymbol{r}) = \left\langle \sum_{i=1}^{N} \delta(\boldsymbol{r} - \boldsymbol{r}_i) \right\rangle. \tag{4.32}$$

We also define the pair distribution as

$$\rho^{(2)}(\boldsymbol{r}, \boldsymbol{r}') = \langle \sum_{i \neq j} \delta(\boldsymbol{r} - \boldsymbol{r}_i)\delta(\boldsymbol{r}' - \boldsymbol{r}_j) \rangle. \tag{4.33}$$

Using the identity

$$\begin{aligned} \int_V \mathrm{d}\boldsymbol{r} \int_V \mathrm{d}\boldsymbol{r}' \sum_{i \neq j} \delta(\boldsymbol{r} - \boldsymbol{r}_i)\delta(\boldsymbol{r}' - \boldsymbol{r}_j) &= (N-1) \int_V \mathrm{d}\boldsymbol{r} \sum_i \delta(\boldsymbol{r} - \boldsymbol{r}_i) \\ &= N(N-1), \end{aligned} \tag{4.34}$$

we have, for a uniform system or ideal gas,

$$\rho^{(2)}(\boldsymbol{r}, \boldsymbol{r}') = \tfrac{N(N-1)}{V^2} \approx \left(\tfrac{N}{V}\right)^2 = \rho^2, \tag{4.35}$$

since for the uniform system $\rho^{(2)}(\boldsymbol{r}, \boldsymbol{r}')$ does not depend on $\boldsymbol{r}$ and $\boldsymbol{r}'$, namely $\rho^{(2)}(\boldsymbol{r}, \boldsymbol{r}')$ is constant, so that $\int_V \mathrm{d}\boldsymbol{r} \int_V \mathrm{d}\boldsymbol{r}' \rho^{(2)}(\boldsymbol{r}, \boldsymbol{r}') = V^2 \rho^{(2)}(\boldsymbol{r}, \boldsymbol{r}')$. We introduce a pair correlation function $g(\boldsymbol{r}, \boldsymbol{r}')$ defined as

$$g(\boldsymbol{r}, \boldsymbol{r}') = \rho^{(2)}(\boldsymbol{r}, \boldsymbol{r}')/\rho^2. \tag{4.36}$$

This definition implies that $g(\boldsymbol{r}, \boldsymbol{r}') = 1$ for the uniform system and $g(\boldsymbol{r}, \boldsymbol{r}')$ deviates from unity when particle–particle correlations arise. Note that $\rho g(\boldsymbol{r}, \boldsymbol{r}')d\boldsymbol{r}'$ corresponds to the conditional probability that a particle is found in a volume $d\boldsymbol{r}'$ at $\boldsymbol{r}'$ when a particle is given at $\boldsymbol{r}$. For an isotropic system $g(\boldsymbol{r}, \boldsymbol{r}')$ depend on only the distance $r = |\boldsymbol{r} - \boldsymbol{r}'|$ between two particles so that $g(\boldsymbol{r}, \boldsymbol{r}')$ is the same as $g(r)$ in (4.25).

Now, we return to (4.29). The last term in this equation can be expressed in another form as

$$\left\langle \sum_{i \neq j} e^{i\boldsymbol{q}\cdot(\boldsymbol{r}_i - \boldsymbol{r}_j)} \right\rangle = \left\langle \int \mathrm{d}\boldsymbol{r} \int \mathrm{d}\boldsymbol{r}' \sum_{i \neq j} \delta(\boldsymbol{r} - \boldsymbol{r}_i)\delta(\boldsymbol{r}' - \boldsymbol{r}_j) e^{i\boldsymbol{q}\cdot(\boldsymbol{r} - \boldsymbol{r}')} \right\rangle$$
$$= \int \mathrm{d}\boldsymbol{r} \int \mathrm{d}\boldsymbol{r}' \left\langle \sum_{i \neq j} \delta(\boldsymbol{r} - \boldsymbol{r}_i)\delta(\boldsymbol{r}' - \boldsymbol{r}_j) \right\rangle e^{i\boldsymbol{q}\cdot(\boldsymbol{r} - \boldsymbol{r}')}. \tag{4.37}$$

Using (4.33) and (4.36), we have

$$\left\langle \sum_{i \neq j} e^{i\boldsymbol{q}\cdot(\boldsymbol{r}_i - \boldsymbol{r}_j)} \right\rangle = \int \mathrm{d}\boldsymbol{r} \int \mathrm{d}\boldsymbol{r}' \rho^{(2)}(\boldsymbol{r}, \boldsymbol{r}') e^{i\boldsymbol{q}\cdot(\boldsymbol{r} - \boldsymbol{r}')} = \rho^2 \int \mathrm{d}\boldsymbol{r} \int \mathrm{d}\boldsymbol{r}' g(\boldsymbol{r}, \boldsymbol{r}') e^{i\boldsymbol{q}\cdot(\boldsymbol{r} - \boldsymbol{r}')}. \tag{4.38}$$

When $g(\boldsymbol{r}, \boldsymbol{r}')$ depends only on $\boldsymbol{r} - \boldsymbol{r}'$, one of the volume integrals can be done and gives V and (4.38) becomes

$$\left\langle \sum_{i \neq j} e^{i\boldsymbol{q}\cdot(\boldsymbol{r}_i - \boldsymbol{r}_j)} \right\rangle = \rho^2 V \int \mathrm{d}\boldsymbol{r}\, g(\boldsymbol{r}) e^{i\boldsymbol{q}\cdot\boldsymbol{r}} = N\rho g_{\boldsymbol{q}}, \tag{4.39}$$

where $g_{\boldsymbol{q}} \equiv \int d\boldsymbol{r}\, g(\boldsymbol{r}) e^{i\boldsymbol{q}\cdot\boldsymbol{r}}$ is a Fourier component of $g(\boldsymbol{r})$. This relation yields the scattering intensity in (4.29) as

$$I(\theta) = \tfrac{|f(\theta)|^2}{R^2} N S_{\boldsymbol{q}}. \tag{4.40}$$

Here, we have introduced the so-called structure factor defined as

$$S_{\boldsymbol{q}} = 1 + \rho g_{\boldsymbol{q}}. \tag{4.41}$$

Using the spherical coordinates ($r = |\boldsymbol{r}|, q = |\boldsymbol{q}|, \boldsymbol{q}\cdot\boldsymbol{r} = qr\cos\theta$, $\mathrm{d}\boldsymbol{r} = r^2 \sin\theta \mathrm{d}r \mathrm{d}\theta \mathrm{d}\varphi$, with the polar angle θ and the azimuthal angle φ), $g_{\boldsymbol{q}}$ is calculated as

$$g_q = 2\pi \int_0^\infty \mathrm{d}r \int_0^\pi \mathrm{d}\theta r^2 \sin\theta g(r) e^{iqr\cos\theta} = \frac{4\pi}{q} \int_0^\infty \mathrm{d}r r g(r) \sin(qr). \tag{4.42}$$

4.3 Colloidal Systems

4.3.1 Brownian Motion as a Stochastic Process

The equations of motion (4.5) and (4.6) are deterministic, that is, once positions $\boldsymbol{r}_i$ and velocities $\boldsymbol{v}_i$ of molecules for $i = 1, 2, \ldots, N$ are given at $t = 0$, they are uniquely determined for $t > 0$, in principle. However, we often encounter a phenomenon that is not a deterministic but stochastic or random process. Such random processes are often observed in mesoscopic systems. A typical example is a colloidal system in which colloidal particles are in a medium, typically water. When the size of a colloidal particle is of the order of nm to μm, the particle motion is appeared to be random. Such random motion is called Brownian motion. When the particle size is smaller than nm, we cannot distinguish it from the molecular motion of the medium. In this case, we do not call such particle motion Brownian motion. Also, when the particle size is much larger than μm, the particle motion is almost deterministic. We cannot call it Brownian motion. Therefore, it can be said that we observe Brownian motion for mesoscopic degrees of freedom. Indeed, the random motion of the colloid particle of nm—μm is a result of random collisions between the colloidal particle and the surrounding small molecules. On the other hand, when the particle is so huge, the surrounding molecules can be regarded as a continuous fluid, and no Brownian motion is observed. Instead, the frictional force exerts on the particle due to the fluid motion around it.

Brownian motion can be described as a stochastic process or a stochastic differential equation in mathematical physics [7–12]. Equations of motion of a colloidal particle is written as

$$\frac{\mathrm{d}\boldsymbol{R}}{\mathrm{d}t} = \boldsymbol{V}, \quad M\frac{\mathrm{d}\boldsymbol{V}}{\mathrm{d}t} = \boldsymbol{F} - \zeta \boldsymbol{V} + \boldsymbol{f}, \tag{4.43}$$

where $\boldsymbol{R}$, $\boldsymbol{V}$, and M are position, velocity, and mass, respectively, and $\boldsymbol{F}$ is the external force or the force due to interaction with other particles. The last term $\boldsymbol{f}$ is the "random force" due to interactions with small molecules, and $-\zeta \boldsymbol{V}$ is the friction force with the coefficient ζ which is equal to $6\pi\eta a$ for the radius a of the particle in the fluid with shear viscosity η. The type of (4.43) is called the Langevin equation. In the absence of the last two terms in (4.43), this equation is Newton's equation of motion. In many cases including colloids, the acceleration term $(M\mathrm{d}\boldsymbol{V}/\mathrm{d}t)$ in (4.43) is negligible (in the long-time limit $t \gg M/\zeta$), and we have, in the absence of the force $\boldsymbol{F}$,

$$\frac{\mathrm{d}\boldsymbol{R}}{\mathrm{d}t} = \zeta^{-1}\boldsymbol{f}. \tag{4.44}$$

This is the simplest equation for the Brownian motion of a single particle. In one-dimensional system, replacing $\boldsymbol{R}$ and $\boldsymbol{f}$ with their x components X and f, respectively, the above equation can be written as $dX/dt = \zeta^{-1}f$. Discretizing the time t as $t_n = n\Delta t$ $(n = 0, 1, 2, \ldots)$, we can introduce a further simplified model as

$$X_{n+1} = X_n + \xi_n, \tag{4.45}$$

where X_n is the position of the Brownian particle at $t = t_n$ and ξ_n is a random number which is characterized by

$$\langle \xi_n \rangle_P = 0, \quad \langle \xi_l \xi_m \rangle_P = \sigma^2 \delta_{lm}, \tag{4.46}$$

where $\langle \xi \rangle_P$ means the average (or expectation value) of ξ by the probability distribution $P(\xi)$, that is, $\langle \xi \rangle_P = \int \xi P(\xi) \mathrm{d}\xi$, and $\delta_{lm} = 1$ if $l = m$, otherwise $\delta_{lm} = 0$. Equation (4.46) express that the mean and variance of random displacement of the particle are 0 and σ^2 (σ is not the particle diameter), respectively, and there are no correlations between the two displacements at different times. From (4.45) and (4.46), we obtain

$$\langle X_n \rangle_P = 0 \tag{4.47}$$

for $n = 1, 2, \ldots$ (we can always take $X_0 = 0$), and

$$\begin{aligned} \left\langle X_n^2 \right\rangle_P &= \left\langle (\xi_0 + \xi_1 + L + \xi_{n-1})^2 \right\rangle_P \\ &= \left\langle \xi_0^2 + \xi_1^2 + L + \xi_{n-1}^2 \right\rangle_P = n\sigma^2, \end{aligned} \tag{4.48}$$

since it holds in general that $a\xi + b_P = a\xi_P + b$ for the random variable ξ and arbitrary constants a and b. The quantity $\left\langle X_n^2 \right\rangle_P$ is called the mean square displacement of the Brownian particle. Equation (4.48) imply that the mean square displacement $\left\langle X_n^2 \right\rangle_P$ is proportional to the time, since $n\sigma^2$ can be written as $\left(\sigma^2/\Delta t\right)t_n$.

It should also be mentioned that the coefficient $\sigma^2/\Delta t$ is related to the diffusion coefficient of the particle. In order to show the relation between $\sigma^2/\Delta t$ and the diffusion coefficient of the Brownian particle, we return to (4.43). In the absence of the external field $\boldsymbol{F}$ in a one-dimensional system, (4.43) read as a second-order differential equation of the one-dimensional position X of the particle,

$$M \frac{\mathrm{d}^2 X}{\mathrm{d}t^2} = -\zeta \frac{\mathrm{d}X}{\mathrm{d}t} + f. \tag{4.49}$$

Multiplying X in both sides of (4.49) and using the identities

$$\frac{\mathrm{d}}{\mathrm{d}t}\left(\frac{X^2}{2}\right) = X\frac{\mathrm{d}X}{\mathrm{d}t}, \quad \frac{\mathrm{d}^2}{\mathrm{d}t^2}\left(\frac{X^2}{2}\right) = \left(\frac{\mathrm{d}X}{\mathrm{d}t}\right)^2 + X\frac{\mathrm{d}^2 X}{\mathrm{d}t^2}, \tag{4.50}$$

we have

$$M\left[\frac{\mathrm{d}^2}{\mathrm{d}t^2}\left(\frac{X^2}{2}\right) - \left(\frac{\mathrm{d}X}{\mathrm{d}t}\right)^2\right] = -\zeta \frac{\mathrm{d}}{\mathrm{d}t}\left(\frac{X^2}{2}\right) + Xf. \tag{4.51}$$

Taking average of the above equation, we obtain

$$M\frac{\mathrm{d}^2}{\mathrm{d}t^2}\left(\frac{\langle X^2\rangle_P}{2}\right)-M\langle V^2\rangle_P=-\zeta\frac{\mathrm{d}}{\mathrm{d}t}\left(\frac{\langle X^2\rangle_P}{2}\right), \tag{4.52}$$

since $\langle Xf\rangle_P = X\langle f\rangle_P = 0$. According to the equipartition theorem of the classical statistical mechanics, $M\langle V\rangle_P^2 = k_B T$, where k_B is the Boltzmann constant and T is the absolute temperature. Then, we have

$$M\frac{\mathrm{d}^2}{\mathrm{d}t^2}\left(\frac{\langle X^2\rangle_P}{2}\right)+\zeta\frac{\mathrm{d}}{\mathrm{d}t}\left(\frac{\langle X^2\rangle_P}{2}\right)=k_B T. \tag{4.53}$$

In the long-time regime, $t \gg M/\zeta$, the first term of (4.53) is negligible. Solving the first order differential equation, we obtain

$$\langle X^2\rangle_P=\frac{2k_B T}{\zeta}t. \tag{4.54}$$

Here, we have set a constant which should be add to the right-hand side zero because the initial condition $X = 0$ at $t = 0$ is considered. The relationship between the diffusion coefficient D and the friction coefficient ζ of the particle is well known as the Stokes–Einstein relation,

$$D=\frac{k_B T}{\zeta}=\frac{k_B T}{6\pi\eta a}, \tag{4.55}$$

where η is the viscosity of the dispersion medium and a is the particle radius. Using this relation, we have the expression

$$\langle X^2\rangle_P=2Dt. \tag{4.56}$$

Therefore, $n\sigma^2$ in (4.48) is replaced with $2Dt_n$. In three-dimensional system, the above equation becomes

$$\langle |\boldsymbol{R}|^2\rangle_P=6Dt, \tag{4.57}$$

because $\langle |\boldsymbol{R}|^2\rangle_P = \langle X^2+Y^2+Z^2\rangle_P = X_P^2+Y_P^2+Z_P^2 = 3X_P^2$, where $\boldsymbol{R} = (X, Y, Z)$.

In the above discussion, we did not mention a precise form of the probability distribution function $P(\xi)$ of the random variable ξ rather than its first and second moments given as (4.46). To determine the distribution $P(\xi)$, we utilize a theorem in statistics called the "central limit theorem," which is as follows. Let $H_1, H_2, \ldots, H_N$ be independent random variables and each distribution function of H_i has mean 0 and variance s_i^2. Define a new random variable Z_N as

$$Z_N=\frac{H_1+H_2+\cdots+H_N}{\sqrt{\sigma_N^2}}, \tag{4.58}$$

where

$$\sigma_N^2=s_1^2+s_2^2+\cdots+s_N^2. \tag{4.59}$$

The distribution function of Z_N converges, in the limit $N \to \infty$, to

$$P(Z_N) = \frac{1}{\sqrt{2\pi}} \exp\left(-\frac{Z_N^2}{2}\right). \tag{4.60}$$

This is a well-known distribution function called normal or Gaussian distribution with zero mean and unit variance. Strictly speaking, we require some conditions for the convergence of the above distribution. However, if we restrict the distribution functions of H_i $(i = 1, 2, \ldots, N)$ to the same distribution for all H_i, (4.60) always holds. In that case, (4.58) is given as

$$Z_N = \frac{H_1 + H_2 + \cdots + H_N}{\sqrt{Ns^2}}, \tag{4.61}$$

where $s^2 \equiv s_1^2 = s_2^2 = L = s_N^2$. The central limit theorem gives a reason why errors in some experimental measurements follow the Gaussian distribution, implying that an observed error is a sum of many unobserved small random events.

In the present case, a Brownian particle collides with surrounding small molecules many times in microscopic view. Since the time scale of the big particle, that is the Brownian particle, is in general much longer than that of the small molecules, the random displacement of the big particle observed in macroscopic measurement can be regarded as a result of many microscopic collisions. This intuitive picture enables us to apply the central limit theorem to the probability distribution of random displacement ξ_n of the Brownian particle in (4.45). The distribution function $P(\xi_n)$ is given by the normal distribution,

$$P(\xi_n) = \frac{1}{\sqrt{2\pi}\sigma} \exp\left(-\frac{\xi_n^2}{2\sigma^2}\right), \tag{4.62}$$

with the variance $\sigma^2 = 2D\Delta t$. Putting $\hat{\xi}_n \equiv \xi_n/\sigma$ in the above equation, we have

$$P(\xi_n)d\xi_n = \frac{1}{\sqrt{2\pi}\sigma} \exp\left(-\frac{\xi_n^2}{2\sigma^2}\right)d\xi_n = \hat{P}\left(\hat{\xi}_n\right)d\hat{\xi}_n. \tag{4.63}$$

Here, we have introduced the standard normal distribution

$$\hat{P}(\xi) = \frac{1}{\sqrt{2\pi}} \exp\left(-\frac{\xi^2}{2}\right), \tag{4.64}$$

with mean 0 and variance 1. Equation (4.63) means that random numbers sampled from the distribution $P(\xi_n)$ are statistically the same as random numbers sampled from $\hat{P}\left(\hat{\xi}_n\right)$ multiplied by $\sigma = \sqrt{2D\Delta t}$. Therefore, our model (4.45) can be expressed as

$$X_{n+1} = X_n + \hat{\xi}_n \sqrt{2D\Delta t}. \tag{4.65}$$

This expression is available to the numerical scheme for (4.44) or its one-dimensional version, $\mathrm{d}X/\mathrm{d}t = \zeta^{-1} f$, with the random force f. Since the random displacement ξ_n can be written as

$$\xi_n = \int_{t_n}^{t_n+\Delta t} \frac{\mathrm{d}X}{\mathrm{d}t}\mathrm{d}t = \int_{t_n}^{t_n+\Delta t} \zeta^{-1} f(t)\mathrm{d}t, \tag{4.66}$$

the mean square displacement $\langle \xi_n^2 \rangle_P$ is expressed as

$$\langle \xi_n^2 \rangle_P = \zeta^{-2} \int_{t_n}^{t_n+\Delta t} \mathrm{d}t \int_{t_n}^{t_n+\Delta t} \mathrm{d}t' f(t) f(t') \rangle_P = 2D\Delta t. \tag{4.67}$$

Here, we have assumed that $\langle \xi_n \rangle_P = 0$ and $\langle f(t) f(t') \rangle_P = C\delta(t - t')$ with a constant C which should be determined after the following calculations. First, we formally solve the differential Eq. (4.49) for the velocity $V(t) = \mathrm{d}X/\mathrm{d}t$ and obtain the time-correlation function of the velocity $\langle V(t) V(t') \rangle_P = \zeta^{-2} \langle f(t) f(t') \rangle_P$. We also need the equipartition theorem $M\langle V^2 \rangle_P = k_B T$ and the relation (4.55). We finally obtain $C = 2k_B T\zeta$ and $\langle \xi_n^2 \rangle = 2D\Delta t$. These results imply the validity of (4.65).

It should be noted that the above discussion can be understood more clearly by considering the Brownian motion as a diffusion process of particles. Introducing the probability distribution of particles $P(X, t)$ at position X and time t, we can find that the distribution $P(X, t)$ obeys the diffusion equation, $\partial P/\partial t = D\partial^2 P/\partial X^2$. This equation has a solution $P(X, t) = \exp(-X^2/4Dt)/\sqrt{4\pi Dt}$ for the initial condition $P(X, t) = \delta(X)$ at $t = 0$. This solution is identical to (4.62) with $\sigma = \sqrt{2Dt}$. This type of analysis was first done by Einstein, see Column 4.1 below.

Column 4.1 Brownian Motion And Molecular Reality

Scottish botanist Robert Brown (1773–1858) discovered that while observing pollen floating in a liquid under a microscope, the fine particles from the pollen were moving in an irregular and vivid motion, referred to as Brownian motion nowadays. This observation by R. Brown was published under the title "A brief account of microscopical observations made in the months of June, July and August 1827, on the particles contained in the pollen of plants; and on the general existence of active molecules in organic and inorganic bodies," [The Philosophical Magazine, 4:21, 161–173 (1828)]. He initially attributed the particle motion to "active Molecules" with life. However, he observed that every particle, whether organic or inorganic, was in an irregular motion. He concluded that such motion is determined by the size and shape of the particles, and that the most active particles, "active Molecules," are spherical, with diameters ranging from 0.85μm to 1.3μm (Brown, "Additional remarks on active molecules," [The Philosophical Magazine, 6:33, 161–166 (1829)]). This conclusion may indicate that Brownian motion is a universal phenomenon.

The discovery of Brownian motion was followed by detailed experiments by L. G. Gouy and several other researchers. In 1905, Albert Einstein provided a theoretical analysis of Brownian motion with the aid of the molecular kinetic theory of heat. The relevant papers are compiled in the book: "INVESTIGATIONS ON THE THEORY OF THE BROWNIAN MOVEMENT" by Albert Einstein, Edited with Notes by R. Furth, Translated by A. D. Cowper, (Dover, New York, 1956). Einstein discussed the relationship between the irregular motion of fine particles and the diffusion of particles and derived (4.55) and (4.56). This work on the diffusion of Brownian particles contributed significantly to the later development of the theory of stochastic processes [12]. His theory attracted the interest of mathematicians as well as researchers in physics and chemistry because it suggested the existence of molecules. He presented several methods for calculating Avogadro's number N_A and the size of particles a from experimentally observable quantities, for example, $N_A = \mathrm{RT}/6\pi\eta a D$ from (4.55) with the gas constant $R = N_A k_B$, and a can be calculated using the expression for the suspension viscosity $\eta^* = \eta(1 + 2.5\phi)$ with the volume fraction $\phi = \frac{4}{3}\pi a^3 n$ of the particles (n is the particle number density).

However, Einstein's theory was not immediately accepted by researchers in physics and chemistry. This is partly because his theory seemed to contain inappropriate assumptions. In fact, he used van't Hoff's equation of osmotic pressure for the forces acting on the particles and used the Stokes's law, derived from macroscopic hydrodynamics for the resistance of fine particles. A more serious situation, however, was that some energetics theorists refused to acknowledge the existence of molecules. It was Perrin's experiment that put an end to this controversy. French physicist Jean Baptiste Perrin trusted Einstein's theory and performed several precise experiments. He observed the Brownian motion of uniformly sized fine particles and recorded particle positions $X(t)$ at regular time intervals. Einstein's theory predicts $\sqrt{\langle X^2(t)\rangle} = b\sqrt{t}$, where $b = \sqrt{k_B T/3\pi\eta a}$. Thus, the value of b is obtained as the slope of a straight line plotting the root-mean-square $\sqrt{\langle X^2(t)\rangle}$ against $\sqrt{t}$. Then, using the value of b, the Boltzmann constant $k_B = R/N_A$ or Avogadro's number N_A can be obtained. Perrin repeated the experiment under various conditions and obtained approximately the same value of Avogadro's number in each experiment. In this way, he experimentally proved that matter is made of molecules, i.e., that molecules are real. Perrin was awarded the Nobel Prize in Physics in 1926 for this work.

4.3.2 Brownian Dynamics

The numerical scheme (4.65) is available for many-particle systems in which the particles interact with each other and undergo Brownian motion. The simulation method of such kinds of systems is called Brownian dynamics. Here, we assume that all the Brownian particles are identical for simplicity. From (4.43) for the overdamped system, equations of motion of N particles in three dimensions are given as

$$\frac{d\boldsymbol{R}_i}{dt} = \zeta^{-1}\big(\boldsymbol{F}_i + \boldsymbol{f}_i\big), (i = 1, 2, \ldots, N), \tag{4.68}$$

where $\boldsymbol{R}_i = (X_i, Y_i, Z_i)$ is the position of particle *i* and $\boldsymbol{F}_i$ and $\boldsymbol{f}_i$ are the interaction force and the random force exerting on particle *i*, respectively. Since in the absence of the random force $\boldsymbol{f}_i$, the particle motion is deterministic, we can apply the simple finite difference method for numerical calculations as

$$\boldsymbol{R}_i(n+1) = \boldsymbol{R}_i(n) + \zeta^{-1}\boldsymbol{F}_i(n)\Delta t, \tag{4.69}$$

where $\boldsymbol{R}_i(n)$ and $\boldsymbol{F}_i(n)$ are $\boldsymbol{R}_i$ and $\boldsymbol{F}_i$ evaluated at $t = n\Delta t$, respectively. In the presence of random forces, we can add the random displacement $\boldsymbol{\xi}_i(n)$, as discussed above, to the right-hand side of (4.69). Then, we have

$$\boldsymbol{R}_i(n+1) = \boldsymbol{R}_i(n) + \zeta^{-1}\boldsymbol{F}_i(n)\Delta t + \boldsymbol{\xi}_i(n), \tag{4.70}$$

where $\boldsymbol{\xi}_i(n) = \big(\xi_i^x(n), \xi_i^y(n), \xi_i^z(n)\big)$ should satisfy the conditions,

$$\big\langle \xi_i^\alpha(n) \big\rangle_P = 0, \quad (i = 1, 2, \ldots, N; \alpha = x, y, z), \tag{4.71}$$

$$\Big\langle \xi_i^\alpha(l)\xi_j^\beta(m) \Big\rangle_P = \sigma^2 \delta_{ij}\delta_{\alpha\beta}\delta_{lm}, \quad (i = 1, 2, \ldots, N; \alpha, \beta = x, y, z), \tag{4.72}$$

where $\sigma^2 = 2D\Delta t$ with the diffusion constant *D*. As we have already shown in (4.65), the α-component of $\boldsymbol{\xi}_i(n)$ is given by

$$\xi_i^\alpha(n) = \hat{\xi}_i^\alpha(n)\sqrt{2D\Delta t}, \tag{4.73}$$

where an independent random number $\hat{\xi}_n^\alpha(n)$ should be numerically generated using a pseud random number generator for the normal distribution with zero mean and unit variance.

4.3.3 Monte Carlo Method

The Monte Carlo method is another kind of stochastic methods of numerical simulations. Before describing the numerical methods of Monte Carlo simulations [2], we make introductory remarks on stochastic processes [9–12].

In the Monte Carlo method, by stochastically generating sequential states of a system, an equilibrium state is found. In a colloidal system with fixed particle number *N*, volume *V*, and temperature *T*, a state of the system is prescribed by a configuration of particles $\boldsymbol{R}^N = (\boldsymbol{R}_1, \boldsymbol{R}_2, \ldots, \boldsymbol{R}_N)$. Successively adding random displacements of particles, we have a chain of states, $\boldsymbol{S}_0 \to \boldsymbol{S}_1 \to \cdots \to \boldsymbol{S}_n \to \cdots$, which forms a stochastic process, where $\boldsymbol{S}_n = \big(\boldsymbol{R}^N, t_n\big)$ is a state of the system at step *n*, or we can imagine a snapshot of randomly moving particles at a discrete time t_n ($t_n > t_{n-1} > \cdots > t_0$). In particular, when $\boldsymbol{S}_n$ depends only on the state of one step before $\boldsymbol{S}_{n-1}$ ($n = 1, 2, \ldots$), such kind of stochastic process is called the Markov process. Let $p(\boldsymbol{S}_n)d\boldsymbol{S}_n$ be the probability that each particle i is found in the infinitesimally small volume $d\boldsymbol{R}_i = dX_i dY_i dZ_i$ at position $\boldsymbol{R}_i = (X_i, Y_i, Z_i)$ at time t_n, where $\mathrm{d}\boldsymbol{S}_n = \mathrm{d}\boldsymbol{R}^N = \mathrm{d}\boldsymbol{R}_1 \mathrm{d}\boldsymbol{R}_2 \cdots \mathrm{d}\boldsymbol{R}_N$ at t_n. For the Markov process, we consider the joint probability density $p(\boldsymbol{S}_n, \boldsymbol{S}_{n-1}, \ldots, \boldsymbol{S}_0)$, that is, $p(\boldsymbol{S}_0)d\boldsymbol{S}_0$ is the probability of finding a state $\boldsymbol{S}_0$ of the system at t_0; $p(\boldsymbol{S}_1, \boldsymbol{S}_0)d\boldsymbol{S}_1 d\boldsymbol{S}_0$ is the probability of finding a state $\boldsymbol{S}_0$ of the system at t_0 and finding a state $\boldsymbol{S}_1$ of the system at t_1; and so on. Since the normalization condition, $\int \mathrm{d}\boldsymbol{S}_0 p(\boldsymbol{S}_0) = 1$, holds, the probability density of finding $\boldsymbol{S}_1$ regardless of $\boldsymbol{S}_0$ can be expressed as $\int \mathrm{d}\boldsymbol{S}_0 p(\boldsymbol{S}_1, \boldsymbol{S}_0)$. In general, following relations are always valid,

$$\int \mathrm{d}\boldsymbol{S}_n p(\boldsymbol{S}_n, \boldsymbol{S}_{n-1}, \ldots, \boldsymbol{S}_0) = p(\boldsymbol{S}_{n-1}, \ldots, \boldsymbol{S}_0), \tag{4.74}$$

$$\int \mathrm{d}\boldsymbol{S}_n \int \mathrm{d}\boldsymbol{S}_{n-1} p(\boldsymbol{S}_n, \boldsymbol{S}_{n-1}, \ldots, \boldsymbol{S}_0) = p(\boldsymbol{S}_{n-2}, \ldots, \boldsymbol{S}_0), \tag{4.75}$$

$$\int \mathrm{d}\boldsymbol{S}_n \int \mathrm{d}\boldsymbol{S}_{n-1} \ldots \int \mathrm{d}\boldsymbol{S}_0 p(\boldsymbol{S}_n, \boldsymbol{S}_{n-1}, \ldots, \boldsymbol{S}_0) = 1, \tag{4.76}$$

and so on, where the integral $\int \mathrm{d}\boldsymbol{S}_n$ is taken over all possible regions of $\boldsymbol{S}_n$. The last equation is the normalization condition of the probability density. Another important notion of probability is the conditional probability $P(X|Y)$ which means the probability of X a for given Y. The conditional probability is related to the joint probability $P(X, Y)$ as $P(X, Y) = P(X|Y)P(Y)$. In the present case, the conditional probability density, for example, $p(\boldsymbol{S}_n|\boldsymbol{S}_{n-1})$ satisfies $p(\boldsymbol{S}_n, \boldsymbol{S}_{n-1}) = p(\boldsymbol{S}_n|\boldsymbol{S}_{n-1})p(\boldsymbol{S}_{n-1})$. For the Markov process in which $\boldsymbol{S}_n$ depends only on $\boldsymbol{S}_{n-1}$,

$$p(\boldsymbol{S}_n|\boldsymbol{S}_{n-1}, \boldsymbol{S}_{n-2}, \ldots, \boldsymbol{S}_0) = p(\boldsymbol{S}_n|\boldsymbol{S}_{n-1}), \tag{4.77}$$

is valid. The above property of the Markov process leads to the following relation

$$\begin{aligned} p(\boldsymbol{S}_n, \boldsymbol{S}_{n-1}, \cdots, \boldsymbol{S}_0) &= p(\boldsymbol{S}_n|\boldsymbol{S}_{n-1}, \ldots, \boldsymbol{S}_0)p(\boldsymbol{S}_{n-1}, \ldots, \boldsymbol{S}_0) \\ &= p(\boldsymbol{S}_n|\boldsymbol{S}_{n-1})p(\boldsymbol{S}_{n-1}, \ldots, \boldsymbol{S}_0), \end{aligned} \tag{4.78}$$

and we have, in general,

$$p(\boldsymbol{S}_n, \boldsymbol{S}_{n-1}, \cdots, \boldsymbol{S}_0) = p(\boldsymbol{S}_n|\boldsymbol{S}_{n-1})p(\boldsymbol{S}_{n-1}|\boldsymbol{S}_{n-2}) \cdots p(\boldsymbol{S}_1|\boldsymbol{S}_0)p(\boldsymbol{S}_0). \tag{4.79}$$

Equation (4.79) implies that the conditional probability densities are fundamental quantities in the Markov process. The conditional probability density $p(\boldsymbol{S}_n|\boldsymbol{S}_{n-1})$ is often called transition probability from $\boldsymbol{S}_{n-1}$ to $\boldsymbol{S}_n$. From the definition $p(\boldsymbol{S}_n|\boldsymbol{S}_{n-1}) = p(\boldsymbol{S}_n, \boldsymbol{S}_{n-1})/p(\boldsymbol{S}_{n-1})$, we have the normalization condition for the transition probability,

$$\int \mathrm{d}\boldsymbol{S}_n p(\boldsymbol{S}_n|\boldsymbol{S}_{n-1}) = 1. \tag{4.80}$$

Using the identities

$$\int \mathrm{d}\boldsymbol{S}_{n-1} p(\boldsymbol{S}_n, \boldsymbol{S}_{n-1}, \boldsymbol{S}_{n-2}) = p(\boldsymbol{S}_n, \boldsymbol{S}_{n-2}) = p(\boldsymbol{S}_n|\boldsymbol{S}_{n-2})p(\boldsymbol{S}_{n-2}), \tag{4.81}$$

and

$$p(\boldsymbol{S}_n, \boldsymbol{S}_{n-1}, \boldsymbol{S}_{n-2}) = p(\boldsymbol{S}_n|\boldsymbol{S}_{n-1})p(\boldsymbol{S}_{n-1}|\boldsymbol{S}_{n-2})p(\boldsymbol{S}_{n-2}), \tag{4.82}$$

we obtain

$$p(\boldsymbol{S}_n|\boldsymbol{S}_{n-2}) = \int \mathrm{d}\boldsymbol{S}_{n-1} p(\boldsymbol{S}_n|\boldsymbol{S}_{n-1})p(\boldsymbol{S}_{n-1}|\boldsymbol{S}_{n-2}). \tag{4.83}$$

This is called the Chapman-Kolmogorov equation, which means that the transition probability from $\boldsymbol{S}_{n-2}$ to $\boldsymbol{S}_n$ is given as a sum of probabilities for all possible paths $\boldsymbol{S}_{n-2} \to \boldsymbol{S}_{n-1} \to \boldsymbol{S}_n$.

In order to understand (4.83) more intuitively, we recast the integral equation into a differential equation. For simplicity, we assume that only two types of transition occur: no transition or a transition with finite displacements of particles during a small time Δt. The transition probability density $p(\boldsymbol{S}_n|\boldsymbol{S}_{n-1})$ can be expressed as

$$p\left(\boldsymbol{R}_3^N, t_3|\boldsymbol{R}_2^N, t_2\right) = \alpha\delta\left(\boldsymbol{R}_3^N - \boldsymbol{R}_2^N\right) + \Delta t w\left(\boldsymbol{R}_3^N|\boldsymbol{R}_2^N\right). \tag{4.84}$$

Here, we have expressed $\boldsymbol{S}_n$ as $\left(\boldsymbol{R}_n^N, t_n\right)$, where $\boldsymbol{R}_n^N = (\boldsymbol{R}_1, \boldsymbol{R}_2, \ldots, \boldsymbol{R}_N)$ at $t = t_n$, and put $n = 3$. In (4.84), we have introduced the transition probability density per unit time $w\left(\boldsymbol{R}_3^N|\boldsymbol{R}_2^N\right)$ from $\boldsymbol{R}_2^N$ to $\boldsymbol{R}_3^N$, and α in front of the delta function is determined by the normalization condition, $\int \mathrm{d}\boldsymbol{R}_3^N p\left(\boldsymbol{R}_3^N|\boldsymbol{R}_2^N\right) = 1$, as

$$\alpha = 1 - \Delta t \int \mathrm{d}\boldsymbol{R}_3^N w(\boldsymbol{R}_3^N|\boldsymbol{R}_2^N). \tag{4.85}$$

Note that α is a function of $\boldsymbol{R}_2^N$, and (4.85) can be written as $\alpha\left(\boldsymbol{R}_2^N\right) = 1 - \Delta t \int d\overline{\boldsymbol{R}}^N w\left(\overline{\boldsymbol{R}}^N|\boldsymbol{R}_2^N\right)$, where $\overline{\boldsymbol{R}}^N$ is a dummy variable. Inserting (4.85) into (4.84), we have

$$\begin{aligned}\frac{p\left(\boldsymbol{R}_3^N, t_3|\boldsymbol{R}_1^N, t_1\right) - p\left(\boldsymbol{R}_3^N, t_2|\boldsymbol{R}_1^N, t_1\right)}{\Delta t} &= -\int \mathrm{d}\overline{\boldsymbol{R}}^N w\left(\overline{\boldsymbol{R}}^N|\boldsymbol{R}_3^N\right)p\left(\boldsymbol{R}_3^N, t_2|\boldsymbol{R}_1^N, t_1\right) \\ &+ \int \mathrm{d}\boldsymbol{R}_2^N w\left(\boldsymbol{R}_3^N|\boldsymbol{R}_2^N\right)p\left(\boldsymbol{R}_2^N, t_2|\boldsymbol{R}_1^N, t_1\right).\end{aligned} \tag{4.86}$$

Writing $t_3 = t_2 + \Delta t$, $t_2 = t$ and taking a limit $\Delta t \to 0$, we obtain a time differential equation of the transition probabilities as

$$\frac{\partial}{\partial t}p(\boldsymbol{R}_3^N,t|\boldsymbol{R}_1^N,t_1)=\int \mathrm{d}\boldsymbol{R}_2^N w(\boldsymbol{R}_3^N|\boldsymbol{R}_2^N)p(\boldsymbol{R}_2^N,t|\boldsymbol{R}_1^N,t_1)$$
$$-\int \mathrm{d}\overline{\boldsymbol{R}}^N w\left(\overline{\boldsymbol{R}}^N|\boldsymbol{R}_3^N\right)p(\boldsymbol{R}_3^N,t|\boldsymbol{R}_1^N,t_1). \tag{4.87}$$

Using the relation $p(\boldsymbol{R}_3^N,t|\boldsymbol{R}_1^N,t_1) = p(\boldsymbol{R}_3^N,t;\,\boldsymbol{R}_1^N,t_1)/p(\boldsymbol{R}_1^N,t_1)$ and integrating (4.87) multiplied by $p(\boldsymbol{R}_1^N,t_1)$ over $\boldsymbol{R}_1^N$, we finally obtain

$$\frac{\partial}{\partial t}p(\boldsymbol{R}^N,t)=\int d\overline{\boldsymbol{R}}^N\left[w\left(\boldsymbol{R}^N|\overline{\boldsymbol{R}}^N\right)p\left(\overline{\boldsymbol{R}}^N,t\right)-w\left(\overline{\boldsymbol{R}}^N|\boldsymbol{R}^N\right)p(\boldsymbol{R}^N,t)\right]. \tag{4.88}$$

Here, $\boldsymbol{R}_3^N$ has been replaced by $\boldsymbol{R}^N$, which is a configuration of particles at time t. Equation (4.88) is called the Master equation. The meaning of (4.88) is now clear. The first and second terms on the right-hand side of (4.88) express the gain of the probability $p(\boldsymbol{R}^N,t)$ by the transition from $\overline{\boldsymbol{R}}^N$ to $\boldsymbol{R}^N$ and the loss by the transition from $\boldsymbol{R}^N$ to $\overline{\boldsymbol{R}}^N$, respectively.

If the following condition,

$$w\left(\boldsymbol{R}^N|\overline{\boldsymbol{R}}^N\right)p\left(\overline{\boldsymbol{R}}^N,t\right)=w\left(\overline{\boldsymbol{R}}^N|\boldsymbol{R}^N\right)p(\boldsymbol{R}^N,t), \tag{4.89}$$

is satisfied in (4.88), $\partial p(\boldsymbol{R}^N,t)/\partial t=0$, that is, the probability density is stationary or in equilibrium. Equation (4.89) is a fundamental condition known as the detailed balance condition that determines an equilibrium state. It should be noticed that the stationary state in which $\partial p(\boldsymbol{R}^N,t)/\partial t = 0$ is satisfied does not mean the state $\boldsymbol{R}^N$ itself is stationary but means the distribution of probability is stationary. For example, a thermodynamic quantity $A=\int \mathrm{d}\boldsymbol{R}^N A(\boldsymbol{R}^N)p(\boldsymbol{R}^N,t)$ is independent of time in equilibrium, whereas the dynamical variable $A(\boldsymbol{R}^N)$ fluctuates in time even in equilibrium.

Now, we describe the Monte Carlo method for our particle systems. Physically, we can expect that the state $\boldsymbol{S}_n$ of the system in an appropriate process $\boldsymbol{S}_0\to\boldsymbol{S}_1\to\cdots\to\boldsymbol{S}_n\to\cdots$ will approach an equilibrium state in the limit $n\to\infty$. According to the statistical mechanics, a classical (N,V,T) system, in which the number of particles N, the volume V, and the temperature T are constant, reaches an equilibrium state in the long-time limit $n\to\infty$ for any initial state $\boldsymbol{S}_0$. In the equilibrium, the probability density distribution is independent of time, $p(\boldsymbol{S}_n)\to p_e(\boldsymbol{S}_n)$ as $n\to\infty$, where $p_e(\boldsymbol{S}_n)$ is the equilibrium probability density distribution of states and given by the Boltzmann distribution $p_e(\boldsymbol{S}_n)\propto \exp(-H/k_BT)$. Here, H is the total energy of the system, that is, $H=K+U$, where K is the kinetic energy and U is the potential energy. Since in the present colloidal systems at a given temperature, the kinetic part of the distribution contributes only to a constant factor, the equilibrium probability density distribution $p_e(\boldsymbol{S}_n)$ depends only on the coordinates of particles $\boldsymbol{S}_n=\boldsymbol{R}^N$ at step n. Hence, the equilibrium probability density is given as

$$p_e(\boldsymbol{S}_n)=\frac{1}{Q_N}\exp[-U(\boldsymbol{S}_n)/k_BT], \tag{4.90}$$

where

$$Q_N = \int \mathrm{d}\boldsymbol{S}_n \exp[-U(\boldsymbol{S}_n)/k_B T] \tag{4.91}$$

is the normalization constant.

How can we generate a Markov chain to get an equilibrium state where the probability density $p_e(\boldsymbol{S}_n)$ is stationary? In equilibrium, the detailed balance condition (4.89), or

$$\frac{w(\overline{\boldsymbol{S}}|\boldsymbol{S}_n)}{w(\boldsymbol{S}_n|\overline{\boldsymbol{S}})} = \frac{p_e(\overline{\boldsymbol{S}})}{p_e(\boldsymbol{S}_n)} = \exp\left[-\frac{U(\overline{\boldsymbol{S}})-U(\boldsymbol{S}_n)}{k_B T}\right] \tag{4.92}$$

should be satisfied. Because of (4.84), we can write the left-hand side of the above equation as $w(\overline{\boldsymbol{S}}|\boldsymbol{S}_n)/w(\boldsymbol{S}_n|\overline{\boldsymbol{S}}) = p(\overline{\boldsymbol{S}}|\boldsymbol{S}_n)/p(\boldsymbol{S}_n|\overline{\boldsymbol{S}})$, which is equal to 1 when $\overline{\boldsymbol{S}} = \boldsymbol{S}_n$. Here, $\overline{\boldsymbol{S}}$ is a candidate for the next step state $\boldsymbol{S}_{n+1}$ under the given $\boldsymbol{S}_n$. $\overline{\boldsymbol{S}}$ can be generated stochastically, for example, by choosing a particle $\boldsymbol{R}_i$ in $\boldsymbol{S}_n$ and adding a random displacement $\overline{\boldsymbol{R}}_i = \boldsymbol{R}_i + \Delta\boldsymbol{R}_i$, where $\overline{\boldsymbol{R}}_i$ is the new position of particle i in $\overline{\boldsymbol{S}}$. The candidate $\overline{\boldsymbol{S}}$ is accepted as $\boldsymbol{S}_{n+1}$ with probability $r \equiv p_e(\overline{\boldsymbol{S}})/p_e(\boldsymbol{S}_n)$. Repeating these procedures, we can expect to obtain the most probable or equilibrium state $\boldsymbol{S}_n$ as $n \to \infty$. The above algorithm is called the Metropolis algorithm of Monte Carlo methods.

In the following, we show a formal procedure of the Metropolis algorithm.

(0) Generate a random initial state $\boldsymbol{S}_0 = \boldsymbol{R}_1, \boldsymbol{R}_2, \ldots, \boldsymbol{R}_N, t_0$ and set $n = 0$. See the pseudocode "Initial" in 4.4.3.

(1) Calculate the total interaction energy $U(\boldsymbol{S}_n)$. Usually, it is calculated as a sum of pair interactions under appropriate boundary conditions, assuming that

$$U(\boldsymbol{S}_n) = \sum_{\langle ij \rangle} u(R_{ij}), \tag{4.93}$$

where $u(R_{ij})$ is the pair interaction potential as a function of $R_{ij} = |\boldsymbol{R}_i - \boldsymbol{R}_j|$, and the sum is taken for all pairs of particles i and j $(i \neq j)$.

(2) Generate a candidate $\overline{\boldsymbol{S}}$ for the next state $\boldsymbol{S}_{n+1}$. Choosing a particle $i = 1 + \mathrm{mod}(n, N)$, we have the particle position $\overline{\boldsymbol{R}}_i = \boldsymbol{R}_i + \Delta\boldsymbol{R}_i$ with a random displacement $\Delta\boldsymbol{R}_i = (\Delta X_i, \Delta Y_i, \Delta Z_i)$. The easiest way to get $\Delta\boldsymbol{R}_i$ is

$$\Delta X_i = (1 - 2\xi_x)\delta, \quad \Delta Y_i = (1 - 2\xi_y)\delta, \quad \Delta Z_i = (1 - 2\xi_z)\delta, \tag{4.94}$$

using uniform random numbers ξ_x, ξ_y, ξ_z each of which is in the range (0, 1), where δ is a constant parameter.

(3) Calculate

$$r = \exp\left[-\frac{U(\overline{\boldsymbol{S}})-U(\boldsymbol{S}_n)}{k_B T}\right]. \tag{4.95}$$

If $r \geq 1$ (or $U(\overline{S}) \leq U(S_n)$), then $\overline{S}$ is accepted as S_{n+1}, that is, $S_{n+1} = \overline{S}$. Otherwise, generate a uniform random number ξ in the range (0, 1), and

if $\xi < r$ then $S_{n+1} = \overline{S}$ ($\overline{S}$ is accepted),
else $r \leq \xi$ then $S_{n+1} = S_n$ ($\overline{S}$ is rejected),

return to **(2)** after updating *n*.

The above procedure shows just an outline of Monte Carlo simulations using the Metropolis method. In practice, we must set a value of δ, which might determine a rate of convergence to an equilibrium state, and we need to evaluate whether the state obtained is in equilibrium or not. For such kind of techniques in detail, see [2].

4.4 Examples of Numerical Studies of Colloidal Systems

Here, we show typical examples of numerical studies of colloidal systems. The first one is the numerical simulation of the crystallization of charged colloids using the Brownian dynamics method. The second is the simulation of cluster formation in a system that consists of oppositely charged colloidal particles. The final example is the simulation of aggregation processes of colloidal particles due to the depletion interaction.

4.4.1 General Description of the Numerical Model

Before showing some practical numerical schemes, we describe our numerical model within a framework of Brownian dynamics. Consider an isothermal system of volume *V* which contains *N* colloidal particles dispersed in a liquid. We assume that these particles interact in a pairwise manner. That is, the total interaction energy *U* is a sum of pair interaction u_{ij} between *i* and *j* particles for all pairs $\langle ij \rangle$ as

$$U = \sum_{\langle ij \rangle} u_{ij}, \tag{4.96}$$

where $\sum_{\langle ij \rangle} u_{ij}$ is equivalent to $\sum_{i<j} u_{ij}$, and $u_{ij} = u_{ji}$ is required. Once the interaction *U* is given as a function of positions $\{\boldsymbol{R}_1, \boldsymbol{R}_2, \ldots, \boldsymbol{R}_N\}$ of particles, time evolutions of the particles are described by (4.68), since the force $\boldsymbol{F}_i$ is determined as

$$\boldsymbol{F}_i = -\frac{\partial U}{\partial \boldsymbol{R}_i}. \tag{4.97}$$

It should be noted that a colloidal system in general consists of big colloidal particles and small liquid molecules. Equation (4.68) describes the time evolution of the system with only the degrees of freedom of the particles. There are no degrees of freedom of the liquid molecules, but instead, the fluctuation force $\boldsymbol{f}_i$ appears in (4.68). Therefore, it is implicitly assumed that the liquid molecules are rapidly equilibrated for a fixed configuration $\{\boldsymbol{R}_1, \boldsymbol{R}_2, \ldots, \boldsymbol{R}_N\}$ of particles. Also, the hydrodynamic interactions among particles are ignored.

The time evolution Eq. (4.68) or

$$\zeta \frac{d\boldsymbol{R}_i}{dt} = -\frac{\partial U}{\partial \boldsymbol{R}_i} + \boldsymbol{f}_i \tag{4.98}$$

should be expressed in the dimensionless form. Introducing the dimensionless variables as

$$\hat{\boldsymbol{R}}_i \equiv \boldsymbol{R}_i/a, \quad \hat{t} = t/\tau, \quad \hat{U} \equiv U/\varepsilon, \tag{4.99}$$

where a, τ, and ε are units of length, time, and energy, respectively, and choosing the particle radius as a, $\tau = a^2/D$, $\varepsilon = k_B T$, we have the dimensionless form of (4.61) as

$$\frac{\mathrm{d}\hat{\boldsymbol{R}}_i}{\mathrm{d}\hat{t}} = -\frac{\partial \hat{U}}{\partial \hat{\boldsymbol{R}}_i} + \hat{\boldsymbol{f}}_i, \tag{4.100}$$

where $\hat{\boldsymbol{f}}_i \equiv \boldsymbol{f}_i/(\varepsilon a^{-1})$ and (4.55) has been used. The finite difference equation of (4.100) is given, corresponding to (4.70), as

$$\hat{\boldsymbol{R}}_i(n+1) = \hat{\boldsymbol{R}}_i(n) + \hat{\boldsymbol{F}}_i(n)\Delta\hat{t} + \hat{\boldsymbol{\xi}}_i(n)\sqrt{2\Delta\hat{t}}, \tag{4.101}$$

where $\hat{\boldsymbol{F}}_i = -\partial\hat{U}/\partial\hat{\boldsymbol{R}}_i$ and a component of $\hat{\boldsymbol{\xi}}_i(n)$ is a random number sampled from the standard normal distribution and $\Delta\hat{t}$ is the dimensionless time step.

For multicomponent systems in which there are particles of different sizes, (4.101) should be modified since the friction constant $\zeta = 6\pi\eta a$ depends on the particle size. In (4.98), ζ should be replaced with $\zeta_i = 6\pi\eta a_i$, where a_i is the radius of particle i. Using the relation $D = k_\mathrm{B}T/\zeta = a^2/\tau$, (4.100) is replaced with

$$\frac{\mathrm{d}\hat{\boldsymbol{R}}_i}{\mathrm{d}\hat{t}} = -\frac{1}{\hat{a}_i}\left(\frac{\partial \hat{U}}{\partial \hat{\boldsymbol{R}}_i} - \hat{\boldsymbol{f}}_i\right), \tag{4.102}$$

where $\hat{a}_i \equiv a_i/a$. Then, (4.101) is given as

$$\hat{\boldsymbol{R}}_i(n+1) = \hat{\boldsymbol{R}}_i(n) + \hat{\boldsymbol{F}}_i(n)\frac{\Delta\hat{t}}{\hat{a}_i} + \hat{\boldsymbol{\xi}}_i(n)\sqrt{\frac{2\Delta\hat{t}}{\hat{a}_i}}. \tag{4.103}$$

4.4.2 Charged Colloids

In a charged colloidal system, each particle is charged on its surface with charge density $Ze_0/(4\pi a^2)$, where Z is the charge number and e_0 is the elementary charge.

There are many ions in the continuous phase or solvent so that the system is charge neutral as a whole. Since a positively (negatively) charged particle attracts negative (positive) ions, which undergo rapid thermal motions, the charge distribution near the particle will be non-uniform. This non-uniform charge distribution reduces the Coulomb interaction between particles. The effective interaction potential between the charged colloidal particles is known as the screened Coulomb or Yukawa potential, which has the following form (see Appendix D),

$$u_{ij} = \frac{(Z' e_0)^2}{4\pi\epsilon} \frac{\exp(-\kappa R_{ij})}{R_{ij}}, \tag{4.104}$$

where $R_{ij} \equiv \left| \boldsymbol{R}_i - \boldsymbol{R}_j \right|$ and

$$Z' = Z\left(\frac{e^{\kappa a}}{1+\kappa a}\right) \tag{4.105}$$

is the effective charge number, and κ is the inverse of screening length given as

$$\kappa^2 = 8\pi l_B n_r \tag{4.106}$$

with $l_\mathrm{B} \equiv e_0^2/(4\pi \in k_\mathrm{B} T)$ the Bjerrum length, $2n_r$ the total ion concentration (here, we are considering 1:1 electrolyte only), and ϵ the dielectric constant of the solvent. In the limit of $\kappa \to 0$, (4.104) expresses the pure Coulomb interaction. The particle system in which the particle–particle interaction is described by (4.104) is often called the "Yukawa system." In this system, there are three characteristic lengths, a, κ^{-1}, and l_B. The dimensionless form of (4.104) is given as

$$\hat{u}_{ij} \equiv \frac{u_{ij}}{k_B T} = \left(Z'\right)^2 \hat{l}_\mathrm{B} \frac{\exp\left(-\hat{\kappa} \hat{R}_{ij}\right)}{\hat{R}_{ij}}, \tag{4.107}$$

where $\hat{l}_\mathrm{B} \equiv l_\mathrm{B}/a$, $\hat{\kappa} \equiv \kappa a$, and $\hat{R}_{ij} \equiv R_{ij}/a$. The dimensionless quantity κa, called the Debye parameter, determines the range of particle–particle interaction and is experimentally controllable by changing the salt concentration since $2n_r = n_c + 2n_s$, where n_c is the counterion concentration, and n_s is the concentration of added salt. There is another length scale $l_p \equiv (V/N)^{\frac{1}{3}}$, the mean interparticle distance, in this system. This length scale is related to the volume fraction of particles ϕ, which is another controllable parameter, given by $\phi = \frac{4}{3}\pi a^3 N/V = \frac{4}{3}\pi \left(a/l_p\right)^3$.

4.4.3 Numerical Simulation: Crystallization of Charged Colloids

Charged colloids composed of particles with the same charge will crystallize when the repulsive interactions between particles are high enough. The charge number Z of a particle, the Debye parameter κa or the salt concentration n_s, and the volume fraction ϕ of particles will be relevant parameters to the crystallization of colloids. Such phenomenon, the liquid crystal phase transition of charged colloids, can be understood via numerical simulations using the Brownian dynamics introduced in the previous section. The time evolutions of the particles are given by (4.101) with the pair potential (4.107) for the Yukawa system. The force $\widehat{\boldsymbol{F}}_i$ in (4.101) is given explicitly as

$$\hat{\boldsymbol{F}}_i = -\frac{\partial \hat{U}}{\partial \hat{\boldsymbol{R}}_i} = (Z')^2 \hat{l}_{\mathrm{B}} \sum_j \left(\frac{\hat{\kappa}}{\hat{R}_{ij}} + \frac{1}{\hat{R}_{ij}^2} \right) \exp\left(-\hat{\kappa} \hat{R}_{ij} \right) \frac{\hat{\boldsymbol{R}}_{ij}}{\hat{R}_{ij}}. \tag{4.108}$$

Hereafter, we omit the hat "^" attached to the symbols of dimensionless quantities as long as no confusion arises.

Below, we will show practical numerical algorithms for the Brownian dynamics of the Yukawa system using pseudocodes. In the pseudocodes, an equation, for example, $A = B$ means that the evaluated value of B is substituted into A. An expression $A == B$ in a conditional means that it is true if A is equivalent to B. Comments are followed by "//."

A typical procedure for the numerical simulation of the particle system is as follows. First, we set various parameters to specify a physical system, such as the number of particles N, the volume fraction of particles ϕ, the charge number of a particle Z, the salt concentration n_s, and so on. We should also set numerical parameters, such as a time step Δt, the system dimensions L_x, L_y, L_z (the system volume $V = L_x L_y L_z$), the cutoff radius of interaction between particles R_c, and so on.

```
Main // main procedure
        SET parameters //set physical and numerical parameters
        CALL Initial //create an initial configuration of particles
        CALL Output //print the present state into a file
        FOR n = 1 to maximum time step
                CALL Update //one time step forward
                IF mod(n, interval)=0 THEN // mod(n, d) returns the remainder of n divided by d,
                                                    // that is, mod(n, d) = n – d*int(n/d)
                        CALL Output          // output at regular intervals
                END IF
        END FOR
        EXIT
```

Next, we generate an initial configuration of particles (only the particles positions are needed in the Brownian dynamics).

```
Initial // create an initial configuration R_i = (X_i, Y_i, Z_i) in (0, L_x)×(0, L_y)×(0, L_z)
// RN_U() : external function of random number generator
        START
        FOR all particle i = 0 to N − 1
                λ_x = RN_U( ) // uniform random number (0,1)
                λ_y = RN_U( )
                λ_z = RN_U( )
                R_i = (L_x λ_x, L_y λ_y, L_z λ_z) // initial position of particle i
                FOR particle j = 0 to i //check for overlap of particles
                        IF R_ij^2 < R_ex^2 THEN //R_ex^3 is excluded volume
                                GOTO START
                        END IF
                END FOR
        END FOR
        EXIT
```

The last part of the main procedure is the time evolution process. The Update procedure consists of three processes: calculation of forces between particles, updating particle positions, and relocating the particles to those positions satisfying the periodic boundary conditions.

The calculation of forces is the most time-consuming process. The force $\boldsymbol{F}_i$ on the particle $i = 0, 1, \ldots, N-1$ is calculated for each i as the sum of $N-1$ pairs of interaction forces $\sum_j \boldsymbol{F}_{ij}$ $(i \neq j)$. Because of $\boldsymbol{F}_{ij} = -\boldsymbol{F}_{ji}$, the total $N(N-1)/2$ interaction forces should be calculated. However, when the interaction is short range, we can set a cutoff radius R_c and restrict the interaction range to the inside of the sphere of radius R_c, that is, a particle i interact only the particles j that satisfy $\left|\boldsymbol{R}_i - \boldsymbol{R}_j\right| < R_c$. This method greatly reduces the computation time, although it depends on R_c. The value of R_c is usually set to about 3 times the mean interparticle distance.

A commonly used method to implement the above procedure in a numerical code is to divide the system into several cells and to create a list of particles in each cell. For a particle i in a cell, a particle j which interacts with the particle i can be found from the lists of particles in the same cell and the neighboring cells under the periodic boundary conditions. The lists of particles should be renewed at every time step. Here, we show an alternative way to calculate the force rather straightforwardly, although it takes the computation time slightly more. More excellent schemes can be found in advanced textbooks, for example, [1, 2].

```
Force // calculation of inter-particle forces
// Position of a particle i: R_i = (X_i, Y_i, Z_i) in the simulation box (0, L_x)×(0, L_y)×(0, L_z)
// The simulation box is divided into MX × MY × MZ cells (volume of each cell: L_c^3 )
// Index of the cell that a particle i belongs: M_i = (l_i, m_i, n_i),
//          0 ≤ l_i < MX, 0 ≤ m_i < MY, 0 ≤ n_i < MZ,
    FOR all particle i = 0 to N − 1
        F_i = 0 // initialize the force F_i on particle i
    END FOR
    FOR all particle i = 0 to N − 1
        M_i = ([X_i/L_c], [Y_i/L_c], [Z_i/L_c]) // index of cell ∋ R_i; [x]: integer part of x
        FOR particle j = i + 1 to N − 1
            M_j = ([X_j/L_c], [Y_j/L_c], [Z_j/L_c]) //index of cell ∋ R_j
            IF 0 < l_i < MX − 1 THEN
                IF −1 ≤ l_j − l_i ≤ 1 THEN X_ij = X_i − X_j
                ELSE GOTO NEXTJ
                END IF
            ELSE IF l_i == 0 THEN
                IF l_j == 0 or l_j == 1 THEN X_ij = X_i − X_j
                ELSE IF l_j == MX − 1 THEN X_ij = X_i − X_j + L_x
                ELSE GOTO NEXTJ
                END IF
            ELSE IF l_i == MX − 1 THEN
                IF l_j == MX − 1 or l_j == MX − 2 THEN X_ij = X_i − X_j
                ELSE IF l_j == 0 THEN X_ij = X_i − X_j − L_x
                ELSE GOTO NEXTJ
                END IF
            END IF
            ≪Do the same procedures as above for Y and Z coodinates ≫
            R_ij^2 = X_ij^2 + Y_ij^2 + Z_ij^2
            IF R_ij^2 < R_c^2 THEN
                F_i = F_i + F_ij // sum up F_ij = −(∂/∂R_i) u_ij, cf. (4.108)
                F_j = F_j − F_ij // F_ij = −F_ji
            END IF
        NEXTJ
        END FOR
    END FOR
    EXIT
```

Now the Update procedure is as follows.

```
Update // update a time step using (4.101)

// RN_G() : external function of random number generator
// RN_G(0, 1) generates a normal (Gaussian) random number (zero mean and unit variance)
// Values of parameters Δt and √(2Δt) should be set at the top of the Main procedure.
        CALL Force // calculate force F_i
        FOR all particle i = 0 to N − 1
                ξ_x = RN_G(0,1)
                ξ_y = RN_G(0,1)
                ξ_z = RN_G(0,1)
                R_i = R_i + F_i Δt + ξ√(2Δt) // ξ = (ξ_x, ξ_y, ξ_z)
        END FOR

// Correct new position R_i = (X_i, Y_i, Z_i) to satisfy the periodic boundary conditions

// using the function fmod(a,b) which returns the floating-point remainder of a/b

        FOR all particle i = 0 to N − 1
                X_i = fmod(X_i + L_x, L_x)
                Y_i = fmod(Y_i + L_y, L_y)
                Z_i = fmod(Z_i + L_z, L_z)
        END FOR
        EXIT
```

After running a simulation, we should analyze the data obtained. As a typical example, we show an algorithm for calculating the pair correlation function $g(r)$ defined in Sect. 4.2. In the following pseudocode, it is assumed that the output data of the above pseudocode are given as the input data, that is, time series of particle positions $\boldsymbol{R}_i(n)$, where $n = 0, 1, 2, \ldots$ corresponds to the discrete time.

Pair_correlation

// input data: $\boldsymbol{R}_i(n)$, particle positions $\boldsymbol{R}_i = (X_i, Y_i, Z_i)$, $i = 0, 1, 2, \ldots, N-1$, at time $n\Delta t$

// output data: time averaged pair correlation function $g(r_k)$, $r_k = k\Delta r$, Δr: increment of r

```
FOR k = 0 to K
        g_k = 0 // initialize g_k = g(r_k)
END FOR
FOR n = N_s to N_e // range of time average of radial distribution G_k(n)
        INPUT DATA R_i(n)
        CALL Gr(n) // radial distribution G_k(n) at time n
        FOR k = 0 to K
                g_k = g_k + G_k(n)
        END FOR
END FOR
FOR k = 0 to K
        g_k = g_k/(N_e − N_s + 1) // time average of g_k
END FOR
OUTPUT DATA g_k
EXIT
```

Gr(n)

```
SET ρ = N/(L_x L_y L_z) // number density of particles
SET ΔV = (4π/3)Δr^3 // volume of the spherical shell
FOR k = 0 to K
        G_k(n) = 0 // initialization
END FOR
FOR i = 0 to N − 1
        FOR j = i + 1 to N − 1
                X_ij = X_i − X_j
                X_a = X_ij + L_x
                X_b = X_ij − L_x
                IF |X_a| < |X_ij| THEN X_ij = X_a
                IF |X_b| < |X_ij| THEN X_ij = X_b
```

```
            Y_ij = Y_i − Y_j
            Y_a = Y_ij + L_y
            Y_b = Y_ij − L_y
            IF |Y_a| < |Y_ij| THEN Y_ij = Y_a
            IF |Y_b| < |Y_ij| THEN Y_ij = Y_b
            Z_ij = Z_i − Z_j
            Z_a = Z_ij + L_z
            Z_b = Z_ij − L_z
            IF |Z_a| < |Z_ij| THEN Z_ij = Z_a
            IF |Z_b| < |Z_ij| THEN Z_ij = Z_b
            r = sqrt(X_ij^2 + Y_ij^2 + Z_ij^2)
            k =[r/Δr] // [x]: integer part of x
            G_k(n) = G_k(n) + 2
        END FOR
    END FOR
    FOR k = 0 to K
        N_k = (N − 1)[3k(k + 1) + 1]ΔVρ // normalization factor
        G_k(n) = G_k(n)/N_k
    END FOR
    EXIT
```

Here, we show some numerical results obtained by using the Brownian dynamics method for charged colloids. The system consists of N identical particles with charge number Z and the dispersion medium of a base solution. Since each particle is charged in the process of dissociation reaction on the surface of its particle, the charge number s is determined by the reaction. Assuming the adsorption reaction of Langmuir-type on the surface with a finite number of reaction sites, such as

$$\equiv \mathrm{Si-OH} + \mathrm{B} \rightarrow^{k_1} \equiv \mathrm{Si-O^-} + \mathrm{BH^+}, \quad \equiv \mathrm{Si-OH} \overset{k_2}{\leftarrow} \equiv \mathrm{Si-O^-} + \mathrm{H^+},$$

we have $Z = Z_m kC_b/(1 + kC_b)$ in equilibrium, where Z_m is the maximum charge number, $k \equiv k_1/k_2$ is the ratio of the forward to the backward reaction rates, and C_b is the base concentration in the medium. Note that Z_m and kC_b are dimensionless parameters in addition to the number of particles N and the volume fraction of the particles ϕ.

Below we show a few results of the Brownian dynamics simulation. In Fig. 4.4, two snapshots of the system for (a) $kC_b = 2.0$ and (b) $kC_b = 5.0$ are shown. We can see that there is no crystalline order in (a) but crystalline order in (b) of Fig. 4.4. We also present the radial distribution function $g(r)$ in Fig. 4.5a, b corresponding to Fig. 4.4a, b, respectively. Since there are no sharp peaks in Fig. 4.5a, the system is in a fluid phase. On the other hand, there are several sharp peaks in Fig. 4.5b, which means that the system shows crystalline order. Because $r^{(1)}/r^{(1)} = 1$, $r^{(2)}/r^{(1)} \approx 2/\sqrt{3}$, $r^{(3)}/r^{(1)} \approx 2\sqrt{2/3}$, and $r^{(4)}/r^{(1)} \approx 2$, where $r^{(n)}$ is the nth peak position of $g(r)$, this crystalline structure can be estimated as BCC structure, although some of these peaks are invisible because of large fluctuations of particle positions. Another quantity to characterize the crystalline order is the bond-orientational order parameter introduced in (4.27). We plot the value of Q_6 as a function of kC_d in Fig. 4.6. We find a jump of Q_6 at $kC_b \approx 3.0$ in this figure, which implies that the fluid to crystal phase transition occurs with increase of kC_b (or charge number Z).

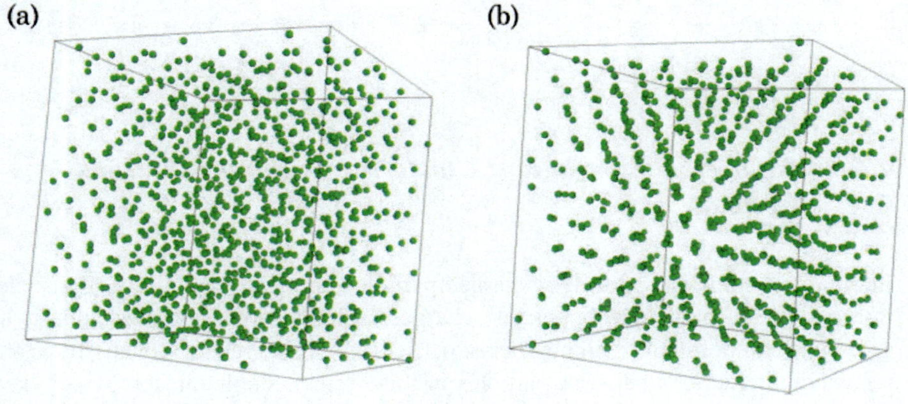

Fig. 4.4 Snapshots of charged colloidal systems obtained by the Brownian dynamics simulations for **a** $kC_b = 2.0$ ($\kappa a = 1.18$) and **b** $kC_b = 5.0$ ($\kappa a = 1.32$). Other parameters are the total number of particles $N = 1000$, the volume fraction of particles $\phi = 0.2$, and the maximum charge number $Z_m = 250$ for both systems (**a**) and (**b**)

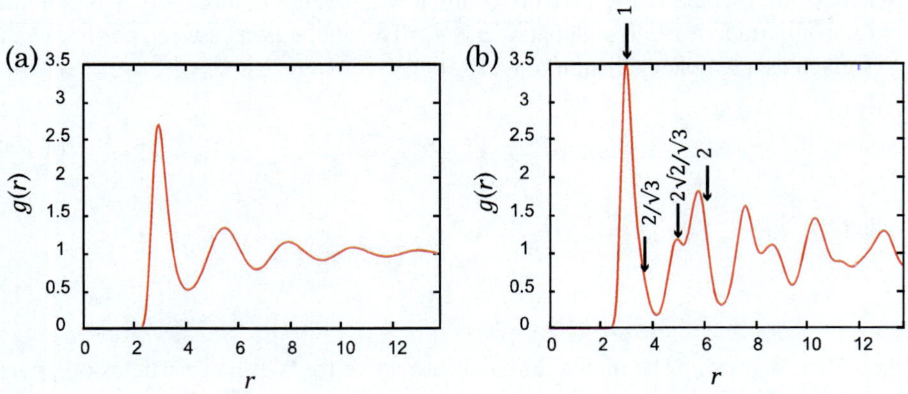

Fig. 4.5 Radial distribution functions for the two systems **a** in a liquid state and **b** in a crystal state, corresponding to Fig. 4.4a, b, respectively. The center-to-center distance r of particles is scaled by the particle radius a

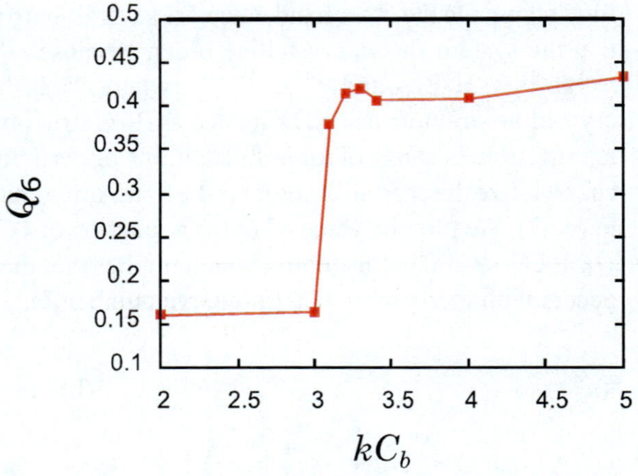

Fig. 4.6 Averaged bond-orientational order parameter Q_6 as a function of kC_b

4.4.4 Numerical Simulation: Clustering in Binary Charged Colloids

In binary colloids of opposingly charged particles, there are two kinds of particles in a system. In particular, oppositely charged colloids form a "cluster" that is an aggregate composed of small number of particles. Since there are both attractive and repulsive interactions between particles in this system, stable clusters which have symmetric structures to one another, are expected to appear.

Here, we outline our simulation procedures for the binary charged colloidal systems [13]. The simulation method is not so different from the Brownian dynamics introduced in the previous sections. The system consists of two kinds of colloidal particles, A and B. The particle A is negatively charged with charge number $Z_A < 0$, whereas the particle B is positively charged with charge number $Z_B > 0$. And the radius of particle A is a_A, and that of B is a_B. The interaction between particles i and j is given as a simple extension of (4.107)

$$\hat{u}_{ij} \equiv \frac{u_{ij}}{k_B T} = Z'_i Z'_j \hat{l}_{\mathrm{B}} \frac{\exp\left(-\hat{\kappa}\hat{R}_{ij}\right)}{\hat{R}_{ij}}, \tag{4.109}$$

where

$$Z'_i = Z_i\left(\frac{e^{\hat{\kappa}}}{1+\hat{\kappa}}\right), \tag{4.110}$$

and $Z_i = Z_A \mathrm{or} Z_B$. The interaction $\hat{u}_{\mathrm{ij}}$ is attractive for A-B pair particles otherwise repulsive. The Debye parameter κ is given as $\kappa^2 = 4\pi l_B\left(\sum_i |Z_i|/V + 2n_s\right)$ with the volume of V the solution and n_s is the salt concentration. The expression (4.109) for the pair potential is not always valid, in particular, for a small distance between particles. However, we adopt this expression because it is simple and still expected to conserve qualitative features of the system.

Another problem arises when there is an attractive pair potential in the particle systems. Since a colloidal particle has a finite size of radius a_i, two particles would occupy the same spatial region when the attractive force acts on these particles. To avoid such unphysical situations, we must introduce a core interaction, that is, a strong repulsive interaction between particles. Here, we use the Weeks-Chandler-Andersen (WCA) potential as

$$v_{ij} = \begin{cases} v_0\left[\left(\frac{a_i+a_j}{R_{ij}}\right)^{12} - \left(\frac{a_i+a_j}{R_{ij}}\right)^{6} + \frac{1}{4}\right] & \text{for } R_{ij} < 2^{\frac{1}{6}}(a_i + a_j) \\ 0 & \text{otherwise} \end{cases}, \tag{4.111}$$

where v_0 is a positive constant. We determine the value of v_0 such that $u_{ij} + v_{ij}$ have a minimum at $R_{ij} = a_i + a_j$ when $Z_i Z_j < 0$. From the condition $\frac{\partial}{\partial R_{ij}}(u_{ij} + v_{ij}) = 0$, we obtain in dimensionless form,

$$\hat{v}_0 \equiv \frac{v_0}{k_B T} = \frac{1}{6}\left|Z_i' Z_j'\right| \hat{l}_B\left(\hat{\kappa} + \frac{1}{\hat{a}_i + \hat{a}_j}\right) \exp[-\hat{\kappa}(\hat{a}_i + \hat{a}_j)], \tag{4.112}$$

where $\hat{a}_i \equiv a_i/a$ and $\hat{a}_j \equiv a_j/a$.

The total interaction potential is now given as $\hat{U} = \sum_{ij}(\hat{u}_{ij} + \hat{v}_{ij})$ so that the force $\hat{\boldsymbol{F}}_i = -\partial\hat{U}/\partial\hat{\boldsymbol{R}}_i$ exerting on particle i now consists of two parts:

$$\hat{\boldsymbol{F}}_i = \hat{\boldsymbol{F}}_i^{(u)} + \hat{\boldsymbol{F}}_i^{(v)}, \tag{4.113}$$

where $\hat{\boldsymbol{F}}_i^{(u)}$ is given as (4.108) and $\hat{\boldsymbol{F}}_i^{(v)}$ is given as

$$\hat{\boldsymbol{F}}_i^{(v)} = \sum_{ij} \frac{\hat{v}_0}{\hat{R}_{ij}}\left[12\left(\frac{\hat{a}_i + \hat{a}_j}{\hat{R}_{ij}}\right)^{12} - 6\left(\frac{\hat{a}_i + \hat{a}_j}{\hat{R}_{ij}}\right)^{6}\right]\frac{\hat{\boldsymbol{R}}_{ij}}{\hat{R}_{ij}}, \tag{4.114}$$

for $\hat{R}_{ij} < 2^{\frac{1}{6}}(\hat{a}_i + \hat{a}_j)$ and $\hat{\boldsymbol{F}}_i^{(v)} = 0$ for $\hat{R}_{ij} \geq 2^{\frac{1}{6}}(\hat{a}_i + \hat{a}_j)$. Using these expressions and (4.103), we can carry out the Brownian dynamics simulation of binary charged colloids.

Note that in the binary system where the radii of particle A and B are different ($a_A \neq a_B$), we have several choices for the unit of length a, that is, $a = a_A, a_B, \frac{1}{2}(a_A + a_B), a_A + a_B$, and so on.

Here, we show some numerical results of the binary charged colloids by using the Brownian dynamics method. The system contains $N = N_A + N_B$ particles of different kinds A and B of charge numbers $Z_A < 0$ and $Z_B > 0$, respectively. All particles have the same size, $a_A = a_B = a$, and the volume fraction ϕ of the particles is given as $\phi = \frac{4\pi}{3}a^3 N/V$ with the total volume V of the system. And the system is salt-free.

In this system, there are three types of interactions between particles $u_{ij} + v_{ij}$, i.e., U_{AA}, U_{BB}, and U_{AB} corresponding to the pairing of particles. Both U_{AA} and U_{BB} are repulsive with different magnitudes, whereas the interaction U_{AB} is attractive. In Fig. 4.7, the attractive interaction U_{AB} and repulsive interaction U_{BB} are plotted as a function of the distance between particles R_{ij} for the typical parameters shown in the caption. Using these interaction potentials and parameters, the Brownian dynamics simulations described above have been carried out and interesting results have been obtained. Figure 4.8 shows a snapshot of the system, which is almost in equilibrium. The parameters used in this simulation are the same as those described in the caption of Fig. 4.7. As shown in Fig. 4.8, a network-like structure is formed in which A and B particles are alternately connected. This is because the dimers of A and B particles are easy to appear under the condition $N_A/N_B = 1$. Such an A-B dimer has an electric dipole, which attracts another particle A or B to form a trimer of A-B-A or B-A-B. Thus, a linear structure grows. As a result, the network structures appear.

On the other hand, in the case that the number ratio is small ($N_A/N_B = 1/12$) and the charge ratio is relatively high ($Q = |Z_A/Z_B| = 11$), we obtain many small aggregates or clusters instead of the network-like structures. In Fig. 4.9, we show a snapshot of the system obtained by the Brownian dynamics simulation. Analyzing the data obtained in this simulation, we find that there are many clusters whose association number is 12, and each of these clusters has a symmetry of icosahedron. In order to examine the relationship between the association number n_a and the charge ratio Q, other simulations have been carried out in extreme situations, that is, $N_A = 1$, $N_B = 999$, and $1 \leq Q \leq 11$ with $Z_B = 100$ (all particles have the same size). In Fig. 4.10, the association number n_a is plotted as a function of the charge ratio Q, and typical shapes of the clusters are also depicted [13]. From these data, we clearly find the relation $n_a = Q + 1$ is valid at least for $1 \leq Q \leq 11$. This result implies that a cluster that consists of binary charged particles is stable in its overcharged state. The cluster in the neutral state is unstable since the neutral cluster will be deformed in its shape, and the electric dipole will be induced in the cluster by approaching of other charged particles. Hence, the neutral cluster becomes unstable due to the attractive interaction between the induced dipole and the charged particle. The overcharged clusters are stable because of the repulsive interaction between them.

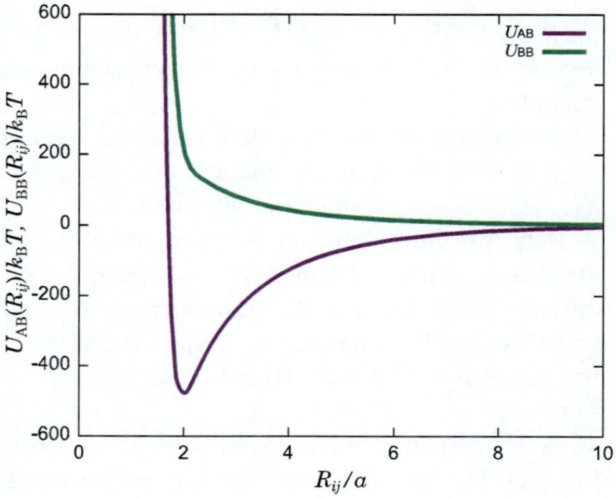

Fig. 4.7 Yukawa plus WCA potential as a function of R_{ij}/a. Here, U_{AB} and U_{BB} are the potentials between different sign of particles AB and same sign of particles BB, respectively. Parameter values are $N_A = N_B = 500$, $Q = |Z_A/Z_B| = 3$ ($Z_A = -300$, $Z_B = 100$), $\phi = 0.01$, and $a_A = a_B = a$

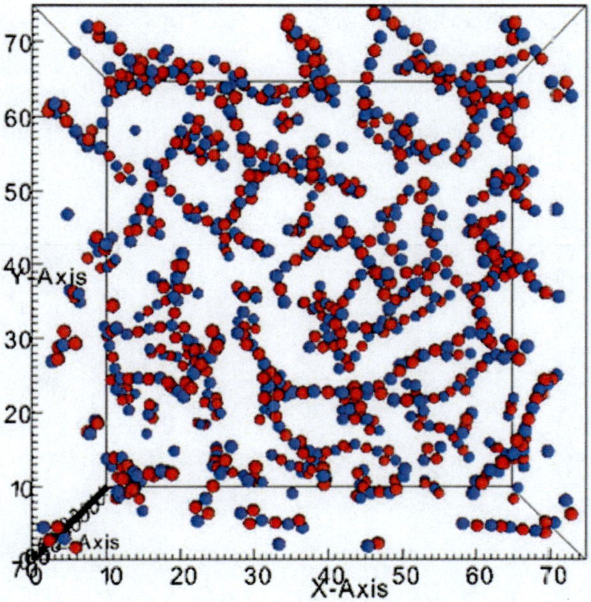

Fig. 4.8 Snapshot of the binary charged colloidal system obtained by the Brownian dynamics simulation with the interaction potentials and the parameters shown in Fig. 4.7. Blue and red particles correspond to A and B particles, respectively

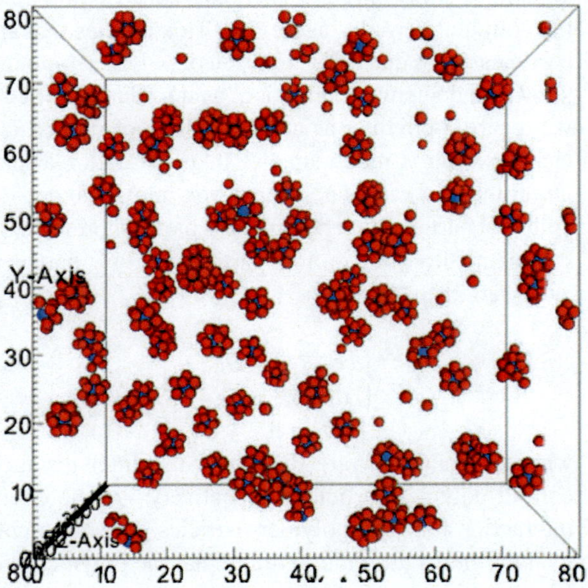

Fig. 4.9 Snapshot of the binary charged colloidal system obtained by the Brownian dynamics simulation for the parameters, $N_A = 100$, $N_B = 1200$, $Z_A = -1100$, $Z_B = 100$, and $\phi = 0.01$

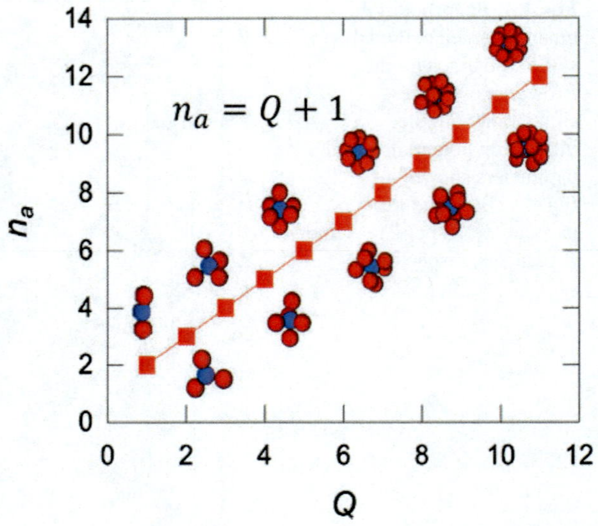

Fig. 4.10 Relation between the association number n_a of a cluster and the charge ratio $Q = |Z_A/Z_B|$ obtained by the Brownian dynamics simulations [13]

4.4.5 Numerical Simulation: Colloids with Added Polymers

When non-adsorbing polymers are added into a colloid system, it is observed that the colloidal particles aggregate. This implies that an attractive interaction between particles arises due to the addition of polymers into the colloid. Such an interaction is known as depletion interaction, as introduced in Sect. 2.2.5 [14, 15]. For simplicity, we regard a polymer as a small sphere of radius R_g (gyration radius of a polymer chain) which is much smaller than the colloidal particle of radius a and is much larger than the radius a_m of a solvent molecule: $a_m \ll R_g \ll a$. We also regard both colloidal particles and polymers as hard spheres. That is, the interaction potentials v_{cc} between colloid and colloid particles and v_{cp} between colloid and polymer particles are given, respectively, as

$$v_{cc}(R_{ij}) = \begin{cases} \infty & R_{ij} \leq 2a \\ 0 & \text{otherwise} \end{cases}, \quad v_{cp}(R_{i\alpha}) = \begin{cases} \infty & R_{i\alpha} \leq a + R_g \\ 0 & \text{otherwise} \end{cases}, \tag{4.115}$$

where R_{ij} and $R_{i\alpha}$ are the center-to-center distance between colloid-colloid and colloid-polymer particles, respectively. On the other hand, we assume there is no interaction between polymer particles. This assumption might be acceptable when the polymer solution is so dilute that the polymer particles behave like an ideal gas. We should point out that the interaction v_{cp} in (4.115) causes an empty layer with thickness R_g around a colloidal particle. That is, there is a no polymer particle α in the region $a < R_{i\alpha} < a + R_g$ due to the colloid-polymer interaction v_{cp}, where $R_{i\alpha} = |\boldsymbol{R}_i - \boldsymbol{R}_\alpha|$ is the center-to-center distance between the colloidal particle located at $\boldsymbol{R}_i$ and the polymer particle located at $\boldsymbol{R}_\alpha$. This empty layer is called the depletion layer.

Now, we consider an isothermal colloidal system in contact with a reservoir of the polymer solution. A fundamental thermodynamic potential in such a system is the grand potential $\Omega(V, T, \mu)$ which is a function of the volume V, the temperature T, and the chemical potential μ of the polymers. Because the size of colloidal particles is much larger than that of polymer particles ($a \gg R_g$) , the motion of colloidal particles is much slower than that of polymer particles. Hence, we can evaluate the grand potential $\Omega(V, T, \mu; R_i)$ of the system for fixed configuration of colloidal particles $\boldsymbol{R}_i$. Using the thermodynamic relation $\Omega = -PV$ with the pressure P and the equation of state for ideal gas $P = n_p k_{\mathrm{B}} T$, where n_p is the number density of polymer particles, we have

$$\Omega = -n_p k_{\mathrm{B}} T V_f(\{\boldsymbol{R}_i\}), \tag{4.116}$$

where $V_f(\{\boldsymbol{R}_i\})$ is the volume of the maximum region where the polymer particles are reachable for given $\boldsymbol{R}_i$. Note that $n_p k_{\mathrm{B}} T$ is the osmotic pressure of polymers in the present case and n_p is equal to the reservoir concentration, which is constant. Therefore, the grand potential Ω depends only on $\boldsymbol{R}_i$. When the particles are far enough from each other, the volume $V_f(\{\boldsymbol{R}_i\})$ is given as $V - \frac{4\pi}{3}\left(a + R_g\right)^3 N$, where N is the total number of particles. However, if there are overlap regions among depletion layers, $V_f(\{\boldsymbol{R}_i\})$ is larger than the above value, that is, $V_f(\{\boldsymbol{R}_i\}) > V - \frac{4\pi}{3}\left(a + R_g\right)^3 N$. When the overlap region is caused by the superposition of the two depletion layers, the volume $V_f(\{\boldsymbol{R}_i\})$ can be expressed as

$$V_f(\{\boldsymbol{R}_i\}) = V_f^{\infty} + \sum_{ij} \Delta V\left(R_{ij}\right), \tag{4.117}$$

where $V_f^{\infty} \equiv V - \frac{4\pi}{3}\left(a + R_g\right)^3 N$ and $\Delta V\left(R_{ij}\right)$ is the volume of the overlap region between depletion layers of particles i and j. It is easy to calculate $\Delta V\left(R_{ij}\right)$ if all particles have the same radius a. In this case, we obtain

$$\Delta V\left(R_{ij}\right) = \begin{cases} \frac{4\pi}{3} R_d^3 \left[1 - \frac{3}{4}\frac{R_{ij}}{R_d} + \frac{1}{16}\left(\frac{R_{ij}}{R_d}\right)^3\right] & 2a < R_{ij} \leq 2R_d \\ 0 & 2R_d < R_{ij}, \end{cases} \tag{4.118}$$

where $R_d \equiv a + R_g$. Equations (4.116) and (4.117) imply that the interaction potential u_{ij} between particles i and j is given as

$$u_{ij} = -n_p k_{\mathrm{B}} T \Delta V\left(R_{ij}\right). \tag{4.119}$$

Here, we have subtracted V_f^{∞} from (4.117) since this constant term does not contribute to any force between particles. From (4.118) and (4.119), we find the interaction between particles attractive. Thus, we should add the hard-sphere potential $v_{ij} = v_{cc}\left(R_{ij}\right)$ to the interaction potential, where $v_{cc}\left(R_{ij}\right)$ is given by (4.115). The total interaction potential $U(\boldsymbol{R}_i)$ is now expressed as

$$U(\{\boldsymbol{R}_i\}) = \sum_{ij} (u_{ij} + v_{ij}) \tag{4.120}$$

with (4.118), (4.119), and (4.115). When the particle sizes are not uniform, we can find the following expression,

$$u_{ij} + v_{ij} = \begin{cases} \infty & 0 < R_{ij} \le a_i + a_j \\ -n_p k_B T \frac{\pi}{12}\left[K_1 - K_2 R_{ij} - \frac{K_3}{R_{ij}} + R_{ij}^3\right] & a_i + a_j < R_{ij} \le R_d^i + R_d^j \\ 0 & R_d^i + R_d^j < R_{ij} \end{cases} \tag{4.121}$$

with

$$\begin{aligned} K_1 &= 8\left[\left(R_d^i\right)^3 + \left(R_d^j\right)^3\right] \\ K_2 &= 6\left[\left(R_d^i\right)^2 + \left(R_d^j\right)^2\right] \\ K_3 &= 3\left[\left(R_d^i\right)^2 - \left(R_d^j\right)^2\right]^2 \end{aligned} \tag{4.122}$$

where a_i is the radius of particle i and $R_d^i \equiv a_i + R_g$.

It is not difficult to use the above expressions of interactions in Monte Carlo simulations, which require the calculations of interaction energy only. However, a difficulty arises in calculating a force for the hard-sphere potential $(-dv_{ij}/d\boldsymbol{R}_i)$ in Brownian dynamics simulations. In order to avoid the difficulty, we must introduce, instead of the hard-sphere potential, a potential with a smooth profile, such as the WCA potential shown in the previous section. Alternatively, here we use the repulsive part of the Morse potential given as

$$v_{ij} = \begin{cases} \varepsilon\left[e^{b(\sigma - R_{ij})}\left(e^{b(\sigma - R_{ij})} - 2\right) + 1\right] & R_{ij} \le \sigma \\ 0 & \text{otherwise,} \end{cases} \tag{4.123}$$

where ε, b, and σ are positive constants. When we impose a condition that $u_{ij} + v_{ij}$ is minimum at $R_{ij} = a_i + a_j$, the parameters should satisfy the relation

$$\sigma = a_i + a_j + \frac{1}{b}\ln\left[\frac{1}{2}\left(1 + \sqrt{1 + \frac{2}{\varepsilon b} u'_{ij}(a_i + a_j)}\right)\right], \tag{4.124}$$

where u_{ij} is given by the attractive part of (4.121) and $u'_{ij}(R)$ is the derivative of $u_{ij}(R)$. In Fig. 4.11, we show the profiles of u_{ij}, v_{ij}, and $u_{ij} + v_{ij}$ as functions of R_{ij} for $\hat{a}_i = \hat{a}_j = 1.0$, $\hat{\varepsilon} \equiv \varepsilon/k_B T = 2.0$, $\hat{b} \equiv ba = 10$, $\hat{R}_g \equiv R_g/a = 0.1$, and the volume fraction of polymer particles $\phi_d \equiv (4\pi/3)R_g^3 n_p = 0.3$.

The Brownian dynamics simulations using the above potential (Asakura-Oosawa plus Morse potential) can reveal the dynamics of aggregation processes of colloidal

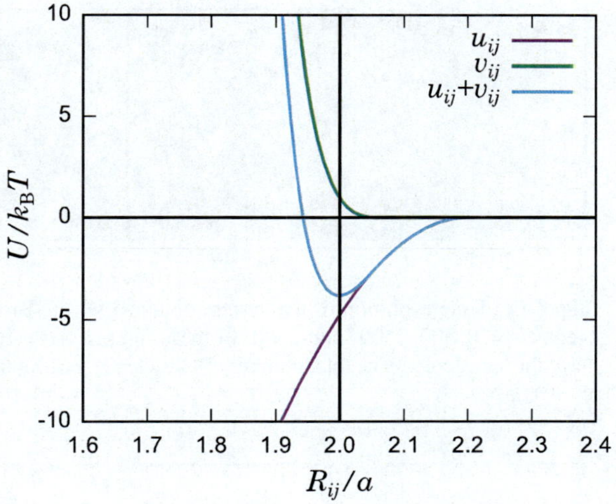

Fig. 4.11 Asakura-Oosawa potential u_{ij} as a function of distance R_{ij} between particle i and j. The total potential U is given as a sum of u_{ij} and the Morse potential v_{ij}

particles due to the depletion attraction. Figure 4.12 shows the time evolution of the system obtained numerically. We can see that the clusters grow: The average cluster size s increases in time, where the size of a cluster is the number of particles that make up the cluster. In the classical nucleation and growth process, one particle (molecule) is incorporated into the cluster at a time, and then the cluster size continuously increases. However, there is another type of process, which is a process of merging and enlargement of clusters and clusters. Such a kind of growing process is called the "cluster–cluster aggregation" (CCA) process. Figure 4.13 shows a CCA process observed in the above simulation. Such a coalescence process was studied extensively [16–19] in the 1980s as one of the growth mechanisms of fractal aggregates. In the diffusion-limited cluster aggregation process, the average cluster size s at time t increases with a power law $s \sim t^z$. The dynamic exponent z is given by [16].

$$z = \frac{1}{1-\gamma+(2-d)/D_f}, \tag{4.125}$$

where γ is an exponent determined by the relation $D_s \sim s^\gamma$, where D_s is the diffusion coefficient of a cluster of size s. In addition, d is the spatial dimension of the system, and D_f is the fractal dimension of the aggregate. Here, we assume $d = D_f = 3$.

It is not straightforward to determine the value of exponent γ. The Stokes–Einstein relation (4.55) implies $D_s \propto 1/a \sim s^{-1/D_f}$, since $s \sim a^{D_f}$. As another way to determine the value of γ, we here discuss the Brownian motion of a cluster of size s. In Sects. 4.3.1 and 4.3.2, we have described the Brownian motion of colloidal particles in a simple manner. The Brownian motion of particle i is described by (4.70)–(4.72) in a discrete-time system. Consider a cluster that consists of s particles each of which undergoes the Brownian motion described as (4.70)–(4.72). The position of the cluster is given as the center of mass of the particles,

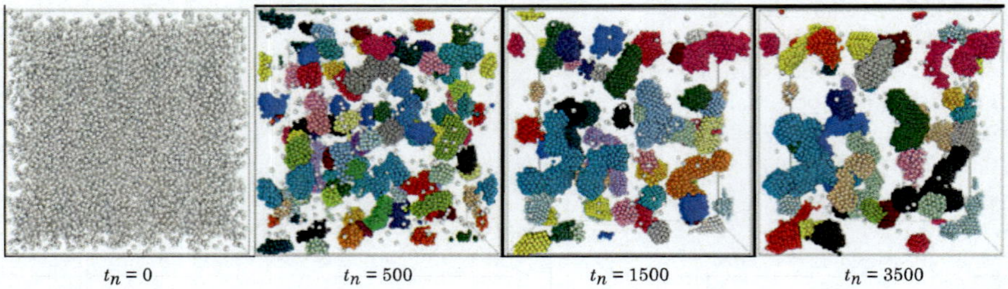

Fig. 4.12 Time evolution of the system obtained by the Brownian dynamics simulation at time steps $t_n = 0, 500, 1500$, and 3500. In these figures, the particles belonging to the same cluster have the same color. The total number of particles $N = 10000$ and the volume fraction of particles $\phi = 0.05$

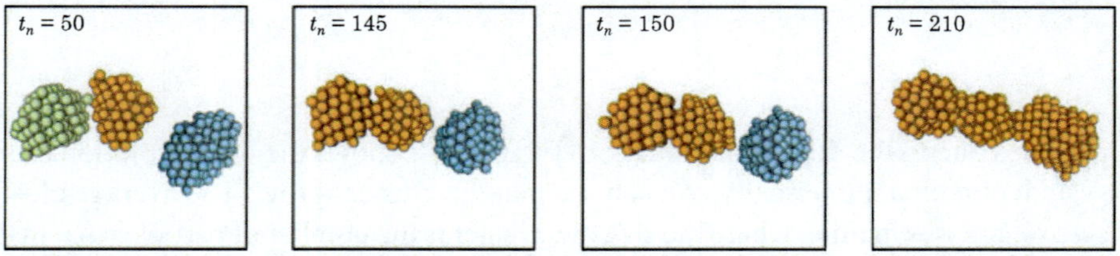

Fig. 4.13 Cluster–cluster aggregation process observed in the Brownian dynamics simulation

$$\boldsymbol{R}_G = \tfrac{1}{s}\sum_{i=1}^{s} \boldsymbol{R}_i. \tag{4.126}$$

This position $\boldsymbol{R}_G$ will undergo a Brownian motion by the Brownian particles whose positions are $\{\boldsymbol{R}_1, \boldsymbol{R}_2, \ldots, \boldsymbol{R}_s\}$. We assume that $\boldsymbol{R}_G$ obeys a similar equation to (4.70) as

$$\boldsymbol{R}_G(n+1) = \boldsymbol{R}_G(n) + X(n), \tag{4.127}$$

for the single cluster system, where $\boldsymbol{X}(n) = (X^x(n), X^y(n), X^z(n))$ is the random displacement at a time step n. From (4.71) and (4.126), we have

$$X^\alpha(n)_P = \left\langle \tfrac{1}{s}\sum_{i}^{s} \xi_i^\alpha(n) \right\rangle_P = \tfrac{1}{s}\sum_{i}^{s} \xi_i^\alpha(n)_P = 0, \tag{4.128}$$

for $\alpha = x, y, z$. We can calculate the time-correlation functions as

$$\langle X^\alpha(l) X^\beta(m) \rangle_P = \left\langle \left(\frac{1}{s}\sum_{i}^{s} \xi_i^\alpha(l) \right) \left(\frac{1}{s}\sum_{j}^{s} \xi_j^\beta(m) \right) \right\rangle_P$$

$$= \frac{1}{s^2} \sum_i^s \sum_j^s \xi_i^\alpha(l) \xi_j^\beta(m)_P$$

$$= \frac{\sigma^2}{s^2} \delta_{\alpha\beta} \delta_{lm} \sum_i^s \sum_j^s \delta_{ij} = \frac{\sigma^2}{s} \delta_{\alpha\beta} \delta_{lm} \tag{4.129}$$

where $\sigma^2 = 2D\Delta t$ with the single-particle diffusion coefficient D. If we write the above equation as

$$\langle X^\alpha(l) X^\beta(m) \rangle_P = 2D_s \Delta t \delta_{\alpha\beta} \delta_{lm}, \tag{4.130}$$

corresponding to (4.72), we find the relation

$$D_s = D/s. \tag{4.131}$$

This relation implies $\gamma = -1$.

In Fig. 4.14, we show the diffusion coefficient D_s of the cluster as a function of the cluster size s obtained by the Brownian dynamics of a single cluster. The slope of the line fitting the data points is $\gamma = -0.94$, which is close to the value -1 discussed above. We also show the average cluster size $\overline{s}$ as a function of time t in Fig. 4.15. The average size of clusters has been calculated as $\overline{s} = \sum_s s^2 n_s / \sum_s s n_s$, where n_s is the number of clusters of size s. We find that the growth of average cluster size obeys the power law $\overline{s} \sim t^z$ with the dynamic exponent z, except for the initial transient regime. The slope of the straight line obtained by fitting the data for $3 < \ln t < 8$ gives the exponent $z \approx 0.69$, which is slightly larger than the value 0.62 for $\gamma = -0.94$ with (4.125) and $D_f = d = 3$.

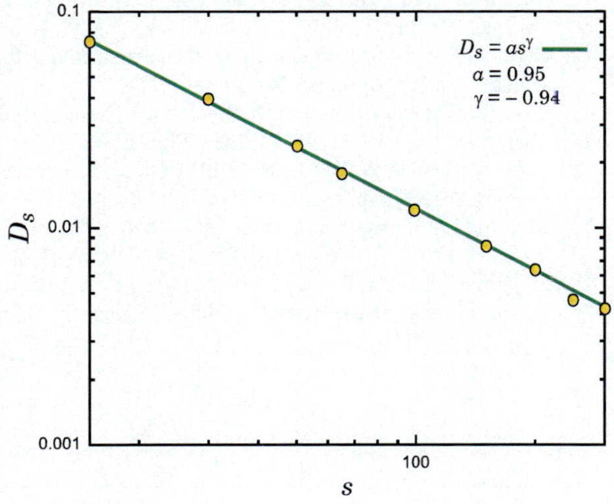

Fig. 4.14 Size dependence of the diffusion coefficient D_s of the cluster obtained by the Brownian dynamics simulation for the single cluster system

Fig. 4.15 Time dependence of the average size of clusters $\overline{s}$ shows the power law behavior $\overline{s} \sim t^z$ with the dynamic exponent $z \approx 0.69$

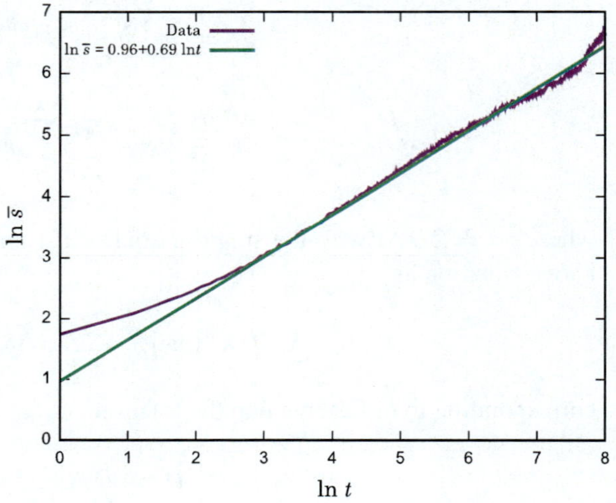

References

1. Allen MP, Tildesley DJ (1987) Computer simulation of liquids. Clarendon Press, Oxford
2. Frenkel D, Smit B (2002) Understanding molecular simulation—from algorithms to applications. 2nd ed. Academic Press, San Diego
3. Hansen J-P, McDonald IR (2013) Theory of simple liquids—with application to soft matter. 4th ed. Academic Press, Oxford
4. Goldstein H (1980) Classical mechanics, 2nd ed. Addison-Wesley, Reading
5. Steinhardt PJ, Nelson DR, Ronchetti M (1983) Phys Rev B 28:784–805
6. Lechner W, Dellago C (2008) J Chem Phys 129:114707
7. Chandrasekhar S (1943) Rev Mod Phys 15:1–89
8. Van Kampen NG (1975) Phys Rep 24:171–228
9. Van Kampen NG (1981) Stochastic processes in physics and chemistry. Elsevier, Amsterdam
10. Kubo R, Toda M, Hashitsume N (1991) Statistical physics II—nonequilibrium statistical mechanics. 2nd ed. Springer, Berlin
11. Risken H (1989) The fokker-plank equation—methods of solution and applications, 2nd ed. Springer, Berlin
12. Gardiner CW (1985) Handbook of stochastic methods for physics, chemistry, and the natural sciences, 2nd ed. Springer, Berlin
13. Okuzono T, Odai K, Masuda T, Toyotama A, Yamanaka J (2016) Phys Rev E 94:012609
14. Asakura S, Oosawa F (1954) J Chem Phys 22:1255
15. Lekkerkerker HNW, Tuinier R (2011) Colloids and the depletion interaction. Lecture Notes in Physics, vol 833. Springer, Dordrecht
16. Kolb M (1984) Phys Rev Lett 53:1653–1656
17. Vicsek T, Family F (1984) Phys Rev Lett 52:1669–1672
18. Meakin P, Vicsek T, Family F (1985) Phys Rev B 31:564–569
19. Lin MY, Lindsay HM, Weitz DA, Ball RC, Klein R, Meakin P (1989) Nature 339:360–362

Chapter 5
Studies on Colloidal Self-Assembly

Abstract In this chapter, we present some of our previous studies on the crystallization of colloids as examples of colloidal self-assembly. In particular, we present the various methods we have developed to produce large colloidal crystals of high quality. After providing an overview of the background in Sect. 5.1, we report the formation of cm-sized charged colloidal crystals by simply adding a small amount of alkali to silica colloids in Sect. 5.2. Based on this finding, we then describe the unidirectional crystal growth of silica colloids under gradients of pH and salt concentration. Section 5.3 describes how crystallization is controlled by temperature. We also present studies of unidirectional crystal growth and zone melting under temperature gradients. Section 5.4 presents the behavior of multicomponent systems. We describe that "impurity" particles, such as particles of different sizes, are spontaneously eliminated from the crystals. Experimental results are reported for the charged colloids and depletion-attraction systems. We also describe eutectics in multicomponent colloids.

Keywords Colloidal crystals · Unidirectional crystal growth · Temperature-induced crystallization · Impurity exclusion · Eutectics

5.1 Introduction

5.1.1 Crystal Growth

A crystal is a structure whose components (atoms, molecules, or particles) are regularly arranged and consist of crystal lattice planes (Sect. 3.4). A crystal whose lattice planes have the same orientation in any part of the sample is called a "single crystal." The artificial production of large single crystals is necessary for practical and theoretical research and is a very important technique in the field of metals and semiconductors. However, actual crystals (whether atomic, molecular, or colloidal crystals) are usually composed of many small crystals of different sizes and orientations. They are called polycrystals. Figure 5.1 shows an example of a confocal laser microscope (LSM) image of a silica colloid crystal (particle diameter $d = 110$ nm, particle volume

© The Author(s), under exclusive license to Springer Nature Singapore Pte Ltd. 2023

J. Yamanaka et al., *Colloidal Self-Assembly*, Lecture Notes in Chemistry 108,
https://doi.org/10.1007/978-981-99-5052-2_5

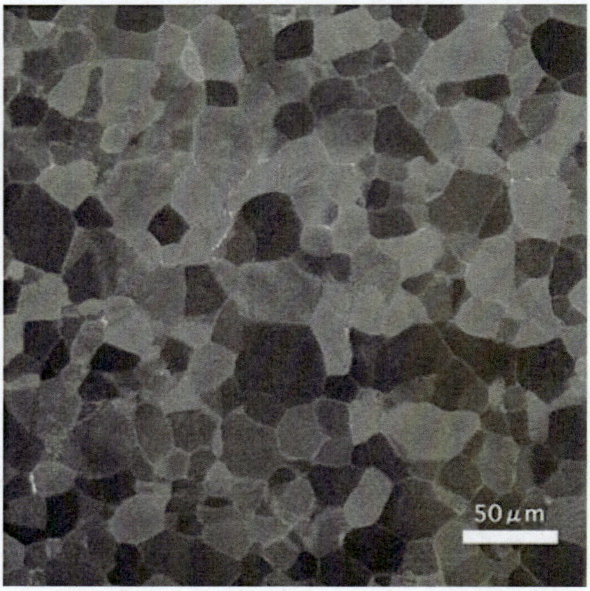

Fig. 5.1 Grain structure of colloidal crystals. Confocal laser scanning micrograph. Silica colloid, diameter $d =$ 110 nm, particle volume fraction $\phi = 0.03$, salt concentration $C_s = 2$ μM, effective charge number Z_{eff} $= 570$, and growth time = 24 h. Taken in reflection mode

fraction $\phi = 0.03$, salt concentration $C_s = 2$ μM, and effective charge number $Z_{eff} =$ 570). The micrograph was taken in reflection mode 24 h after sample preparation. A single compartment in a polycrystalline structure with oriented crystal lattice planes is called a *crystal grain* or *crystal domain*. The crystal grains appear to be different in brightness because the intensity of reflected light differs depending on the orientation of the crystal lattice planes (Fig. 5.1).

Figure 5.2 shows schematically how crystal structures are formed. Crystal nuclei are first generated (nucleation) and then grow in size (crystal growth). The grain structure shown in Fig. 5.1 is the result of the growth and collision of microcrystals. Details of previous studies on colloidal crystal growth are presented in a review by Palberg [1].

The size of crystal grains varies significantly depending on the conditions under which crystals are formed. The driving force for crystallization is weaker for conditions closer to the phase boundary of the crystallization phase diagram. Then nucleation is less frequent, and the growth rate slows down. Therefore, under conditions close to the phase boundary, large crystal grains grow, though it takes a longer time. For an introduction to crystal growth studies, see textbooks on crystal growth science [2–5].

As explained in Chap. 2, according to thermodynamics, the state of the system changes in the direction of decreasing free energy. Under conditions where crystallization occurs, the crystalline state has smaller free energy than the disordered state. At the boundaries between neighboring crystal grains, the local free energy is higher than in the crystalline state because the particle's arrangements are disordered. Thus, the change proceeds so that the total area of grain boundaries decreases with time. The larger the average size of the grain, the smaller the total area of the grain

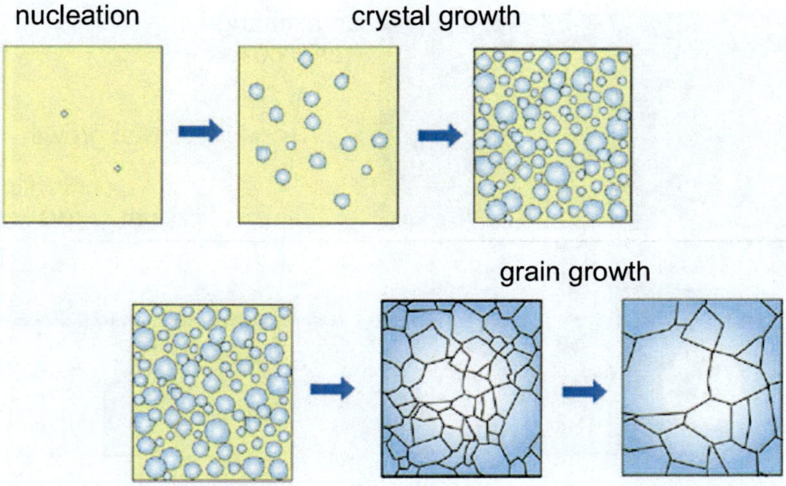

Fig. 5.2 Illustrations of the crystal growth and grain growth processes

boundary is. Therefore, the average size of the grain increases with time, as shown in the lower panel of Fig. 5.2. This process is called "grain growth." Grain structures and their growth processes have been studied in detail in the field of metallurgy [6]. In addition, various structures similar to grain structures exist in nature, including the structure of soap bubbles. These are called cellular structures, also studied in the field of soft matter physics [7].

5.1.2 Overview of This Chapter

As discussed in Chap. 1, colloidal crystals have potential applications as optical and other materials. Since the applications of crystalline materials are significantly dependent on their size and quality, methods have been devised to control crystallization [8–10]. Examples include shear annealing of polycrystals confined in submillimeter gaps [11, 12], epitaxial growth from 2D templates[13], and electrophoresis [14]. The construction of two-dimensional crystals using interfaces has also been actively investigated [15]. These methods have yielded large-area, thin (<100 μm), optically fine film-like colloidal crystals. However, it is difficult to fabricate large 3D colloidal crystals.

In this chapter, we present the results of our work on the fabrication of large, high-quality, 3D colloidal crystalline materials. Section 5.2 introduces the phenomenon of the formation of large cm-sized charged colloidal crystals, simply by adding a base to silica colloids. Next we report on the unidirectional crystal growth of silica colloids that we examined based on this finding. Growing large single crystals is a very important issue in the field of metal and semiconductor crystal growth. For this purpose, unidirectional crystal growth under a temperature gradient is often used. As

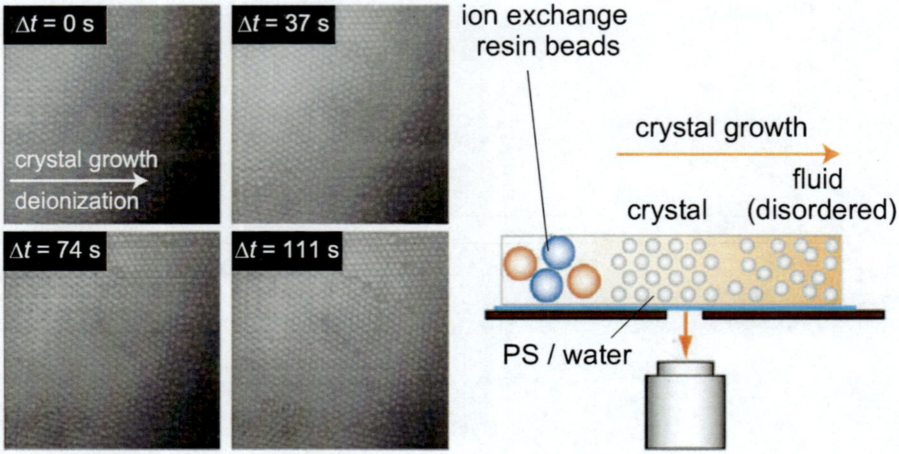

Fig. 5.3 Unidirectional crystal growth by deionization. An aqueous dispersion of charged polystyrene (PS) particles, $d = 430$ nm

discussed in Chap. 2, experimental parameters such as Z, ϕ, and C_s are more effective than the temperature in controlling crystallization in colloidal systems. A gradient of these parameters allows the unidirectional growth of colloidal crystals. As an example, Fig. 5.3 shows a photograph of crystal growth in one direction when deionized by ion exchange resin beads. The sample is an aqueous dispersion of charged polystyrene (hereinafter, PS) particles of $d = 430$ nm. In charged colloidal crystals formed by electrostatic repulsion, the solid–liquid interface is not well defined.

In addition, by using the temperature dependence of Z, Cs, or ϕ, colloidal crystallization can be controlled by temperature. This allows crystal growth under a temperature gradient. Section 5.3 summarizes the effect of temperature on crystallization in charged colloidal systems. We report unidirectional growth and zone melts under temperature gradients.

Finally, Sect. 5.4 introduces impurity exclusion in charged colloids and depletion-attraction systems. Unlike atoms and molecules, it is impossible to create colloidal dispersion systems consisting of particles of exactly equal size. Also, actual colloidal samples contain particle aggregates. Particles or agglomerates that differ significantly in size from the majority of particles act as impurities in the crystal and disrupt the crystal structure. Atomic and molecular crystalline materials often require high purity in their applications. For example, the silicon used in electronic components requires a purity of 99.9999% or higher. Therefore, crystal refining technology is very important. Impurities in materials can be spontaneously eliminated from the crystal structure during the crystallization process. Various methods utilizing this spontaneous exclusion of impurities have been used to purify single crystals of metals and semiconductors [3]. For low-molecular-weight organic and inorganic compounds, crystals of high purity can be obtained by repeatedly crystallizing from a saturated solution, dissolving, and then crystallizing again in a saturated solution. This method is called recrystallization purification. The same impurity elimination phenomenon occurs with colloidal crystals. In colloidal systems, the elimination of impurities can be observed in situ and in real time with an optical microscope.

5.2 Unidirectional Crystallization of Charged Colloids

5.2.1 *Formation of Large Crystals by Addition of Base*

First, we would like to introduce the growth of colloidal silica crystals by the addition of a base, which was found by chance [16]. Figure 5.4 shows a schematic of the crystal growth process. Initially, the colloid was in a fluid phase where particles were randomly arranged. As described in Chap. 2, adding a base to the silica colloid increases the surface charge number of the silica particles, resulting in crystallization. Adding a small drop of concentrated NaOH caused the NaOH to diffuse through the dispersion, and the crystal region increased accordingly. Measurement of the Bragg wavelength of the crystals by fiber spectroscopy (Sect. 3.4.2) showed that the ϕ values in the crystals were slightly higher than in the fluid phase. Thus, the mass density of the crystalline region was greater than its value in the fluid region, and the crystals settled at the bottom of the cell. The excess base in the settled crystals then diffused upward, causing the crystals to grow in a unidirectional manner. Figure 5.5a shows the growth process of the colloidal silica ($\phi = 0.023$, 3 mL) upon the addition of 10 μL of 10 mM NaOH solution. Figure 5.5b presents the appearance of colloidal silica ($\phi = 0.023$, 4 mL) three days after adding NaOH solution (13.3 μL, 10 mM). Colloidal crystals consisting of large crystal grains of about 1 cm^3 were formed. Note that when the NaOH solution of the same concentration and same volume was dropped and immediately shaken well to homogenize the concentration, the resulting crystal grain size was a few 10 μM. Thus, growth by base diffusion increased the crystal size by three orders of magnitude.

5.2.2 *Unidirectional Crystallization of Charged Colloids Under pH Gradients*

Based on the finding described in Sect. 5.2.1 and another related observation (See Column 5.1), we have studied the unidirectional crystal growth of silica colloids. Unidirectional crystallization under a temperature gradient is often used to obtain single crystals of atomic and molecular crystalline materials, including metals and semiconductors. In this process, the material is melted at a high temperature to a fluid state and then cooled from one end to grow large crystals (e.g., the Choklarsky method, Bridgman method, etc.) [3].

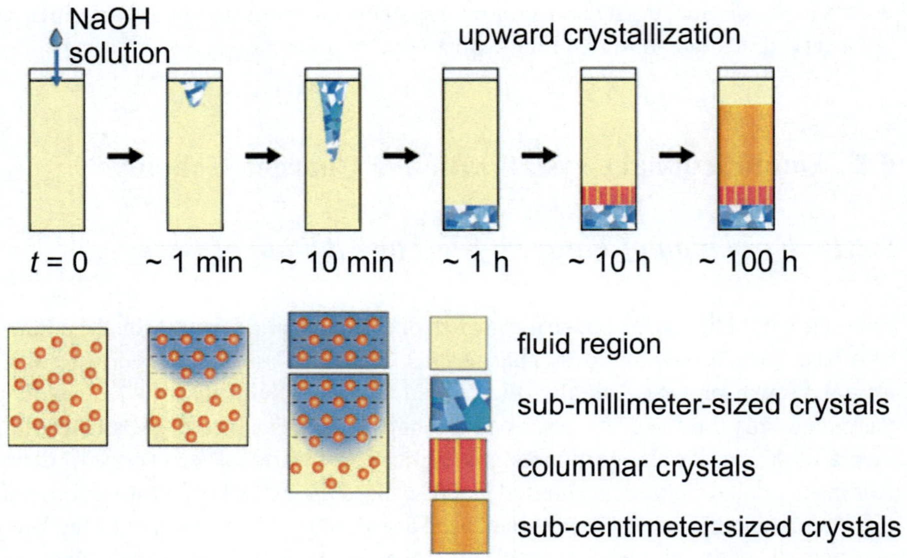

Fig. 5.4 Illustration of the crystal growth process of colloidal silica by additions of concentrated NaOH solution

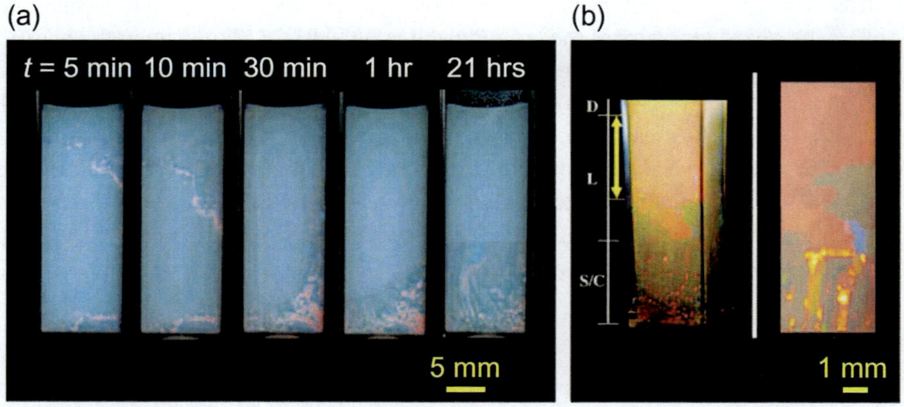

Fig. 5.5 **a** Photographs of the crystal growth process of colloidal silica ($\phi = 0.023$, 3 mL + 10 mM NaOH, 10 μL), **b** appearance of large colloidal crystals three days after adding NaOH ($\phi = 0.023$, 4 mL + 10 mM NaOH, 13.3 μL). Reprinted with permission from Ref. [16]. Copyright (2006) American Chemical Society

As will be discussed in Sect. 5.3, temperature itself has little effect on the crystallization of colloidal systems. Rather, for charged colloidal systems, the important parameters are Z, ϕ, and C_s. Thus, instead of a temperature gradient, a spatial gradient of these three parameters would allow unidirectional growth (Fig. 5.6). Figure 5.7 shows the correspondence of the growth methods with the crystallization phase diagram. The crystal growth described in Sect. 2.5.1 is due to the diffusion of the base, resulting in a gradient in base concentration (in other words, a gradient in pH) and a gradual change in the Z of the silica particles.

Crystal growth by base diffusion was examined in detail with collaborators [17–20]. A schematic of the experimental setup is shown on the left panel in Fig. 5.8. A cell equipped with a semipermeable membrane is filled with a silica colloid sample and brought into contact with a reservoir of an aqueous base solution. The cell has size = 1 × 1 × 4 cm^3, and the reservoir has volume = 500 ml. We used aqueous solutions of weak base pyridine (Py) in the experiments. Diffusion was unlikely to occur with the strong base NaOH at low concentrations. This is because NaOH dissociates completely into Na^+ and OH^- ions in water and reacts almost completely

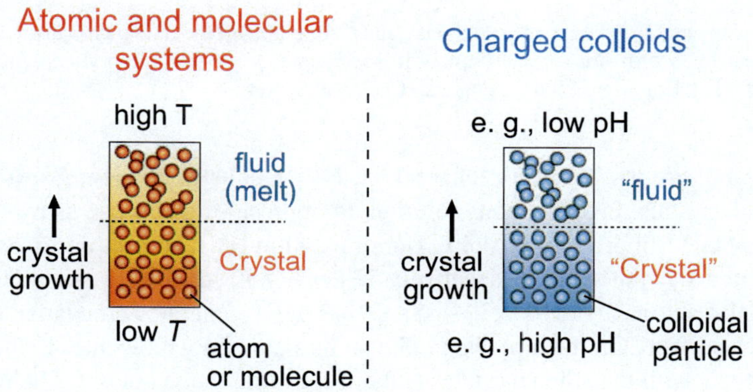

Fig. 5.6 Illustrations of unidirectional crystal growth in atomic/molecular systems and colloidal systems

Fig. 5.7 Unidirectional crystallization of charged colloids under the gradient of the three parameters

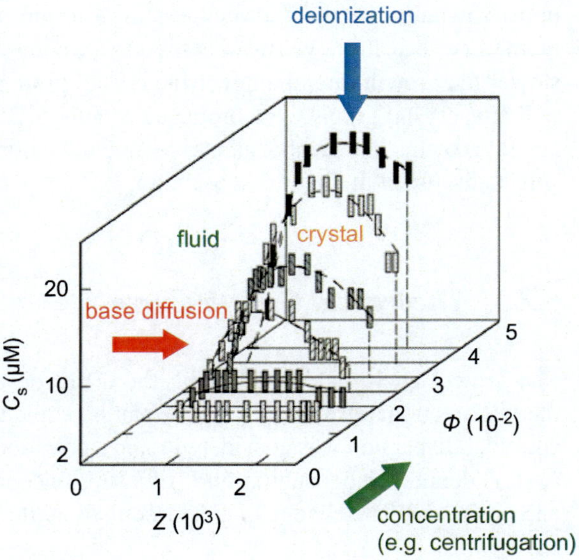

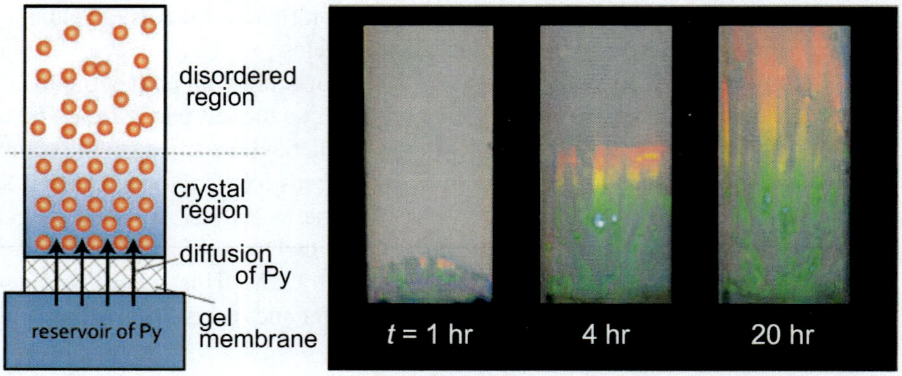

Fig. 5.8 Schematic of the experimental setup and photographs of the crystal growth process of silica colloid due to Py diffusion ($d = 110$ nm, salt-free system, $\phi = 0.034$). Reprinted with permission from Ref. [17]. Copyright (2004) American Chemical Society

with silanol groups. Diffusion of weak base Py results in efficient growth of silica colloidal crystals. In an aqueous solution, Py molecules dissociate as py + $H_2O \rightarrow$ $pyH^+ + OH^-$, indicating basicity. Here py and pyH^+ represent undissociated and dissociated Py molecules, respectively. When Py is added to a silica colloid, the chemical reaction Si–OH + py $\rightarrow$ $Si–O^-$ + pyH^+ generates a negative charge on the silica surface. On the other hand, the undissociated Py molecules in the medium do not react with the silica particles and diffuse through the colloid. The right panel of Fig. 5.8 shows a photograph of the crystal growth process due to Py diffusion ($d = 110$ nm, salt-free system, $\phi = 0.034$). The Py concentration in the reservoir $[Py]_0$ was 100 mM. The crystalline region shows Bragg diffraction of visible light. Figure 5.9 presents some other examples of the growth process. The images in the upper row are a result of a thickness of 4~6 mm polymer gel as a semipermeable membrane, and the lower row is the case where a Teflon membrane is used. The slower the growth rate, the larger the crystal grain size, and with the largest, $1 \times 1 \times 3$ cm^3 crystal grains were produced. Figure 5.10 shows the $[Py]_0$ dependence of crystal growth. The symbols are experimental values, and the curves are theoretical values discussed in the next subsection.

5.2.3 *Theoretical Growth Curve*

The growth curve calculated using the diffusion equation [17, 18] indicates that the unidirectional growth of crystals can be explained by base diffusion. Here, the colloidal dispersion was considered to be a continuous, uniform substance. Let $C = C(x, t)$ denote the concentration C of a substance in solution as a function of time t and position x. The change in C is described by the diffusion equation

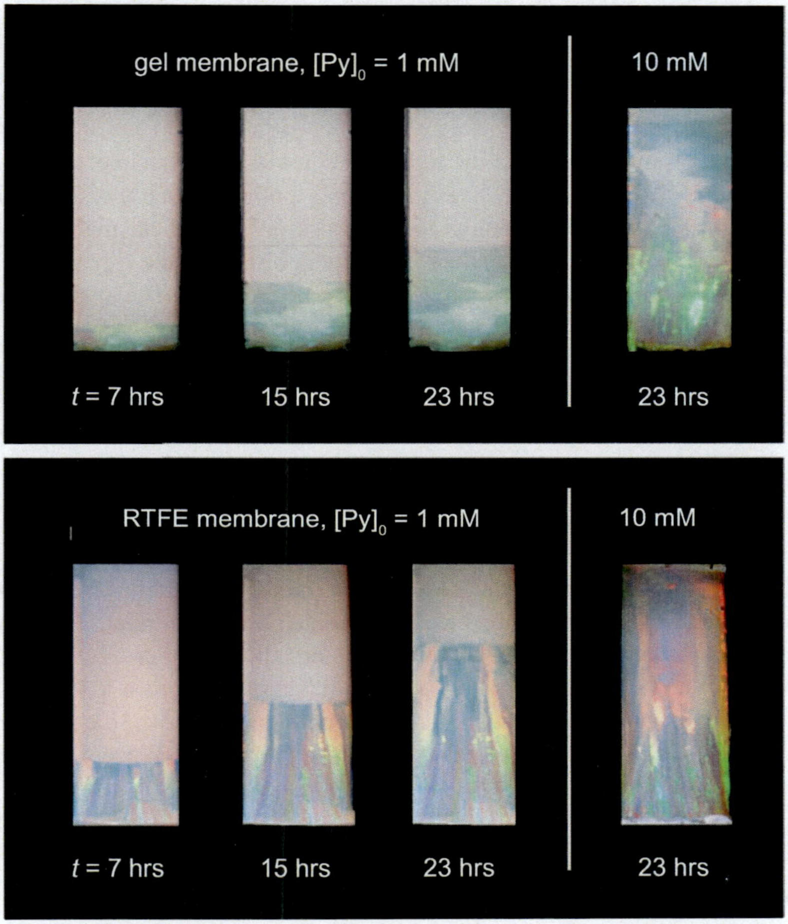

Fig. 5.9 Growth process of silica colloidal crystals by diffusion of the base. The upper part, gel membrane (thickness = 4~6 mm); the lower part, Teflon membrane. Reprinted with permission from Ref. [18]. Copyright (2007) American Chemical Society

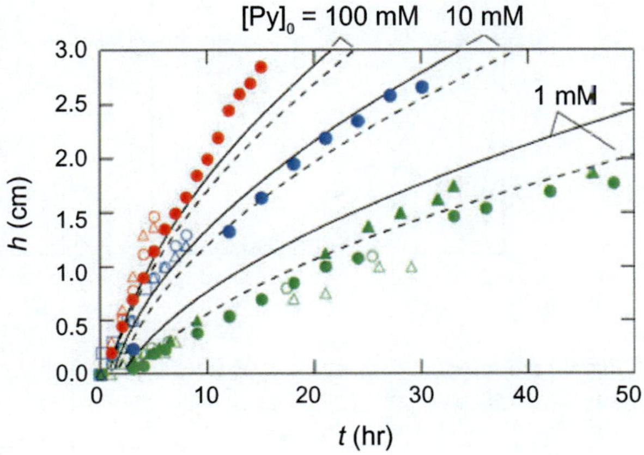

Fig. 5.10 Growth curves of the colloidal silica crystals at three values of $[Py]_0$. Symbols, observed growth; curves, calculated from adsorption-diffusion equation. Solid and dashed curves are those at salt-free conditions and at 5 μM NaCl. Reprinted with permission from Ref. [18]. Copyright (2007) American Chemical Society

$$\partial C/\partial t = D\left(\partial^2 C/\partial x^2\right). \tag{5.1}$$

Equation (5.1) means that for diffusion, the time variation of C (left-hand side) is proportional to the rate of change of the concentration gradient of C (right-hand side). The constant D is called the diffusion constant (for Py molecule in water at 25 °C, $D = 8.75 \times 10^{-6}$ cm^2/s). See Appendix E for the derivation of (5.1). The time variation of the Py concentration at various positions can be calculated using (5.1) if the reaction between Py and silica is negligible. Because the reaction with silica also decreases the Py concentration, (5.1) must be modified. The chemical reaction between Py and silanol groups can be regarded as the electrical adsorption of Py ions on the silica surface. When the concentration of adsorbed Py is denoted as $S = S\ (x, t)$, the concentration of free Py, $C(x, t)$, decreases not only by diffusion but also by adsorption. Therefore, a term $-\partial S/\partial t$ must be added to the right-hand side of (5.1). By noting that $\partial S/\partial t = (\partial S/\partial C)(\partial C/\partial t)$, we have

$$\partial C/\partial t = [D/(1 + \partial S/\partial C)]\partial^2 C/\partial x^2. \tag{5.2}$$

Equation (5.2) is called the adsorption-diffusion equation. The relationship between S and C can be obtained from titration experiments (Sect. 3.2.3) to calculate $\partial S/\partial C$. Equation (5.2) can be solved numerically to obtain $C(x, t)$ and $S(x, t)$. Figure 5.11 shows the profiles at $[\mathrm{Py}]_0 = 1$ mM. Here x is taken as the distance from the reservoir. In calculation, we took into account the thickness of the gel membrane = 5 mm. Note that the charge number Z is proportional to S. When S is sufficiently large, the silica colloid crystallizes.

Figure 5.12 shows the crystallization phase diagram defined by the concentration of NaCl, [NaCl], and base concentrations. The pH titration confirmed that all added

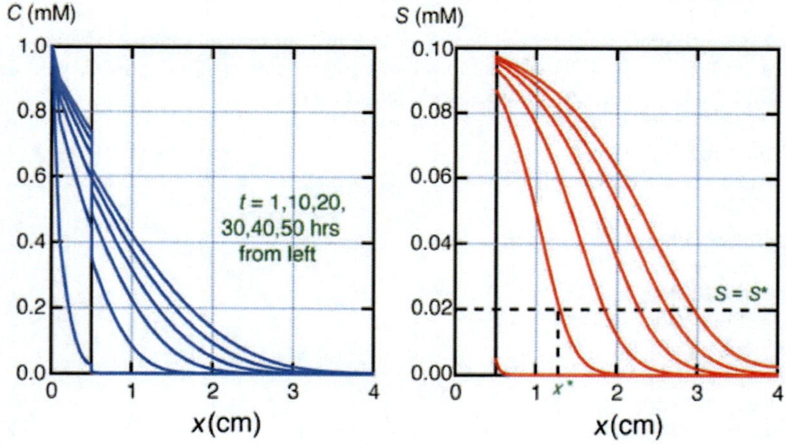

Fig. 5.11 Profiles of $C\ (x, t)$ and $S\ (x, t)$ for $[\mathrm{Py}]_0 = 1$ mM (x is the distance from the reservoir). The thickness of the gel membrane = 5 mm. Reprinted with permission from Ref. [18]. Copyright (2007) American Chemical Society

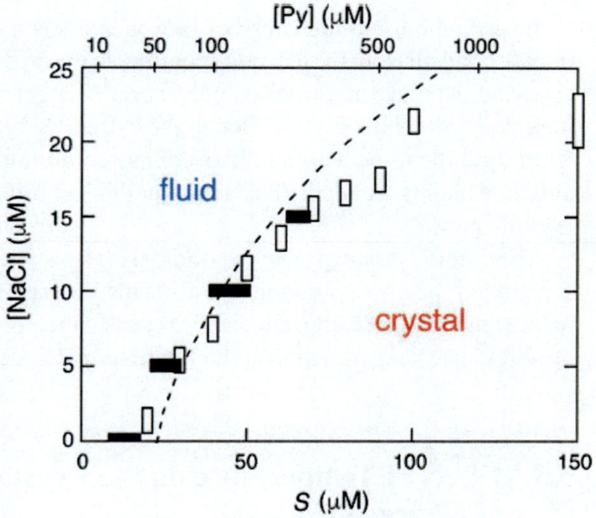

Fig. 5.12 Crystallization phase diagram of silica colloid defined by the concentrations of NaCl and bases. The open rectangular symbols are the phase boundaries for NaOH. Filled symbols are phase boundaries for Py systems. The dashed curve is the theoretical phase boundary for the RGK phase diagram. Reprinted with permission from Ref. [18]. Copyright (2007) American Chemical Society

NaOH reacted with silica under the experimental conditions. Thus, therefore, we can assume [NaOH] = S. The open rectangular symbols in Fig. 5.12 are the crystallization phase boundaries when NaOH is used. We also determined the phase boundary at four [NaCl]s. The relationship between S and [Py] was obtained by the pH titration. The corresponding S values are shown as filled symbols in Fig. 5.12 ([Py] is also shown as the upper axis in the figure). The phase boundaries for the NaOH and Py thus obtained agreed well. The dashed curve is the theoretical phase boundary for the RKG phase diagram (Chap. 2), which is in good agreement with the experiment.

From Fig. 5.12, we see that the S value required for crystallization, S^*, is about 20 μM at salt-free conditions. The crystal growth curves were calculated for x (= x^*) satisfying $S\ (x,\ t) = S^*$ at various times t. The solid curves in Fig. 5.10 are the theoretical growth, which agreed well with the observations over three orders of magnitude. Note that a small amount of ionic impurities may occur and affect the crystal growth. The growth curves for [NaCl] = 5 μM are shown by the dashed curves in Fig. 5.10.

We note that the diffusion of the salt of a strong base and weak acid, e.g., sodium carbonate $NaHCO_3$, also causes a pH gradient and similar unidirectional growth. See the [18] for more details.

Column 5.1

Unidirectional growth under pH gradient

In order to use colloidal crystals as optical materials, we explored a method to fabricate cm-size charged colloidal crystals together with our collaborators. Since the crystal size is usually a few mm at most, a new fabrication method was needed. We then came up with the idea of using gel-fixed colloidal crystals (Chap. 6) as seed crystals. As a trial, we put silica microcrystals fixed with a thin gel film into a purified silica colloid. Then, the crystals grew from the gel pieces in a columnar shape.

Investigation into the cause of this phenomenon revealed that it was due to the release of alkali from the gel, quite contrary to the original assumption that it would act as a seed. At that time, we used polyacrylamide gel obtained by photo-radical polymerization to fix the crystal. Due to the extremely high output of the mercury vapor lamp used then, the acrylamide decomposed during the UV irradiation, producing amide, which is an alkali. The amide in this gel diffused into the silica colloid, and crystals grew.

The unidirectional growth of colloidal crystals was then reproduced by adding the aforementioned base solution and allowing the base to diffuse through the semipermeable membrane. Large cm-sized crystals were produced by this method and were used for space experiments on the International Space Station (Chap. 7).

5.3 Effect of Temperature on the Crystallization of Charged Colloids

In this section, we discuss the effect of temperature on the crystallization of charged colloids. In Sect. 5.3.1, we first show that the effect of temperature on electrostatic interactions is small due to the effect of the temperature dependence on the permittivity of the medium [21]. In Sect. 5.2.2, we discuss the temperature control of crystallization using the temperature dependence of Py dissociation [22]. In Sect. 5.3.3, we report crystallization based on the temperature dependence of surfactant adsorption [23, 24]. Furthermore, in Sect. 5.3.4, unidirectional growth under a temperature gradient will be presented, and in Sect. 5.3.5, zone melts will be described.

5.3.1 Temperature Dependence of Electrostatic Interaction

In their paper on the crystallization phase diagram, Robbins et al. [25] pointed out that temperature T does not significantly affect electrostatic interaction. In both of the parameters k_BT/U and λ (Fig. 2.15) in their theoretical phase diagram, T is included as the product by the permittivity ε, i.e., εT. It is known that ε is approximately proportional to $1/T$. Therefore, εT does not change significantly with temperature. In other words, the state of the colloid does not change significantly even if the temperature changes. In aqueous systems, εT rather decreases slightly as T increases, so it is predicted that charged colloidal systems may "crystallize as the temperature rises" [25].

We investigated the effect of temperature on various colloidal systems under experimental conditions near phase boundaries [21]. By calculating the temperature changes of k_BT/U and λ under various conditions near the phase boundary, it is expected that (i) the crystals melt due to the temperature rise if λ is significant and that (ii) the system crystallizes due to the temperature rise if λ is small. Figure 5.13

shows some examples of the temperature dependence of the reflection spectrum. Figure 5.13a is the reflection spectra for an aqueous dispersion of PS particles, with high C_s and large λ (d = 96 nm, Z = 735, ϕ = 0.01, C_s = 10.8 μM) and the crystals melted on heating. On the other hand, Fig. 5.13b is data for a sample with low C_s and small λ (silica, d = 108 nm, Z = 314, ϕ = 0.035, C_s = 2.0 μM), which crystalized upon heating. Figure 5.13c shows changes in the reflectance spectra of the sample in Fig. 5.13b on successive variations of T, showing that the crystallization was reversible. The blue and red curves represent the spectra at 5 °C and 50 °C, respectively.

Figure 5.14 shows the correspondence between the experimental results of various colloidal samples and the RKG phase diagram, which are qualitatively consistent [21]. However, except for the conditions near the phase boundary examined here, the influence of temperature on the phase behavior of the charged colloid was small.

Nonetheless, the temperature is a universal external variable and can be easily adjusted experimentally. Therefore, it would be helpful to find out the method to control colloidal crystallization by temperature. As will be described below, the crystallization can be indirectly controlled by using the temperature dependence of

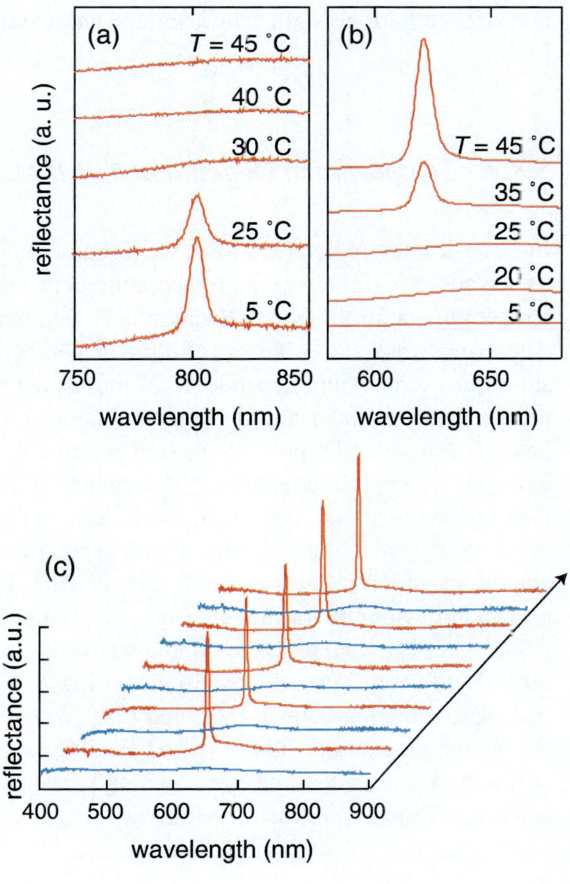

Fig. 5.13 Reflection spectra of aqueous charged colloids undergoing **a** melting and **b** freezing transitions upon heating. Samples: aqueous dispersions of **a** PS (d = 96 nm, Z = 735, ϕ = 0.01, C_s = 10.8 μM), **b** silica (d = 108 nm, Z = 314, ϕ = 0.035, C_s = 2.0 μM). The spectra are shifted vertically for clarity. **c** Changes of the reflectance spectra of the silica samples used in **b** upon successive variations of T. The blue and red curves represent the spectra at 5 °C and 50 °C, respectively. Reprinted with permission from Ref. [21]. Copyright (2011) American Chemical Society

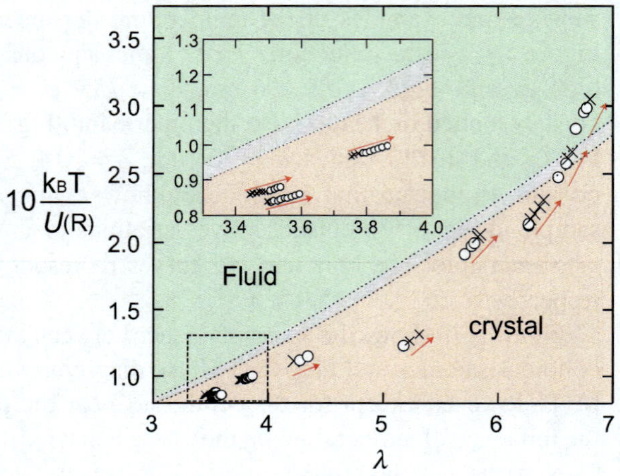

Fig. 5.14 Phase diagram defined by $k_B T/U(a)$ and λ. The circles and crosses correspond to the crystal and liquid states observed upon elevating the temperature from 5 to 45 °C. The inset shows the magnified diagram at low λ. Reprinted with permission from Ref. [21]. Copyright (2011) American Chemical Society

the central variables Z, ϕ, and C_s instead of the temperature itself. In the following, we will introduce methods that utilize the temperature changes of Z and C_s [23, 24]. As a method using the temperature dependence of ϕ, it has been reported that microgel colloids crystallize by a volume phase transition on decreasing temperature [26].

5.3.2 *Temperature Dependence of Base Dissociation*

In Chap. 2 and previous sections, we discussed the crystallization of silica colloids by the addition of bases. The temperature dependence of base dissociation enables temperature control of crystallization [22]. At higher temperatures, the dissociation degree of Py in water is larger, resulting in a higher base concentration. Figure 5.15a shows the temperature dependence of the degree of dissociation of Py. Thus in the presence of Py, under appropriate conditions, Z of silica particles increases upon heating. Figure 5.15b presents the temperature dependence of Z of silica particles having three kinds of counterions, determined from electrical conductivity measurements. For H– or Na-type silica, Z was almost constant regardless of temperature, but Z of Py-type silica increased with temperature. Therefore, under appropriate conditions, silica colloids crystallize on heating. Figure 5.16 is a photograph of the temperature variation of the states of silica samples of two [Py]s (d = 110 nm, ϕ = 0.035). At [Py] = 60 μM, the sample was crystalline at T = 5–40 °C, but at [Py] = 30 μM, the sample crystallized on increasing T. Figure 5.17a shows the crystallization phase diagram defined by T and [Py]. We note that the crystalline phase is on the higher temperature side. It is seen that the crystallization temperature T_c can be adjusted over a wide range by tuning [Py]. The reflection spectra at [Py] = 30 μM at various Ts are represented in Fig. 5.17b.

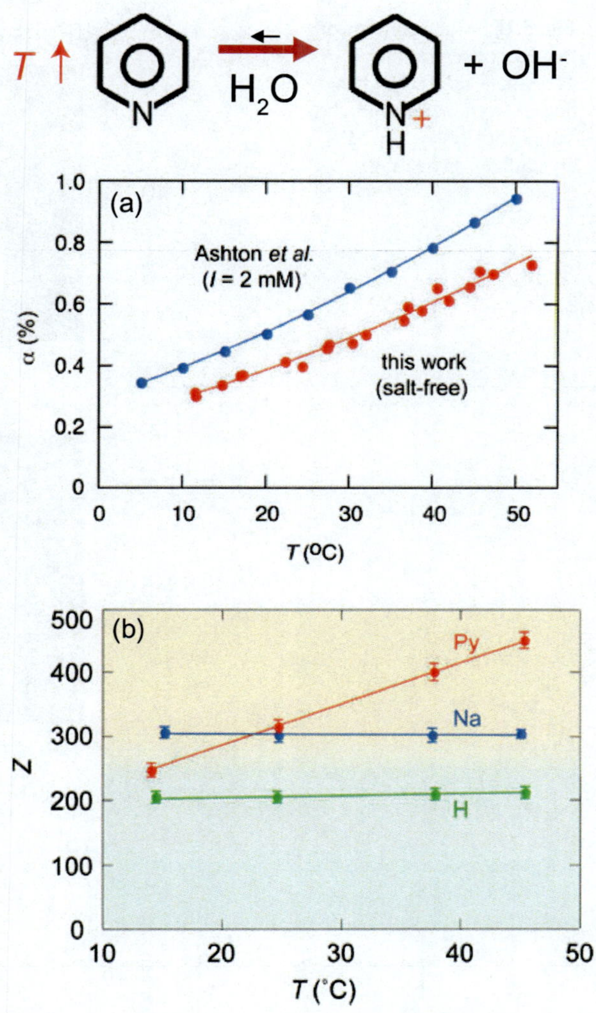

Fig. 5.15 **a** Temperature dependence of the degree of dissociation of Py. **b** Z versus T plot of H, Na, and Py-type silica ($d = 110$ nm). Reprinted with permission from Ref. [22]. Copyright (2013) American Chemical Society

5.3.3 Temperature Dependence of Ionic Surfactant Adsorption

The method described in 5.3.2 is applicable to the particles with weakly acidic dissociating groups. We then examined more generic methods. By using adsorption of ionic surfactants, we were able to crystallize various kinds of particles other than silica [23, 24]. The principle is shown in Fig. 5.18. First, consider an aqueous dispersion of particles with a hydrophobic surface, e.g., charged PS particles. The surfactant concentration was set below the critical micelle concentration. When ionic surfactant molecules with the same charge as the particles are added to the dispersion, some of them adsorb on the particle surface and impart a surface charge, while the remaining non-adsorbed molecules dissolve in the medium as ions.

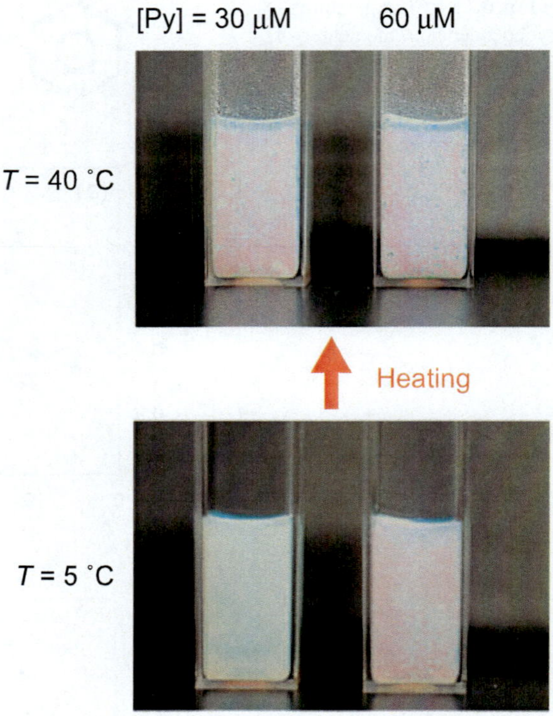

Fig. 5.16 Silica colloids of two Py concentrations (d = 110 nm, $\phi = 0.035$) when heated from 5 to 40 °C. The sample at [Py] = 30 μM crystallizes on heating

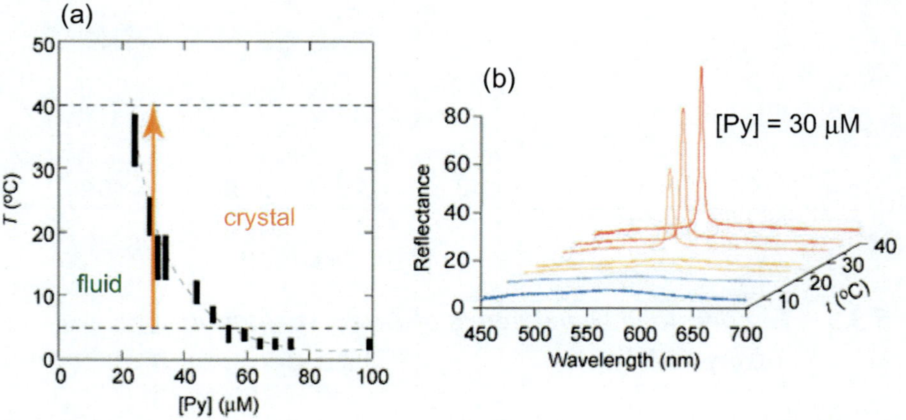

Fig. 5.17 **a** Crystallization phase diagram of silica + Py sample and **b** reflection spectra at various values of temperatures ([Py] = 30 μM). Reprinted with permission from Ref. [22]. Copyright (2013) American Chemical Society

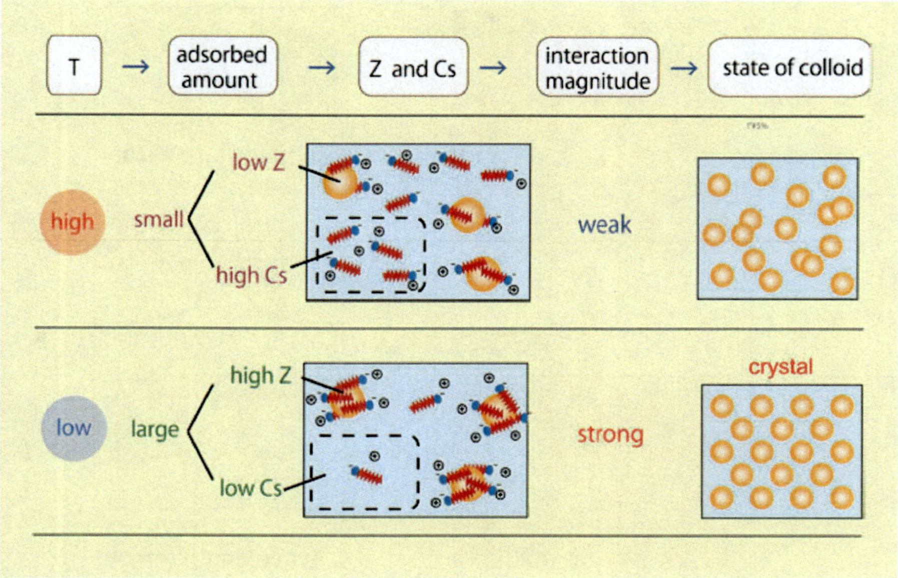

Fig. 5.18 Principle of the control of the crystallization by the addition of ionic surfactants. When a colloidal system containing an ionic surfactant is cooled, Z increases and Cs decrease. Thus, under appropriate conditions, the system crystallizes upon cooling and melts upon heating

Based on thermodynamics, it can be concluded that the amount of adsorbed molecules is larger at lower temperatures. Adsorption occurs due to the attraction between the surfactant molecule and the particle, but molecular entropy is greater when the molecule is unabsorbed. At higher temperatures, thermal motion is more intense, and the concentration of unabsorbed molecules increases. Thus, cooling a colloidal system containing an ionic surfactant increases the adsorption amount, resulting in an increase in the Z of the particles and a decrease in the C_s of the medium. The two changes both result in stronger electrostatic repulsion between particles, and under the appropriate conditions, the colloidal system crystallizes when cooled and melts when heated.

Figure 5.19a presents micrographs of an aqueous dispersion of PS particles in the presence of an ionic surfactant, sodium octadecyl sulfate (SOS), at two temperatures ($d = 430$, $\phi = 0.01$, [SOS] = 4 μM). The reflection spectra of the PS + SOS colloid ($d = 124$, $\phi = 0.01$, [SOS] = 100 μM) are shown in Fig. 5.19b. The PS colloids thus crystallized upon cooling when the ionic surfactants were added. Figure 5.20 shows the crystallization phase diagram of the PS colloid ($d = 132$, $\phi = 0.01$) with sodium alkyl sulfate, having various chain lengths. In Fig. 5.20, n is the number of carbon atoms in the alkyl group. The crystal phase is on the lower temperature side than the rectangular symbol, and the crystallization temperature can be adjusted over a wide range by the carbon number and the surfactant concentration. A phase diagram with sodium chloride is also shown for comparison; in this case, the phase boundaries reflect the effect of the temperature dependence of the electrostatic interactions.

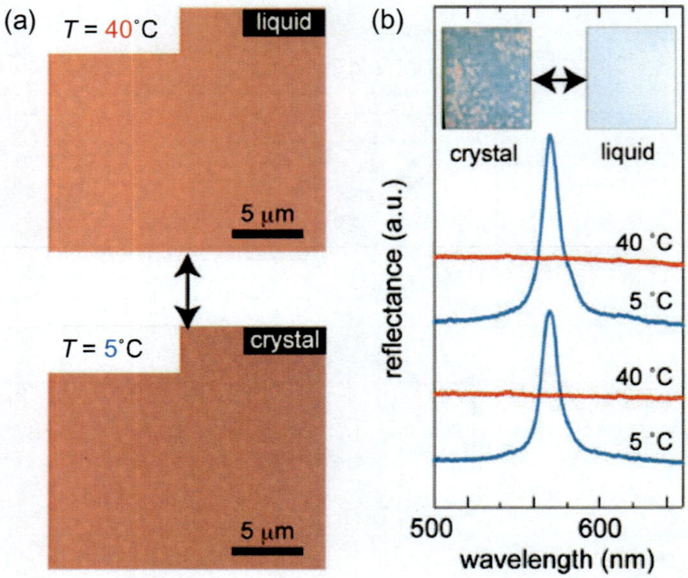

Fig. 5.19 Micrographs of PS dispersions at two temperatures in the presence of sodium octadecyl sulfate ($d = 430$, $\phi = 0.01$, [SOS] = 4 μM). Reprinted with permission from Ref. [23]. Copyright (2014) American Chemical Society

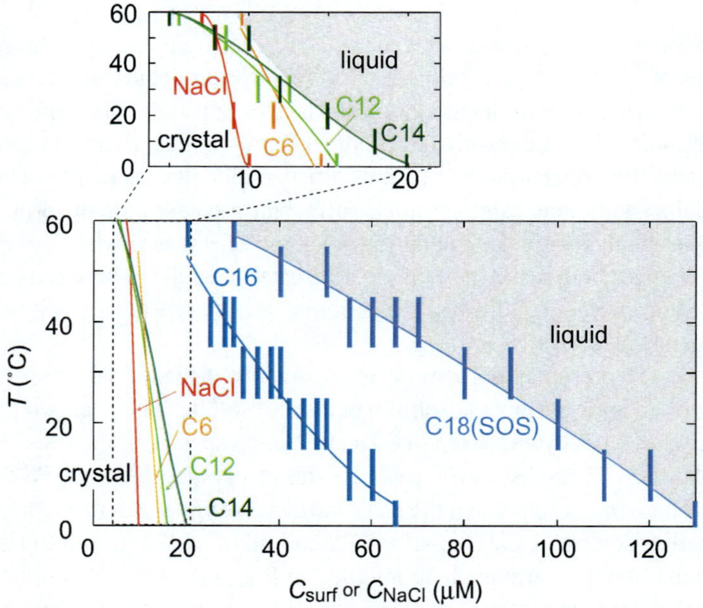

Fig. 5.20 Crystallization phase diagram of the aqueous dispersion of PS particles with various ionic surfactants and sodium alkyl sulfate. *n* is the number of carbon atoms in the alkyl group. For comparison, a phase diagram with sodium chloride is also shown. Reprinted with permission from Ref. [23]. Copyright (2014) American Chemical Society

For hydrophilic particles, the adsorption of ionic surfactants in water is unlikely to occur. Even then, the crystallization can be similarly controlled by adsorbing nonionic surfactants on the particle surface in advance and introducing a hydrophobic group. We used a nonionic surfactant composed of a polyoxyethylene chain and an alkyl chain (polyoxyethylene phenyl ether, EPC) to make the particle surfaces hydrophobic. Then an ionic surfactant, sodium octadecyl sulfate (SOS), is added. In this system, crystallization could be controlled by temperature.

Using this method, we were able to crystallize titania (titanium dioxide, TiO_2) particles with a high refractive index (~2.5), which are expected to be applied to photonic crystals. Micrographs of titania samples (d = 540 nm, ϕ = 0.072) are shown in Fig. 5.21a. The addition of EPC and SOS allowed for temperature response. Figure 5.21b shows the reflection spectrum of titania colloid (d = 210 nm, ϕ = 0.15; [EPC] = 600 μM, [SOS] = 60 μM) as a function of temperature.

Gold particles have attracted much attention in plasmonics. Here, the gold particle surface was modified with hexadecane thiol, which is used to prepare self-assembled monolayers on gold substrates, to introduce hydrophobic alkyl chains. SOS was then added to impart a surface charge; Fig. 5.21c shows micrographs of gold colloidal crystals (d = 100 nm, ϕ = 0.004; [SOS] = 40 μM).

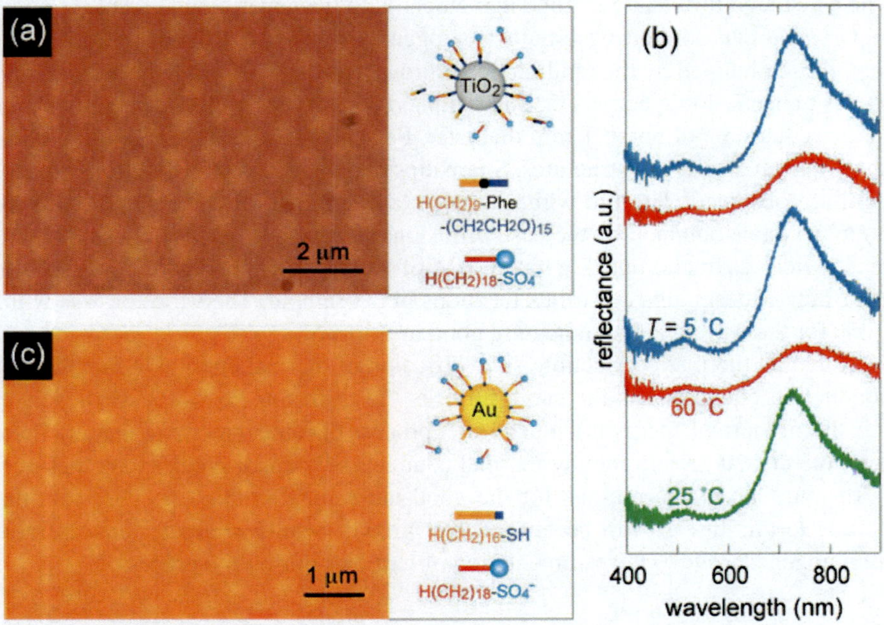

Fig. 5.21 **a** Optical micrograph of charged colloidal crystals of titania colloid (d = 540 nm, ϕ = 0.072) **b** successive change of reflection spectra of titania colloid (d = 210 nm, ϕ = 0.15; [EPC] = 600 μM, [SOS] = 60 μM). Each spectrum is shifted vertically for clarity. **c** A micrograph of the colloidal crystals of gold particles (d = 100 nm, ϕ = 0.004; [SOS] = 40 μM). Reprinted with permission from Ref. [23]. Copyright (2014) American Chemical Society

All of the experiments described above have used negatively charged colloidal particles. For positively charged particles, a similar temperature-induced crystallization occurs by using a cationic surfactant. In this case, cetylpyridinium chloride was used as the cationic surfactant [27].

5.3.4 *Crystallization Under Temperature Gradient*

Using the temperature-induced crystallization described above, we achieved unidirectional growth of colloidal crystals under a temperature gradient. Here we present results for the silica colloid + Py system [28, 29]. Figure 5.22a illustrates an experimental setup used. The sample (d = 110 nm, ϕ = 0.035, [Py] = 27.5 μM) was placed in a quartz cell (cell size = 0.1 × 1 × 4.5 cm^3), and one end of the cell was heated by a heater block. The sample had T_c = 35 °C, and the heater temperature T_0 was 60 °C. An example of the growth process is shown in Fig. 5.22b. Figure 5.23a shows the crystal growth curves at three T_0 values. Figure 5.23b is the temperature profile on the cell surface measured with a thermocouple array. The growth curves (x–t plots) obtained from the relation $T(x, t) = T_c$ are shown for various t values. The calculated results are in good agreement with the experimental results.

Crystals thus formed were quite homogeneous and had a better optical quality than those obtained by the unidirectional growth under a pH gradient. Figure 5.24a shows transmission spectra of crystal samples taken every 5 mm with an optical fiber in a circular area of about 1 mm diameter. For comparison, the broken lines show spectra from disordered structures. Sharp dips due to the Bragg diffraction of visible light are observed. The full width at half maximum (FWHM) was approximately 5 nm. Crystals obtained by the base diffusion, discussed in Sect. 5.2, usually had a much wider diffraction peak with FWHM of 10 nm or more. Figure 5.24a presents the Bragg wavelengths at various locations in the sample. The variation was within 0.1% for the wavelengths, indicating good uniformity.

Note that thermal conduction is the diffusion of thermal energy. In fact, the heat conduction equation has the same form as the diffusion Eq. (5.1). However, the transfer of thermal energy is a much faster process than the transfer of matter. Therefore, the crystal growth rate by thermal conduction is much greater than that of Py diffusion. One of the reasons for the good uniformity in this crystallization may be due to this fast growth because slower growth results in greater concentration inhomogeneity due to the sedimentation of particles during the process.

5.3.5 *Zone Melting*

Zone melting is a method for producing large, high-purity crystals. Its principle is illustrated in Fig. 5.25. When a portion of the crystal is heated by a heater to melt it and then the heater is removed, recrystallization occurs. In the zone-melting method,

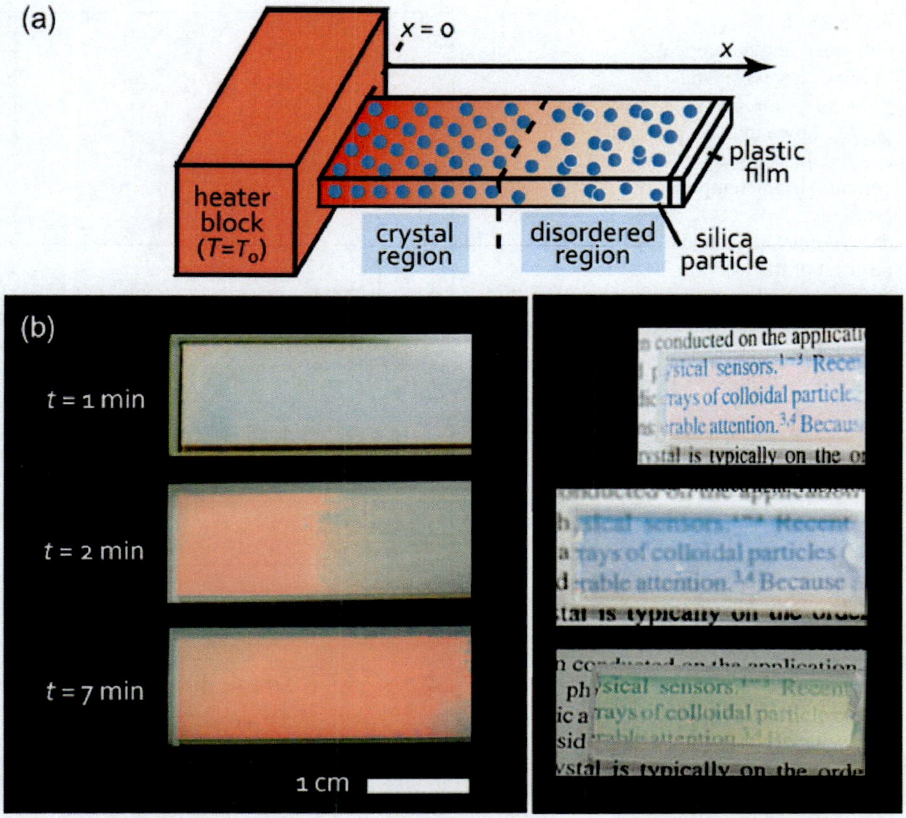

Fig. 5.22 **a** Unidirectional crystallization under a temperature gradient for silica + Py colloid. **b** An example of the growth process. $\phi = 0.035$, [Py] = 35 μM; crystallization temperature T_c = 19 °C; heater temperature T_0 = 40 °C. Reprinted with permission from Ref. [28]. Copyright (2007) American Chemical Society

the recrystallization occurs continuously by moving a narrow heater ribbon along the sample [3]. This method can be used to fabricate single crystals from polycrystals. Furthermore, impurity atoms in the sample can be separated by recrystallization because they tend to segregate during crystallization.

We applied the zone-melting methods to silica + Py colloids [22]. Figure 5.26 shows a schematic drawing of the apparatus and the time variation of the temperature profile. A Peltier device was used to cool and melt the colloidal crystals. Polycrystalline colloidal crystals were used as the starting material. Figure 5.27 shows pictures of the samples recrystallized at various migration velocities. When the migration was sufficiently slow (~10 m/s), crystals of several centimeters in size were formed from submillimeter-sized original polycrystals.

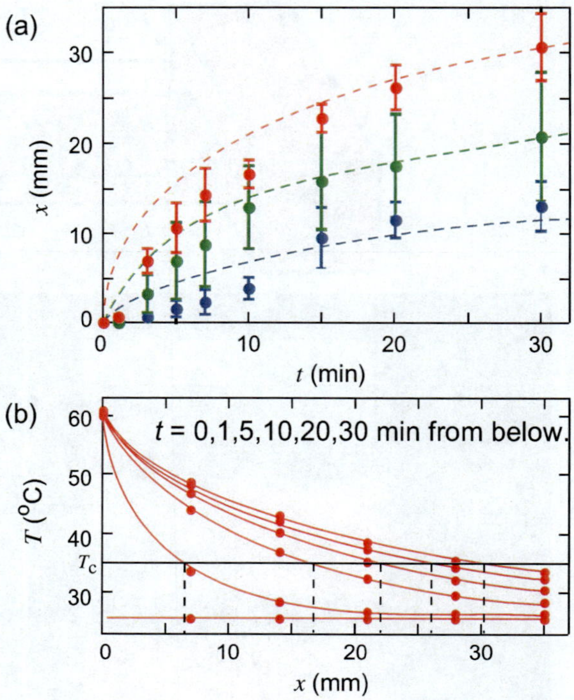

Fig. 5.23 **a** Observed (symbols) and theoretical (dashed curves) crystal growth at $T_0 = 40$, 50, and 60 °C, shown in blue, green, and red, respectively. **b** Time variation of the temperature profile for $T_0 = 60$ °C. The dashed lines indicate the position of the front of crystals at each temperature. Reprinted with permission from Ref. [28]. Copyright (2007) American Chemical Society

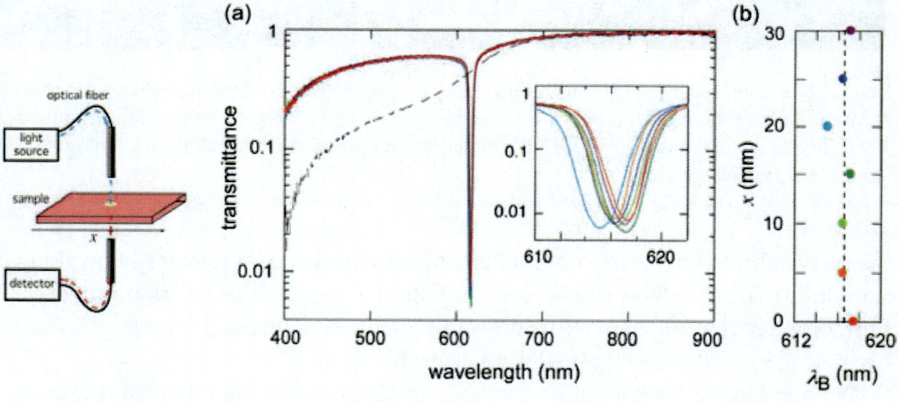

Fig. 5.24 **a** Transmission spectra of the crystals obtained by unidirectional crystallization under temperature gradients. **b** The Bragg wavelength at every 5 mm by an optical fiber in a circular area of about 1 mm diameter. Reprinted with permission from Ref. [28]. Copyright (2007) American Chemical Society

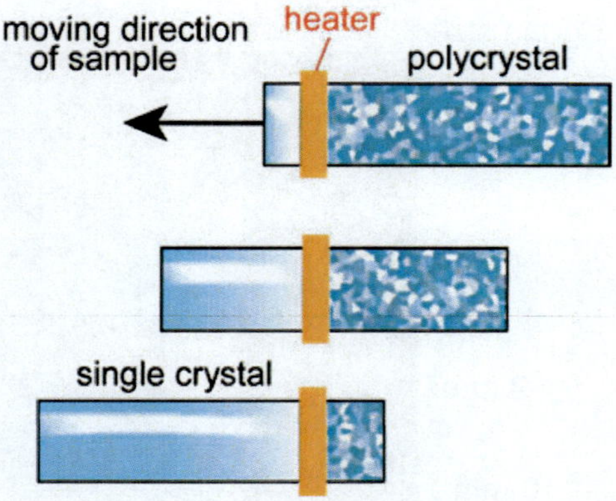

Fig. 5.25 Illustration of zone melting

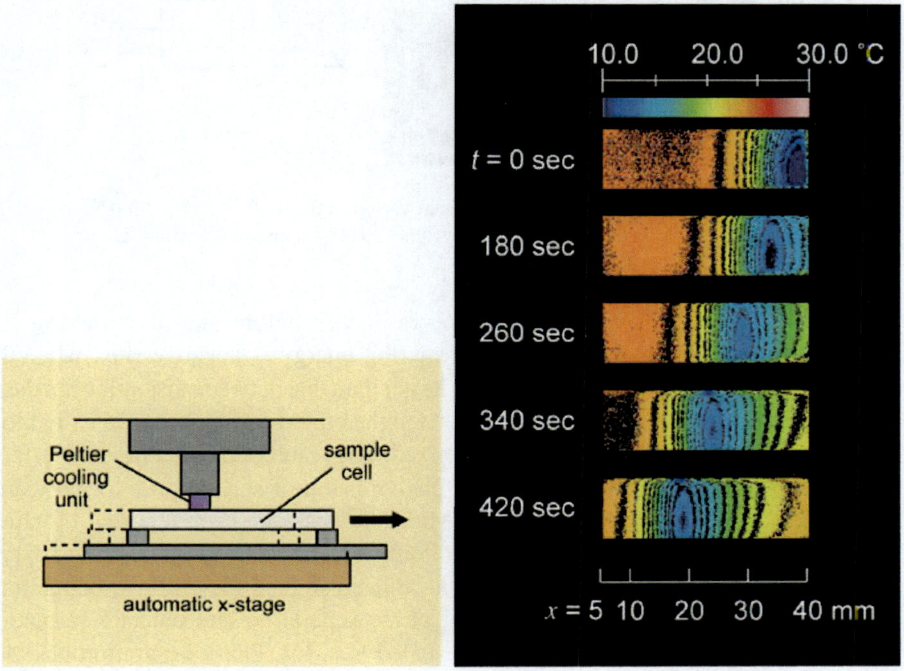

Fig. 5.26 Schematic diagram of the zone-melting apparatus and the time variation of the temperature profiles of the cell surface. Migration velocity $v = 50$ mm/s. Reprinted with permission from Ref. [22]. Copyright (2013) American Chemical Society

5.4 Impurity Exclusions and Phase Separation

5.4.1 Behavior of Multicomponent Colloids

In previous sections, we have discussed the behavior of one-component systems. In multicomponent colloids, phase separation is sometimes observed as common in atomic and molecular systems. In general, whether two substances are miscible

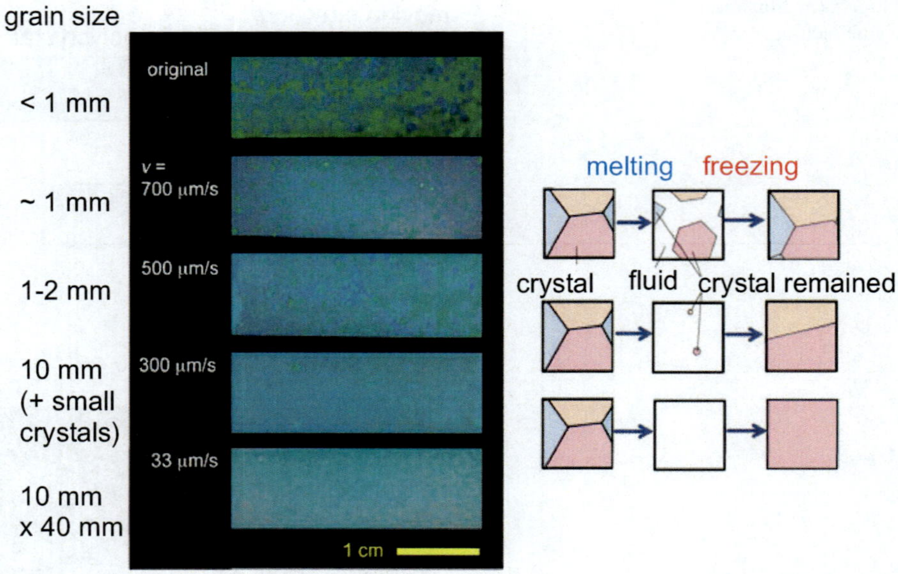

Fig. 5.27 Pictures of the recrystallized samples at various values of v. [Py] = 50 μM, $\phi = 0.05$. Reprinted with permission from Ref. [41]. Copyright (2011) American Chemical Society

or immiscible depends on the difference in energy before and after mixing and the difference in mixing entropy. If the mixing energy is negative (i.e., when the interaction energy between like species is larger than that between dissimilar species), the energetic separation is more favorable. On the other hand, it is entropically more advantageous for each component to be mixed. The result of these competing effects determines the state of the system. The phenomenon of separation into individual components is called phase separation. Many studies have been reported on phase separation in colloidal systems; for HS colloids, crystallization kinetics [30–32] and size classification during crystallization have been studied [33]. In two-component colloidal systems with added polymers, phase separation and eutectic formation due to depletion attraction have been revealed [34–36]. Phase separation has also been observed in charged two-component colloids with asymmetric particle size and charge number [37].

In two-component colloids, if the amount of one component is significantly less than the other, this can be considered an "impurity." Also, unlike atomic materials, colloidal dispersions are composed of particles that are more or less non-uniform in size. These impurities are excluded during crystallization processes. Nozawa et al. studied the distribution of impurities in the crystallization process based on the crystal growth theory [38, 39]. Note that small amounts of impurities have a significant effect on crystal size and crystallization rate [40].

This section describes the results of studies on excluding impurity particles from colloidal crystals in the processes of homogeneous crystallization, controlled crystallization, and grain growth.

5.4.2 *Impurity Exclusions on Crystallization*

Yoshizawa et al. [41, 42] studied the behavior of impurity particles on crystallization. The sample used was a two-component mixture of fluorescently labeled charged PS particles ($d = 333$ nm, $Z = 1680$, $\phi = 10^{-4}$) added to colloidal crystals of charged silica particles ($d = 108$ nm, $Z = 220$, $\phi = 0.05$). A schematic diagram of the exclusion process is shown in the top row of Fig. 5.28.

The sample was observed in the fluorescence mode of LSM. In this observation method, only fluorescent PS particles are observed, not silica particles. The lower row of Fig. 5.28 shows the observed impurity distribution. The uniformly dispersed green PS particles are excluded from the silica colloidal microcrystals during crystallization. Figure 5.29 shows a well-timed sample $t = 1260$ min observed by LSM. From left to right are micrographs of the non-fluorescent mode, the fluorescent mode, and the superposition of the two. The micrograph in the non-fluorescent mode shows that the silica colloids form a microcrystalline structure consisting of numerous grains. Green-colored PS particles are accumulated at the grain boundaries. Also shown in Fig. 5.30 is a 3D construction of the distribution of green PS particles at $t = 780$ min. They accumulated at the crystal grain boundaries to form cellular structures similar to those found in metals and bubbles.

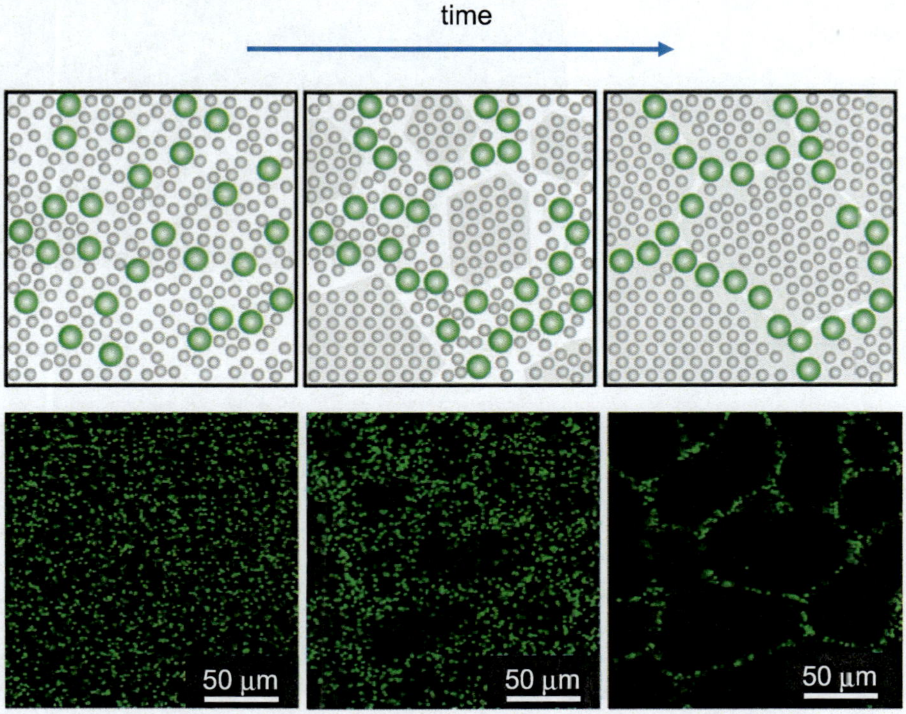

Fig. 5.28 Schematic diagram of the exclusion process and experimental results of the impurity distribution. $t = 0$, 177, and 385 s from left

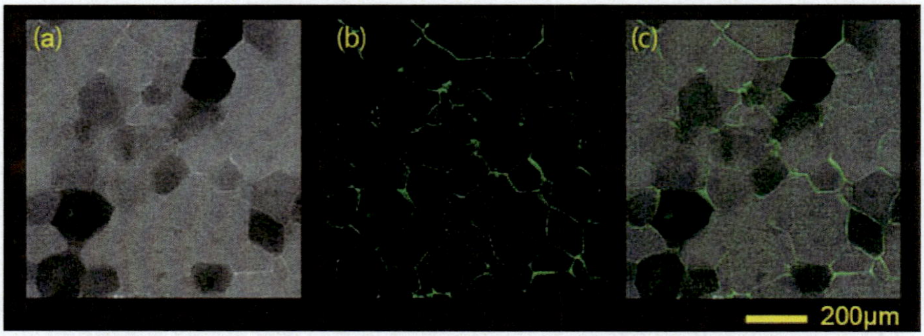

Fig. 5.29 Images of sample $t = 1260$ min after homogenization. **a** Reflection image, **b** fluorescence image, **c** superposition of (**a**) and (**b**) images. Reprinted with permission from Ref. [41]. Copyright (2011) American Chemical Society

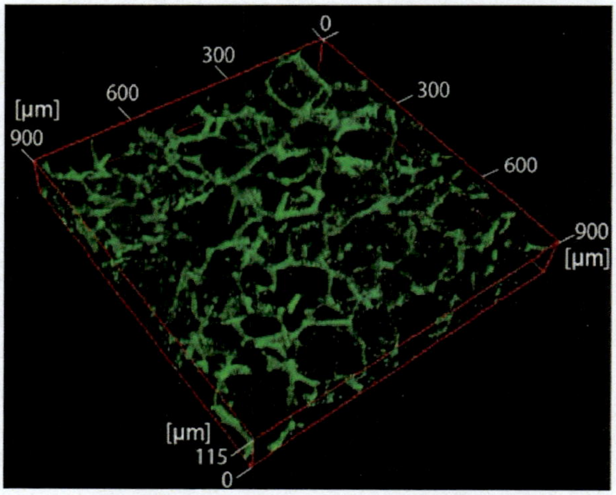

Fig. 5.30 Three-dimensional construction of the distribution of green PS particles at $t = 780$ min. Reprinted with permission from Ref. [41]. Copyright (2011) American Chemical Society

5.4.3 Impurity Exclusions on Grain Growth

As described in Sect. 5.4.1, crystal grains in a polycrystal coalesce and grow over time; Fig. 5.31 shows a micrograph of the grain growth process. The exclusion of impurity particles is observed during grain growth, as shown schematically in Fig. 5.32 [41]. The sample used was a mixture of silica ($\phi = 0.05$) and green-fluorescent PS particles ($\phi = 10^{-4}$), as in the experiment in Sect. 5.4.1. Initially, the impurities were distributed almost uniformly. Figure 5.33a–i shows CLSM images showing the time evolution of the PS particle distribution. Individual PS particles are visible in these images. In (a), an exclusion of PS particles at grain boundaries is already observed. Hereafter, the three crystal grains shown in (a) will be referred to as I, II, and III. (b) and (c) show that grain boundaries begin to move due to grain growth. At $t = 265$ (min), grain III grows larger while grain I shrinks and eventually disappears (d). During this process, PS particles are concentrated and accumulate at the grain boundary of colloidal silica. Grain II then shrinks with time and eventually disappears at 435 min (g). Thus, it can be seen that impurity particles are eliminated from the region pushed away by the grain boundary. This means that the migration of the grain boundary triggered the impurity elimination. A similar exclusion was observed for impurity particles with a diameter of 183 nm, but no exclusion was observed for particles with a diameter of 2420 nm due to their low mobility. PS particles that did not accumulate at the grain boundaries were found to remain near their original positions. This suggests that the migration of PS particles observed in this study was triggered by the migration of grain boundaries.

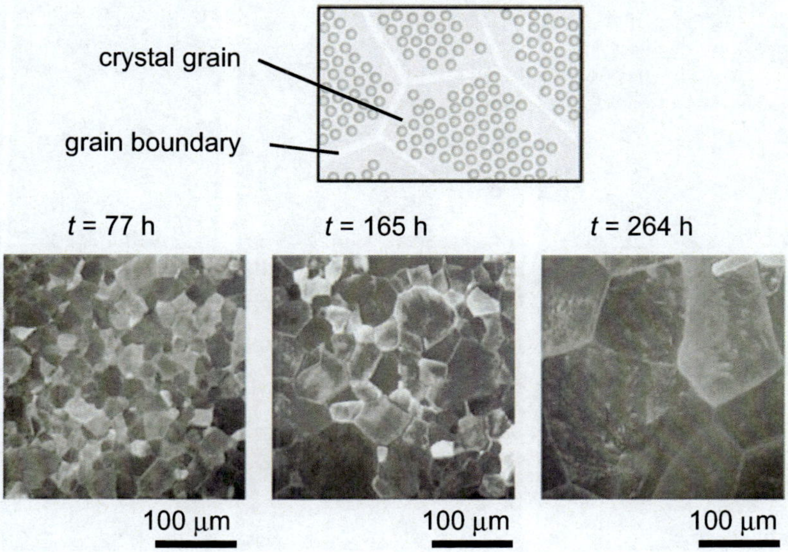

Fig. 5.31 Micrograph of the grain growth process. Silica colloid. $d = 118$ nm, $\phi = 0.03$, $Z = 570$

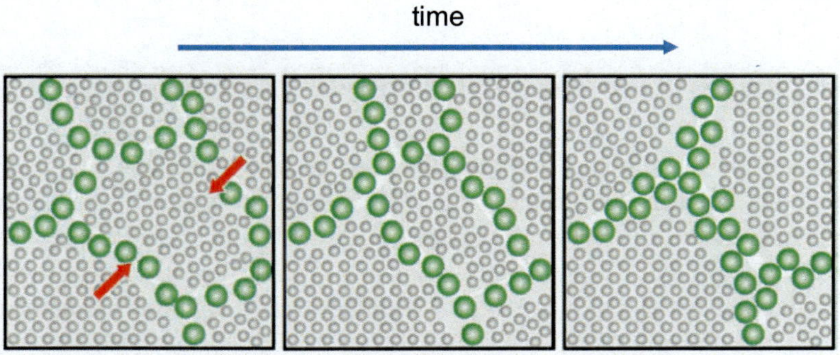

Fig. 5.32 Illustration of the impurity exclusion during the grain growth

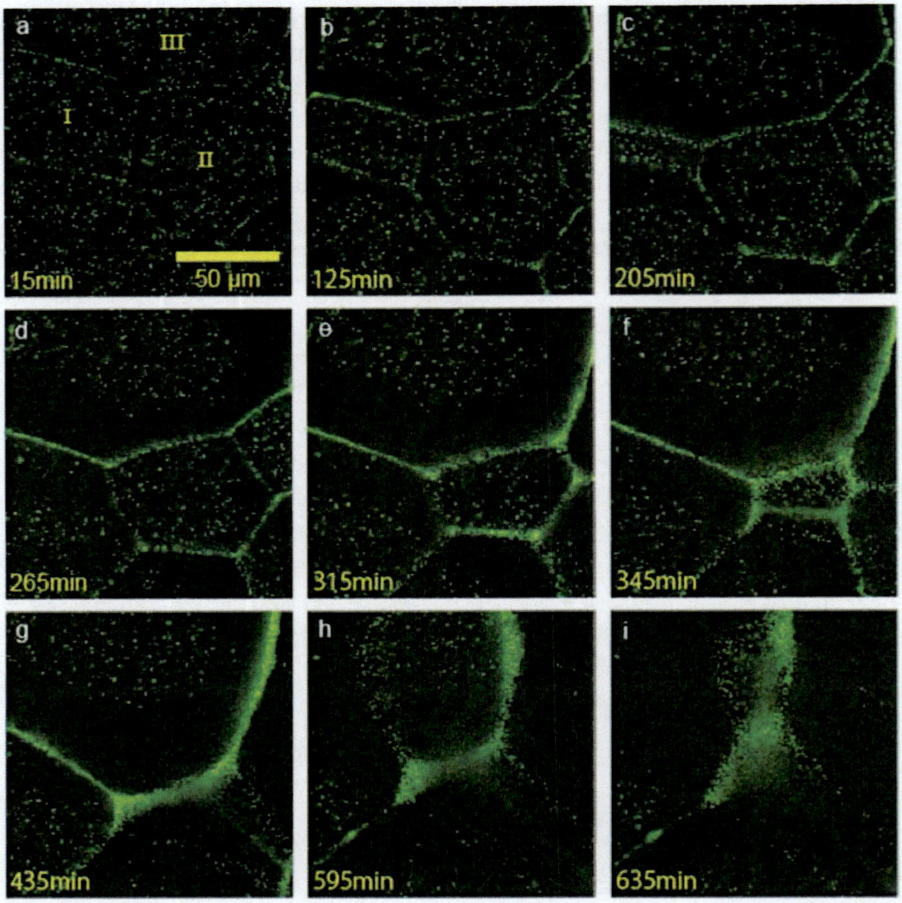

Fig. 5.33 LSM images showing the time evolution of the PS particle distribution. Individual PS particles are visible in these images. Reprinted with permission from Ref. [41]. Copyright (2011) American Chemical Society

5.4.4 Impurity Exclusions on Controlled Crystallization

The exclusion of impurities was also observed in the unidirectional crystal growth process described in this chapter; Fig. 5.34a shows an example of unidirectional growth of colloidal silica crystals under a pH gradient with green PS particles added (silica, $\phi = 0.05$; PS, $\phi = 10^{-4}$. $[\mathrm{Py}]_0 = 1$ mM). At sufficiently slow growth rates, the crystals consist of a thin lamellar layer and grow in a direction along the crystallization plane and orthogonal to the crystal growth [43]. At slow crystallization rates, coexisting impurity grains accumulate at grain boundaries, forming a stripe pattern, as shown in Fig. 5.34b. Pyridine diffuses from the right, and the crystal grows in the direction of the arrow, forming a stripe pattern. As shown in the LSM image on the right, impurities are distributed in stripes along the crystal growth direction. Figure 5.34c is a 3D LSM image.

Figure 5.35a shows the exclusion of impurities by unidirectional crystallization under a temperature gradient [44]. Two-component colloids of silica ($\phi = 0.035$) and green PS ($\phi = 5 \times 10^{-5}$) were used; [Py] = 65 μM and $T_c = 20$ °C. The resulting crystals were partially melted at $T = 5$ °C and kept at $T = 25$ °C for recrystallization. Crystalline regions formed within $t < 10$ min, indicating left-to-right growth. Impurity particles were found to be pushed toward the crystal/melt boundary and accumulated in the boundary region; Fig. 5.35b shows the impurity concentration as a function of position x along the crystal growth direction. From the migration rate of the peak position, the exclusion rate was estimated to be about 0.2 μm/s. At higher crystallization rates, no exclusion was observed. When the heating temperature was varied stepwise, the impurities assembled in a stripe pattern.

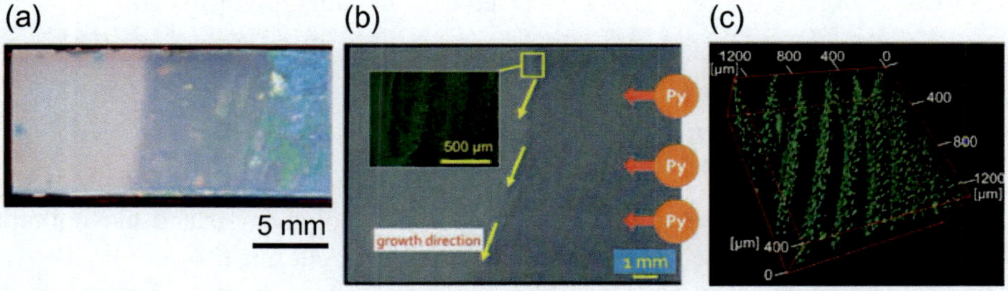

Fig. 5.34 **a** Example of unidirectional growth of silica colloidal crystals under a pH gradient with green PS particles added (silica, $\phi = 0.05$; PS, $\phi = 10^{-4}$. $[Py]_0 = 1$ mM). **b** An image of a stripe pattern. **c** 3D LSM image. Reprinted with permission from Ref. [43]. Copyright (2012) Chemical Society of Japan

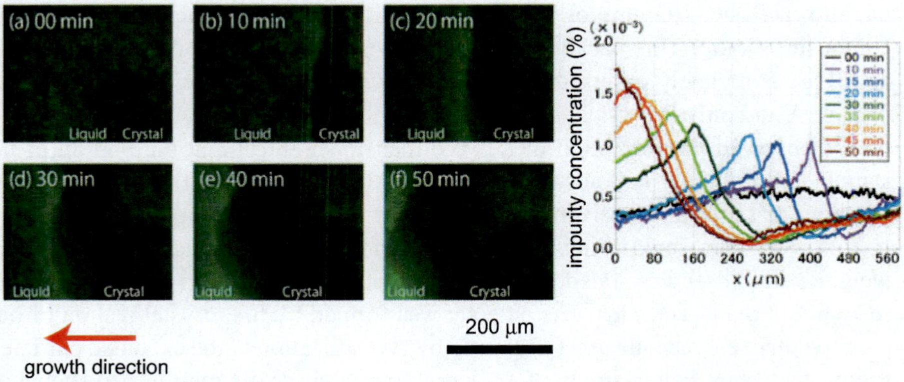

Fig. 5.35 **a** Elimination of impurities by unidirectional crystallization under a temperature gradient silica, $\phi = 0.035$, PS, $\phi = 5 \times 10^{-5}$; [Py] = 65 μM; crystallization temperature = 20 °C) **b** impurity concentration as a function of position x along the crystal growth direction. Reprinted with permission from Ref. [22]. Copyright (2013) American Chemical Society

What are the conditions under which crystallization and elimination occur in colloidal systems, and what is the threshold for such conditions, are very important questions. Phase diagrams of binary charged colloids with various particle sizes and mixing ratios have been reported for charged two-component systems [37, 46].

Nozawa and Uda et al. studied impurity exclusions during colloidal crystallization based on crystal growth science [47]. A small amount of charged fluorescent polystyrene particles were added as impurities to submicron-sized charged polystyrene crystals. They observed that the concentration of impurities in the crystal (C_S) differs from the initial solution (C_0) during the growth of impurity-doped colloidal crystals. The effective distribution coefficients k_{eff} ($= C_S/C_0$) were determined under various growth rates. They found that the impurity distribution in colloidal systems is consistent with the results from the Burton-Prim-Slichter model on the impurity partition and pointed out the analogy between colloidal and atomic systems. Other similarities between colloidal and atomic systems have been reported [48–50]. Future developments are expected from crystal growth studies.

5.4.5 Depletion-Attraction Systems

In depletion-attraction systems with the addition of polymers, island-like colloidal crystals are formed when appropriate conditions are chosen [34–36, 38, 39, 45]. For single-component systems, equilibrium phase diagrams have been reported with polymer and particle concentrations as variables. The phase behavior of multicomponent colloids is similar to that of binary alloys. Under appropriate conditions, eutectics are formed [34–36]. Eutectic is a form of solidification, in which the liquid phase solidifies to form multiple solid phases. For example, when two metals A and B are melted to form an alloy, if the ratio of A to B is not within the miscible range, the alloy will be a mixture of two different kinds of crystals, each with a different composition ratio. This is called eutectic microstructure. When the components A and B are completely immiscible (zero miscibility eutectics), the alloy is a mixture of pure A and only B crystals.

We observed the formation of a two-dimensional eutectic at the bottom of the sample cell in binary colloids [36]. The sample used was a zero-miscible eutectic, meaning that crystals consisting only of each component were formed. The experiments were performed using a system in which sodium polyacrylate (molecular weight ~ 800,000) was added to a dispersion of PS particles of two different sizes ($d = 600$ nm and 500 nm). The eutectic was formed by first crystallization of one component to exclude the other, followed by crystallization of the excluded particles. Figure 5.36 shows an example of a micrograph. A phase diagram is also shown on the left in Fig. 5.37. In phase diagrams of atomic systems, the temperature is usually used as a variable, but here the vertical axis is the reciprocal of the area fraction of particles ϕ_A. The abscissa is the fraction of 500 nm PS particles. Optical microscope images for conditions (a)–(e) in the phase diagram are shown in the right panel of

Fig. 5.37. The obtained phase diagram is similar to the eutectic phase diagram of atoms and molecules.

Eutectic was also observed in the three-component system; Fig. 5.38 shows a mosaic-like two-dimensional eutectic structure formed in a mixture of PS particles of sizes = 216, 245, and 312 nm. The size of the colloidal particles was chosen so that the structural colors from Bragg diffraction are red, blue, and green (RBG colors), the three primary colors of light. As with monitors of TV or PC, varying the RBG intensity ratio causes the color to appear to the human eye as a variety of intermediate colors. Three-component eutectics are expected to be used as light materials in the future.

In this subsection, we reported on various schemes of impurity exclusion in charged colloidal crystals analogous to impurity exclusion in atomic and molecular crystal materials. Colloidal multicomponent systems are useful models for studying the modification of defects in materials and crystals. Unlike atomic crystals, colloidal crystals are composed of particles that are not uniform in size. In addition, colloidal samples inherently contain aggregates of particles that act as impurities in the crystallization process. Therefore, this finding is expected to be useful for the preparation of high-quality colloidal crystals.

Column 5.2

Phase behavior of multicomponent systems

We have sometimes synthesized silica particles in our laboratory for crystallization experiments. On one occasion, the synthesis did not go well, resulting in a sample with a wide size distribution. According to the literature, crystallization should not have occurred with the synthesized particles because the size distribution was too

Fig. 5.36 Example of phase separation of a two-component colloid due to depletion-attraction PS (d = 536 nm) and silica (308 nm, colored with green-fluorescent dye); polymer, sodium polyacrylate (NaPAA), MW = 840,000, polymer concentration C_p = 0.05 wt%

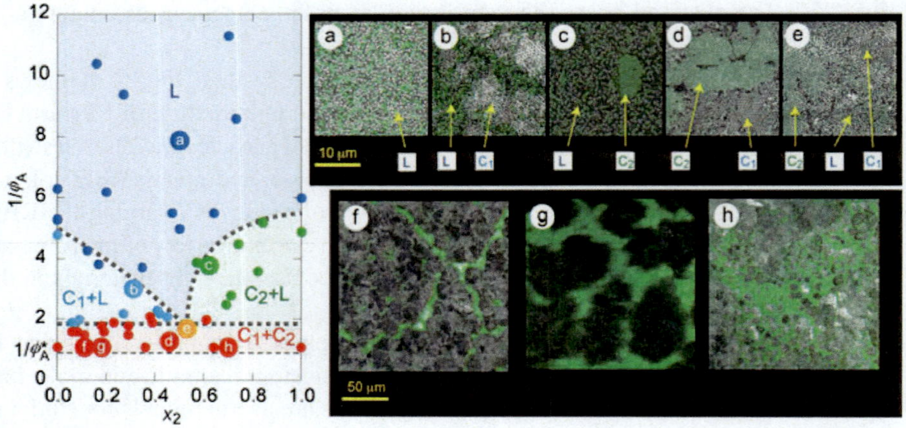

Fig. 5.37 Phase diagram of the binary PS colloid ($d = 500$ nm and 600 nm), defined by the inverse of area fraction, $1/\phi_A$, and volume fraction of PS2, X_2. Optical microscope images for conditions a–e in the phase diagram are shown in the right panel. *L*, liquid phase; C_1 and C_2, crystal phases of PS1 and PS2. $C_p = 0.08$ wt%. Reprinted with permission from Ref. [36]. Copyright (2016) Springer Nature

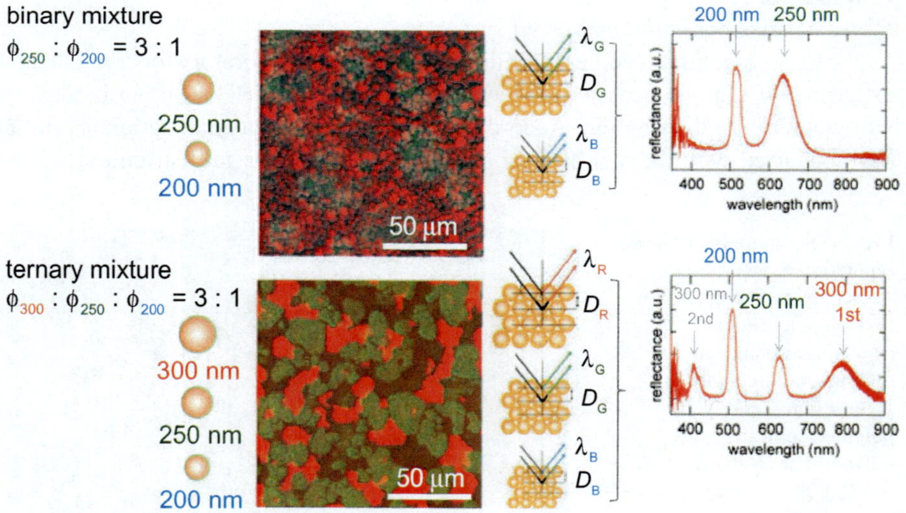

Fig. 5.38 Mosaic-like two-dimensional eutectic structure formed in a mixture of PS particles. **a** The binary system, $d = 200$ and 250 nm. **b** The ternary system, $d = 200$, 250, and 300 nm. Reflection spectra are also shown

wide. However, when we added ion exchange resin beads for deionization and left the sample to stand, only one small microcrystal was formed in the sample. This microcrystal disappeared with time.

This phenomenon was speculated to result from crystallization and fractionation occurring simultaneously. In chemistry, "recrystallization purification" is often used. This is a method of eliminating impurities and increasing purity by crystallizing the target substance in a solution containing impurities and the target substance. We believed that a similar phenomenon occurred in colloidal systems and began our research using two-component colloids, which are easy to analyze. Phase separation phenomena were observed in various two-component systems.

References

1. Palberg T (2014) J Phys Condens Matter 26:333101
2. Mutaftschiev B (2001) The atomistic nature of crystal growth Springer Series in Materials Science 43. Springer, Berlin
3. Markov IV (2017) Crystal growth for beginners, fundamentals of nucleation, crystal growth and epitaxy, 3rd edn. World Scientific, Singapore
4. Dhanaraj G, Byrappa K, Prasad V, Dudley M (eds) (2010) Springer handbook of crystal growth Part B, 1st edn. Springer, Berlin
5. Kashchiev D (2000) Nucleation, basic theory with applications. Butterworth-Heinemann Oxford, 2000
6. Cahn RW, Rivier N (eds) (1996) Physical metallurgy, vol 3, Chapter 28, 4th ed. North-Holland, Amsterdam
7. Weaire D, Hutzler S (2000) The physics of foams. Oxford University Press, Oxford
8. Vogel N et al (2015) Chem Rev 115:6265–6311
9. Li B, Zhou D, Han Y (2016) Nat Rev Mater 1:1–13
10. Cai Z et al (2021) Chem Soc Rev 50:5898–5951
11. Clark N, Hurd A, Ackerson B (1979) Nature 281:57–60
12. Kanai T, Sawada T, Toyotama A, Kitamura K (2005) Adv Funct Mater 15:25–29
13. van Blaaderen A, Ruel R, Wiltzius P (1997) Nature 385:321–324
14. Holgado M et al (1999) Langmuir 15:4701−4704
15. Lotito V, Zambelli T (2017) Adv Colloid Interf Sci 246:217–274
16. Wakabayashi N, Yamanaka J, Murai M, Ito K, Sawada T, Yonese M (2006) Langmuir 22:7936–7941
17. Yamanaka J, Murai M, Iwayama Y, Yonese M, Ito K, Sawada T (2004) J Am Chem Soc 126:7156–7157
18. Murai M, Yamada H, Yamanaka J, Onda S, Yonese M, Ito K, Sawada T, Uchida F, Ohki Y (2007) Langmuir 23:7510–7517
19. Yamada H, Sawada T, Yamanaka J, Yonese M, Uchida F (2008) Chem Lett 37:172–173
20. Murai M, Fujioka M, Toyotama A, Yonese M, Yamanaka J (2011) Chem Lett 40:42–43
21. Toyotama A, Yamanaka J (2011) Langmuir 27:1569–1572
22. Shinohara M, Toyotama A, Suzuki M, Sugao Y, Okuzono T, Uchida F, Yamanaka J (2013) Langmuir 29:9668–9676
23. Toyotama A, Yamamoto M, Nakamura Y, Yamazaki C, Tobinaga A, Ohashi Y, Okuzono T, Ozaki H, Uchida F, Yamanaka J (2014) Chem Mater 26:4057–4059
24. Murakado A, Toyotama A, Yamamoto M, Nagano R, Okuzono T, Yamanaka J (2016) J Colloid Interf Sci 465:200–207
25. Robbins MO, Kremer K, Grest GS (1988) J Chem Phys 88:3286–3312
26. Alsayed AM, Islam MF, Zhang J, Collings PJ, Yodh AG (2005) Science 309:1207–1210
27. Nagano R, Toyotama A, Yamanaka J (2011) Chem Lett 40:1366–1367
28. Toyotama A, Yamanaka J, Yonese M, Sawada T, Uchida F (2007) J Am Chem Soc 129:3044–3045

29. Toyotama A, Yamanaka J, Shinohara M, Onda S, Sawada T, Yonese M, Uchida F (2009) Langmuir 25:589–593
30. Martin S, Bryant G, van Megen W (2005) Phys Rev E 71:021404
31. de Villeneuve VWA et al (2009) Soft Matter 5:2448–2452
32. de Villeneuve VWA, Dullens RPA, Aarts DGAL, Groeneveld E, Scherff JH, Kegel WK, Lekkerkerker HNW (2005) Science 309:1231–1233
33. Bartlett P (1998) J Chem Phys 109:10970
34. Kozina A, Sagawe D, Daz-Leyva P, Bartsch E, Palberg T (2012) Soft Matter 8:627–630
35. Kozina A, Daz-Leyva P, Palberg T, Bartsch E (2014) Soft Matter 10:9523–9533
36. Toyotama A, Okuzono T, Yamanaka J (2016) Sci Rep 6:23292
37. Yoshizawa K, Wakabayashi N, Yonese M, Yamanaka J, Royall CP (2012) Soft Matter 8:11732
38. Nozawa J, Uda S, Suxia G, Hu S, Toyotama A, Yamanaka J, Okada J, Koizumi H (2017) Langmuir 33:3262–3269
39. Nozawa J, Uda S, Guo S, Toyotama A, Yamanaka J, Okada J, Koizumi H (2018) Cryst Growth Des 18:2948–2955
40. Engelbrecht A, Shöpe HJ (2012) Soft Matter 8:11034–11037
41. Yoshizawa K, Okuzono T, Koga T, Taniji T, Yamanaka J (2011) Langmuir 27:13420–13427
42. Yoshizawa K, Toyotama A, Okuzono T, Yamanaka J (2014) Soft Matter 10:3357–3361
43. Yoshizawa K, Onda S, Sawada T, Yamanaka J (2012) Chem Lett 41:322–323
44. Sugao Y, Yoshizawa K, Toyotama A, Okuzono T, Yamanaka J (2012) Chem Lett 41:1163–1164
45. Lekkerkerker HNW, Tuinier R (2011) Colloids and the depletion interaction. Springer, Amsterdam
46. Lorenz NJ, Schöpe HJ, Reiber H, Palberg et al T (2009) J Phys Condens Matter 21:464116
47. Nozawa J, Uda S, Naradate Y, Koizumi H, Fujiwara K, Toyotama A, Yamanaka J (2013) J Phys Chem B 117:5289–5295
48. Burton JA, Prim RC, Slichter WP (1953) J Chem Phys 21:1987–1991
49. Nozawa J, Uda S, Hu S, Fujiwara K, Koizumi H (2016) J Cryst Growth 439:13–18
50. Hu S, Nozawa J, Koizumi H, Fujiwara K, Uda S (2015) Cryst Growth Des 15:5685–5692

Chapter 6
Applied Research on Colloidal Self-Assembly

Abstract This chapter introduces some applied research on the self-assembled structures of colloids. First, we present studies on the preparation of materials in which charged colloidal crystals are immobilized in polymer gels. Next, we report examples on the applications of colloidal crystals in depletion-attraction systems. Since the depletion force does not depend significantly on the salt concentration of the medium, crystals can be generated even at high salt concentrations, in which charged colloids do not crystallize. Therefore, they can be used even at physiological salt concentrations (around 0.1 M), making it suitable for cosmetics, pharmaceutics, and other applications. Gold colloidal crystals, which are expected to be plasmonic materials, can be prepared from dilute dispersions of gold particles by using depletion forces and gravitational sedimentation. We also describe a method to fabricate two-dimensional colloidal crystals by adsorbing three-dimensional charged colloidal crystals onto a substrate. Finally, space experiments on colloidal self-assembly performed on the International Space Station (ISS) will be described. The micro-gravity environment of the ISS is ideal for experiments on colloidal systems because it is not affected by sedimentation. Space experiments on the charged colloidal crystals and colloidal aggregates will be presented.

Keywords Immobilized colloidal crystal · Polymer gel · Microgel colloidal crystal · Gold colloidal crystal · Plasmonic material · Sensing materials · Surface-enhanced Raman scattering · 2D colloidal crystal · The International Space Station · Space experiments · Colloidal clusters

6.1 Introduction

6.1.1 Applications of Colloidal Crystals

In this chapter, we will introduce applied research on colloidal crystals and self-assembled structures. As mentioned in Chap. 1, the diffraction wavelength of colloidal crystals can be set from visible light to the near-infrared region. Thus, their application to various optical materials has been actively investigated in recent

© The Author(s), under exclusive license to Springer Nature Singapore Pte Ltd. 2023

J. Yamanaka et al., *Colloidal Self-Assembly*, Lecture Notes in Chemistry 108,
https://doi.org/10.1007/978-981-99-5052-2_6

years. Structures in which the refractive index changes periodically with the period of light wavelengths are referred to as "photonic crystals" [1]. This name is derived from photons, and as materials that manipulate the way light travels, they are the subject of active research in optics. Photonic crystals, on the other hand, have a stop band of light that prohibits light propagation because it diffracts light of specific energy. This is similar to semiconductor crystals which have a stop band that prohibits the propagation of electrons because it diffracts electron waves of specific energy. As shown in Fig. 6.1, 1D, 2D, and 3D photonic crystals are produced depending on the application. Of these, the 1D and 2D photonic crystals can be built into precise structures by bottom-up methods such as thin-film layering and lithographic techniques. Colloidal crystals are 3D photonic crystals, which are difficult to fabricate by other methods [2, 3]. Because colloidal crystals are relatively easy to prepare and can be mass-produced, their application to various optical elements is currently under active investigation; they include wavelength filters, optical pulse delay elements, photo-detectable physical and chemical sensors, etc. Recently, laser oscillation using colloidal crystals has been realized [4, 5], and it is expected to develop as an advanced material.

In the previous chapters, we introduced charged colloidal crystals formed by electrostatic repulsions between particles. Although large and uniform crystals were obtainable using charged colloids, they are mechanically weak structures that form in water and are easily disrupted by shaking. Therefore, immobilization of the systems, for example, by using polymer gels, has been investigated to maintain their structures [6–11]. Section 6.2 presents the method of gel immobilization of the charged colloidal crystals. They deform elastically when force is applied, changing the Bragg diffraction wavelength of the fixed colloidal crystal. Namely, they work as tunable photonic crystals [11]. Immobilization using flexible elastomers [12] has also been reported.

Conversely, the shape change of the gel can be evaluated from the measurement of the Bragg wavelength of the fixed crystals. Volumes of gels change in response to

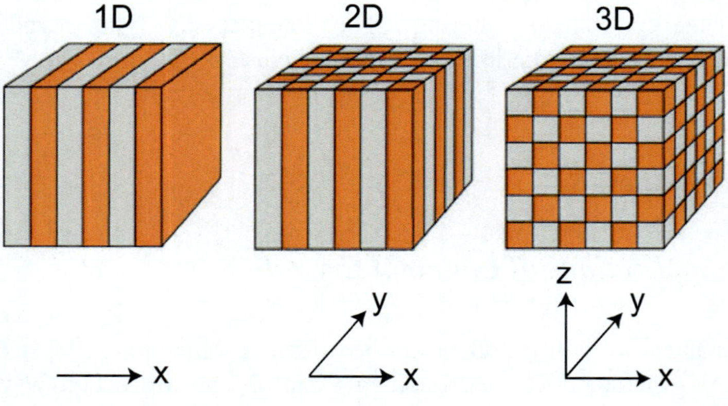

Fig. 6.1 1D, 2D, and 3D photonic crystals. Structures in which the refractive index changes periodically with the period of light wavelengths are called photonic crystals

various external environments (pH, solvent composition, temperature change, etc.). Thus, colloidal crystals fixed in the gel can function as sensors capable of optical monitoring for the external environment [8–10].

Next, we introduce the application of crystals formed by depletion attraction. Unlike electrostatic interaction, depletion attraction is less sensitive to the salt concentration of the medium. For this reason, the crystallization can occur under physiological salt concentrations (around 0.1 M), where charged colloidal crystals cannot be made, making it suitable for pharmaceutical and cosmetic applications. Section 6.3 reports the results of crystallization experiments using uniformly sized polymer microgel spheres [13–16] as particles.

In Sect. 6.4, we report on the crystallization of gold colloids by depletion attraction [17]. When gold is irradiated with light in the visible to near-infrared region, the internal free electrons oscillate, causing a collective oscillation phenomenon of electrons called surface Plasmon resonance (SPR) [17–22]. This property enables highly sensitive analysis, and applied research, such as analytical substrates for Raman spectroscopy, is now being actively conducted. The application of colloidal crystals of gold particles to Raman scattering substrates will also be introduced.

Section 6.5 reports on 2D colloidal crystals, which can be used in various areas, including sensing, photonics, and plasmonics. There has been extensive research on the 2D self-assembly of colloidal particles, including crystallizations at air–liquid interfaces [23] and at liquid–liquid interfaces [24], aggregation under electric fields [25, 26], and colloidal lithography [27]. Here, we show that 2D colloidal crystals can be fabricated by electrostatically adsorbing 3D-charged colloidal crystals onto a substrate with opposite sign charges [28]. Combined with the unidirectional growth of silica colloids (Chap. 5), cm-sized 2D crystals are obtainable.

6.1.2 Space Experiments on Colloids

Colloidal particles in a dispersion settle under the influence of gravity unless the density difference between the medium and the particles is exactly the same. The settling particles are subjected to three forces: gravity, buoyancy, and the drag force from the medium. The drag force is often expressed by Stokes' equation, $F = 6\pi\eta a v$ (η: viscosity of the medium, a: radius of the particle, v: : settling velocity). The sum of buoyancy and gravity is expressed as $(4/3)\pi a^3 \Delta\rho g$, where $\Delta\rho$ is the density difference between the particle and medium and, and g is the gravitational acceleration. When these are balanced and settling, we have

$$v = (2/9\eta)a^2 \Delta\rho g. \tag{6.1}$$

For example, for a titania particle with a diameter of 1 μm in water at 25°C ($\Delta\rho$ ~ 3 g/cm^3), the settling velocity is v ~ 1.8 μm/s ~ 15 cm/day. This is fast enough to be observed in real-time under a microscope and to confirm the effect by visual observation.

However, not all particles necessarily settle and accumulate to the bottom of the sample cell by sedimentation since the concentration of particles varies to become uniform by diffusion (see Appendix E). The balance between sedimentation and diffusion achieves a state called sedimentation equilibrium, where particle concentrations are higher at lower elevations [29–31]. When the particles are sufficiently concentrated at the bottom of the cell, the colloid crystallizes. Davis et al. reported crystallization due to an increase in f during transient sedimentation [32, 33]. They described a detailed theoretical study on sedimentation of the hard-sphere (HS) colloid [34].

For charged colloidal crystals, the crystal lattice spacing becomes non-uniform due to gravitational compression. In a pioneering study, Crandall and Williams used laser light scattering to determine the equilibrium sedimentation profile of ϕ of charged PS colloids [35]. A detailed analysis of the anisotropy of crystal lattices caused by gravity has been reported by using the Kikuchi–Kossel line diffraction [36]. More recently, it has been suggested that a macroscopic electric field is present during the equilibrium sedimentation of charged colloids [37–39]. We have investigated the time evolution of the particle concentration distribution in colloidal silica crystals by detecting the Bragg diffraction wavelengths using fiber-optic reflection spectroscopy [40].

As described above, the effect of particle sedimentation on colloidal self-assembly cannot be ignored on the ground. In addition, when the temperature of the sample is inhomogeneous, the density is also inhomogeneous, and convection due to gravity may occur. A microgravity environment, where the effects of sedimentation and convection can be neglected, is ideal for experiments on the structure formation of colloidal systems. Various experiments have been conducted on the crystallization of colloids in the HS system under microgravity. The experiments using the space shuttle have revealed that particle sedimentation has a significant effect on the nucleation and growth processes [41]. It has also been reported that dendritic growth is the dominant mode of crystal growth under microgravity [42–44]. Our group also reported a decrease in the nucleation rate in the crystallization of charged colloids under microgravity conditions obtained in parabolic flights [45].

In Subsection 6.6, we describe space experiments on colloidal self-assembly performed on the International Space Station (ISS). We participated in two projects on colloidal self-assembly conducted in collaboration with the Japan Aerospace Exploration Agency (JAXA) and a number of other organizations. The microgravity environment of the ISS is ideal for experiments on colloidal systems because it is not affected by precipitation. One of the experimental projects is the 3D Photonic Crystal Formation Project (3DPC project), which creates charged colloidal crystals in space. The other is the Colloidal Clusters Project on colloidal associations. In both experiments, colloidal samples were fixed in polymer gels and returned to the ground for structural analysis.

6.2 Gel Immobilization of Charged Colloidal Crystals

Here, we present a method for immobilizing charged colloidal crystals in polymer gels to fabricate soft and self-standing colloidal crystals. The colloidal crystal deforms elastically when an external force is applied, changing the Bragg diffraction wavelength of the immobilized colloidal crystal. We briefly describe the optical properties of the gel-immobilized crystals as tunable photonic crystals.

6.2.1 Gel Immobilization Methods

A polymer gel is a three-dimensional network of polymer chains in a medium. They can be synthesized by the radical polymerization method described in Chap. 3 as a method for synthesizing polystyrene particles (Fig. 6.2). Acrylamide-based polymer gels, which are often used in experiments, can be synthesized using monomers (~0.1 M) and photoinitiators (~0.1–0.4 mg/mL) as well as cross-linking agents such as methylene-bis-acrylamide (~10 mM). When ultraviolet (UV) light is irradiated, radicals are generated from the initiator, and a polymerization reaction occurs. The monomers are combined in chains to form polymer chains. The cross-linker molecule is incorporated into the growing chain, causing the two polymer chains to bond and form a three-dimensional network. The concentration of polymer chains in the gel is usually around 10 volume percent, and the majority of the gel is a medium.

To make gel-immobilized colloidal crystals, we add the colloidal dispersion to an aqueous solution of the gelator reagent. After the crystals have grown, the entire colloid is fixed by irradiating the medium with UV light for gelation. The immobilized crystals can be removed from the sample cell as a self-standing material. Once fixed in the gel, the colloidal crystals are stable for a long time (more than twenty years), and the crystal structure is retained in the gel. If the crystals are dried by evaporating water, and then the sample is kept in the water again, they will produce the same diffractive color as before drying.

We observed the gel immobilization process under a microscope. Figure 6.3 shows optical microscopy results recording colloidal particles' movement during the gelation. The sample (PS colloidal crystal, $d = 430$ nm) is irradiated with UV light while being observed under the optical microscope. Before gelation, the particles oscillate around the crystal lattice points, but as gelation proceeds, the amplitude of the oscillation gradually decreases and becomes stationary. This indicates the fixation of the crystal structure which was completed. Figure 6.3 bottom figure shows the time variation of the positions of the four particles. In this experiment, the particles are fixed in about 30 s.

The colloidal crystals obtained by various methods described so far could be immobilized in polymer gels. Examples of the immobilized colloidal crystals are shown in Fig. 6.4. From left to right, the crystal samples by unidirectional growth by Py diffusion (Subsection 5.2.2) [46], one using a pH gradient by $NaHCO_3$ diffusion

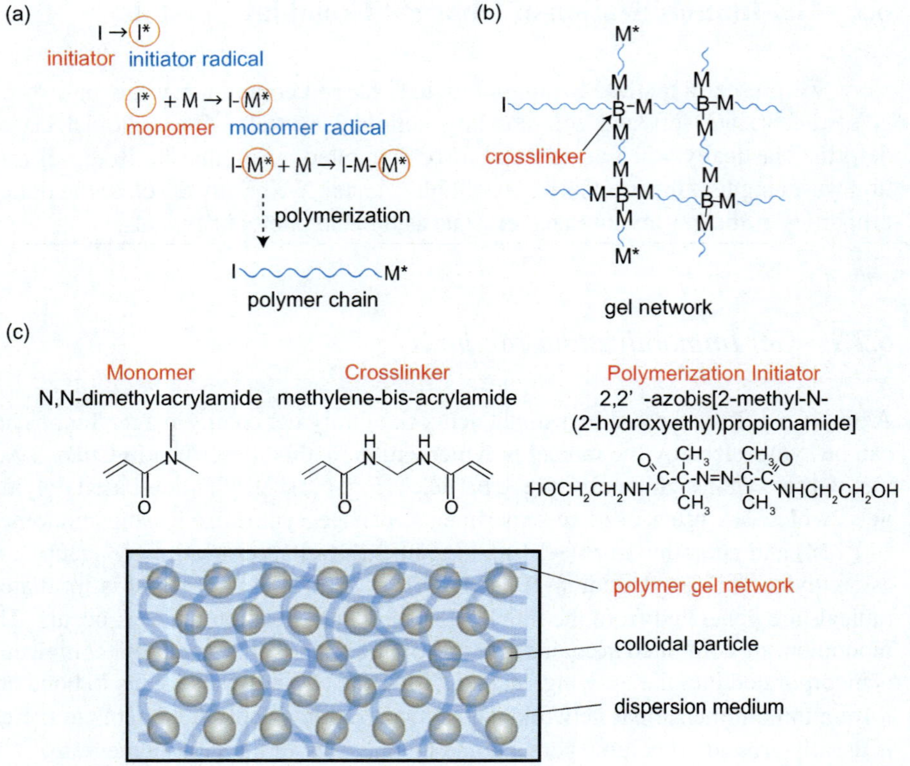

Fig. 6.2 Gelation reagents and illustration of immobilization of colloidal crystals in polymer gels. (a) polymerization reaction, (b) cross-linking reaction, (c) schematic of gelation reagents and colloidal crystals immobilized in the gel

(5.2.3) [47], and one by unidirectional growth under a temperature gradient (5.3.3) [48]. In all cases, silica colloidal particles of about 100 nm in diameter are used, and crystals are prepared in dilute dispersions with $f = 0.03$–0.04.

6.2.2 *Tuning the Diffraction Wavelength Using Gel Deformation*

Colloidal crystals have fixed in the polymer gel deform by compression. Charged colloidal crystals prepared under dilute conditions have a large enough distance between the particles. When a particle size = about 100 nm and $\phi = 0.01$, the particles do not touch each other even when the gel is compressed to half the original thickness. As shown in Fig. 6.5, the lattice spacing of the colloidal crystals becomes narrower by compression, and the diffraction wavelength shifts to the short wavelengths.

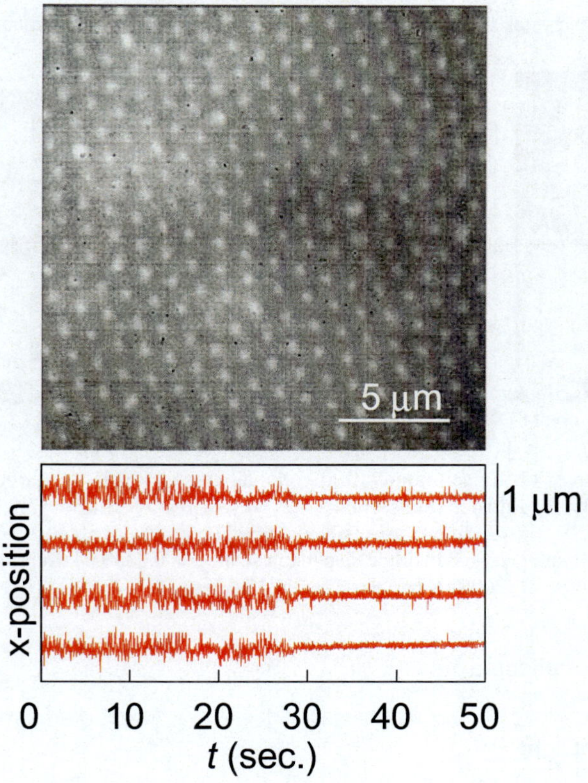

Fig. 6.3 Optical microscopy of the gelled colloidal crystals and motion of the colloidal particles during the gelation (PS particles, $d = 430$ nm)

In the experiment, a gel-immobilized colloidal crystal was compressed between two glass plates. Figure 6.6a shows an appearance of the colloidal crystals fixed in polyacrylamide gel (colloidal silica, $d = 110$ nm, $Z = 510$, $\phi = 0.03$). The crystals are randomly oriented polycrystals of several hundred micrometers in diameter. Figure 6.6b is a reflection microscope image of a single crystal grain before compression, and Figs. 6.6c, d, e are images after compression to 85, 80, and 75% thickness, respectively. The diffraction color changed from orange to blue.

By deforming the gel, the diffraction wavelength can be continuously controlled. Figure 6.7 shows the relationship between the thickness t (reduced by an initial thickness t_0) and diffraction wavelengths λ_m of the three samples. Samples 1 and 2 are colloidal crystals of PS particles ($\phi = 0.008$ and 0.012, respectively), while sample 3 is a colloidal silica crystal shown in the image in Fig. 6.6. The solid line is drawn while assuming that the Bragg wavelength is exactly proportional to t/t_0. The visible light range is approximately 400–800 nm, and Sample 1 covers almost the entire visible light range.

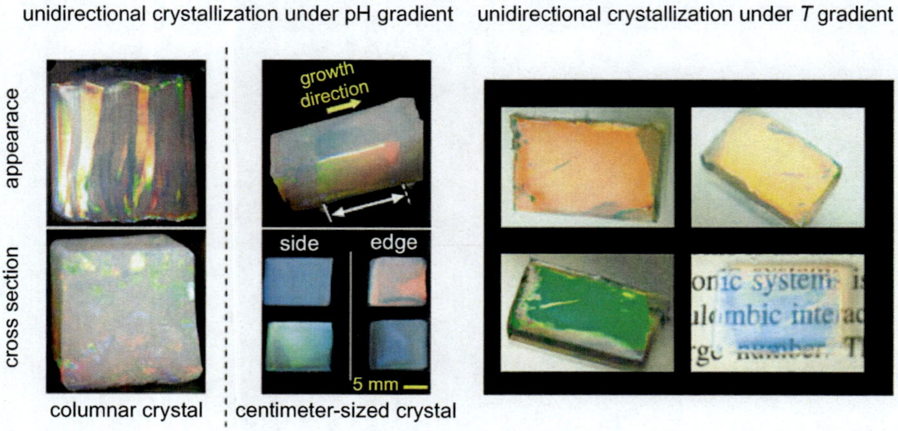

Fig. 6.4 Examples of the gel-immobilized colloidal crystals made by various growth methods. From left to right, Py diffusion (Subsection 5.2.2) [46], $NaHCO_3$ diffusion (5.2.3) [47], and temperature gradient (5.3.3) [48]. In all cases, silica colloidal particles of about 100 nm in diameter are used, and crystals are prepared in dilute dispersions with $\phi = 0.03$–0.04. Reprinted with permission from Refs. [46–48]. © (2004) American Chemical Society

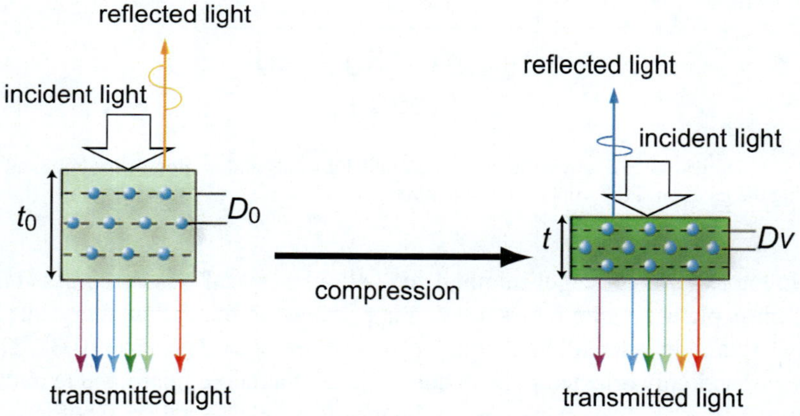

Fig. 6.5 Illustration of the tuning of the Bragg wavelength of gelled colloidal crystal by mechanical compression

6.3 Microgel Crystals

As discussed in Chap. 2, when non-adsorbing polymer chains are added to a colloidal system, colloidal crystals form under appropriate conditions. Crystallization by depletion attraction can be used for a variety of colloids. Here, we describe the crystallization of soft microgel particles [16]. Recently, the synthesis of microgels with submicron and uniform sizes has been reported [10]. We used a copolymer gel of poly-*N*-isopropyl acrylamide (PNIPAM) and acrylamide (AM) as one capable

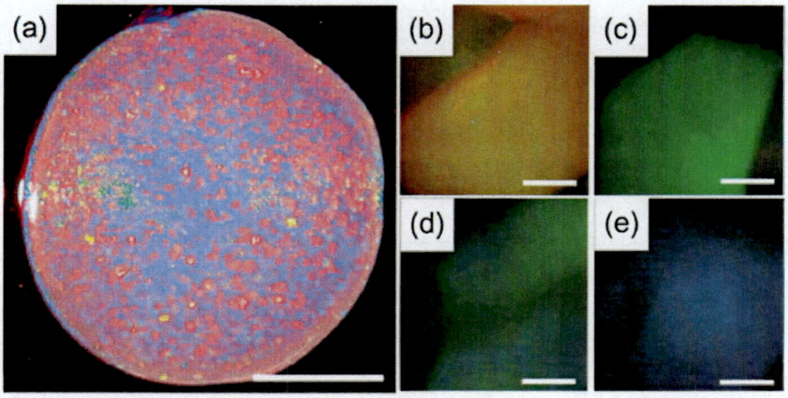

Fig. 6.6 **a** Colloidal crystal immobilized in a disk-shaped poly(acrylamide) gel matrix (silica, $d = 110$ nm, $\phi = 0.03$). Scale bar, 5 mm. **b–e** Reflection microscope images of single grains in the gelled crystal under compression. Scale bar, 50 mm. The thickness of the gel t reduced by the initial thickness $t_0 = 2.15$ mm **b** 1.0, **c** 0.85, **d** 0.80, and **e** 0.75. Reprinted with permission from Ref. [11]. © (2003) American Chemical Society

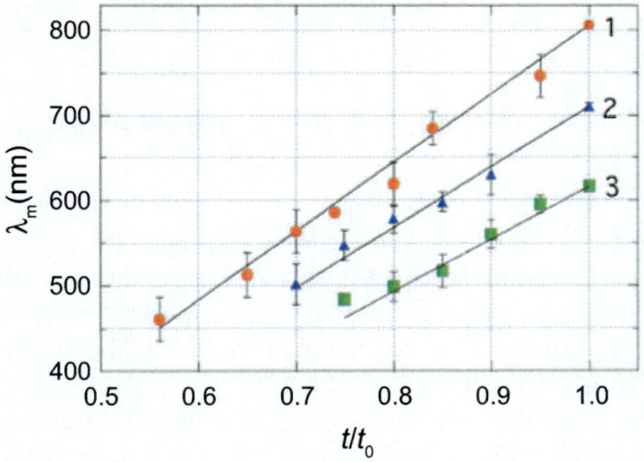

Fig. 6.7 Bragg wavelength λ_m plotted against t/t_0 for three kinds of gelled colloidal crystals. Solid lines were drawn assuming that λ_m is exactly proportional to t/t_0. Samples 1 and 2 are PS particles ($\phi = 0.008$ and 0.012), respectively. Sample 3 is the colloidal silica crystal ($d = 110$ nm, $\phi = 0.03$). Reprinted with permission from Ref. [11]. © (2003) American Chemical Society

of producing particles with a narrow particle size distribution. The PNIPAM-AM microgel is a thermosensitive gel and shrinks at temperatures above 35°C. The crystallization experiments were conducted at 25°C.

We used polyacrylamide with a molecular weight of 4.0×10^5 as a depletant. Colloidal crystals were formed when the gel concentration C_{gel} was 0.5–4 wt% and the polymer concentration $C_p > 0.41$ wt%. The inset figure of Fig. 6.8 shows pictures of the crystals of $d = 231$, 271, 320, and 363 nm (from left to right, $C_p =$

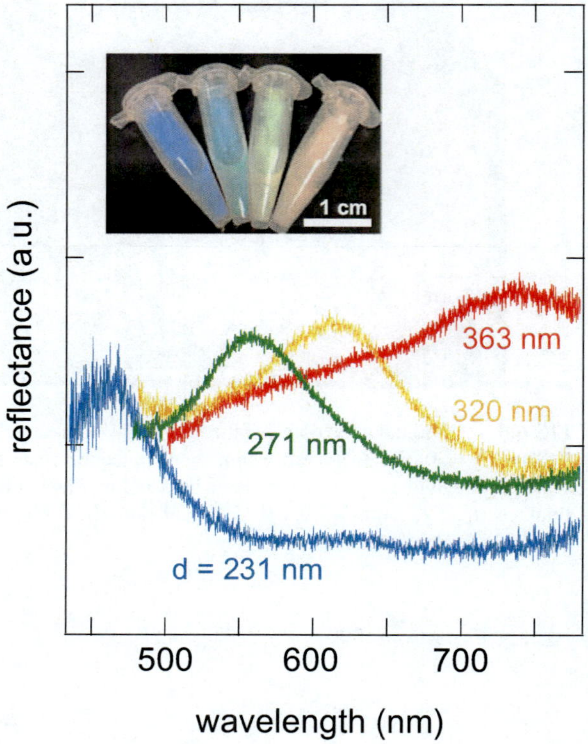

Fig. 6.8 Reflectance spectra of four microgel crystals of different sizes formed in the depletion–attraction system. Inset shows overviews of the samples (from left to right, $d = 363, 320, 271, 231$ nm, polymer concentration $C_P =$ 0.91 wt%, gel concentration $C_{gel} = 2$ wt%)

0.91 wt%, $C_{gel} = 2$ wt%). Reflectance spectra of the samples are shown in Fig. 6.8. A diffraction peak, albeit broad, was observed. The smaller the particle, the shorter the peak wavelength λ_B, which is almost consistent with the Bragg wavelength when assuming an opal structure. The coloration of the microgel colloids is attributable to the Bragg diffraction from the crystal structure.

Figure 6.9a shows the reflection spectra of colloidal crystals ($d = 231$ nm) at various values of C_p (0.5–0.9 wt%, from bottom to top, at 0.045 wt% intervals), indicating that the peak wavelength shifts toward shorter wavelengths as C_p increases. Figure 6.9b presents the lattice spacing of colloidal crystals. The dashed line is the calculated value for a closely packed opal crystal. The observed Bragg wavelength dependence of C_p is reduced from this value, possibly due to the deformation of the softer microgel particles, which reduces the lattice spacing.

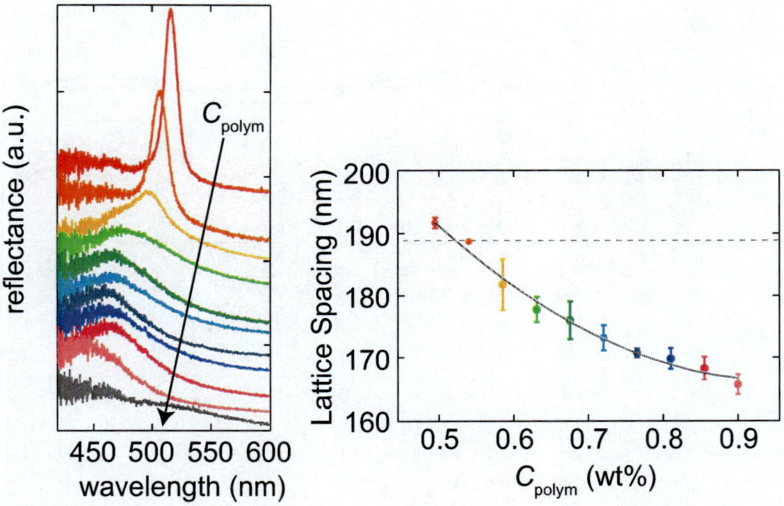

Fig. 6.9 **Left** Reflectance spectra of colloidal crystals at various values of C_p (from the bottom, 0.5–0.9 wt%, with an interval of 0.045 wt%). **Right** The lattice spacing of the colloidal crystal. Reprinted with permission from Ref. [16]. © (2019) Chemical Society of Japan

6.4 Gold Colloidal Crystals and Their Application for SERS

6.4.1 Surface Plasmon Resonance

When the surface of a noble metal such as Au or Ag is irradiated with electromagnetic waves of an appropriate frequency, the free electrons inside the metal vibrate collectively (Fig. 6.10a). This phenomenon is called surface plasmon resonance (SPR) [17, 18].

All physical phenomena have both particle-like and wave-like properties. When light, which is an electromagnetic wave, is viewed as a particle, it is called a photon, while a particle of a sound wave is called a phonon. The term "plasmon" is used when the oscillation of an electron is viewed as a particle.

Since the resonant frequency varies with the dielectric constant of the surrounding medium, SPR is used to quantify molecules attached to metal surfaces and is widely applied in chemistry and bioscience [19, 20]. On bulk metal surfaces, these vibrations are known to be transmitted as waves (Fig. 6.10b). In contrast, in Plasmon resonance using gold colloidal particles, the Plasmon's vibrating of free electrons are localized within the particles and referred to as localized Plasmon (LSPR) (Fig. 6.10c).

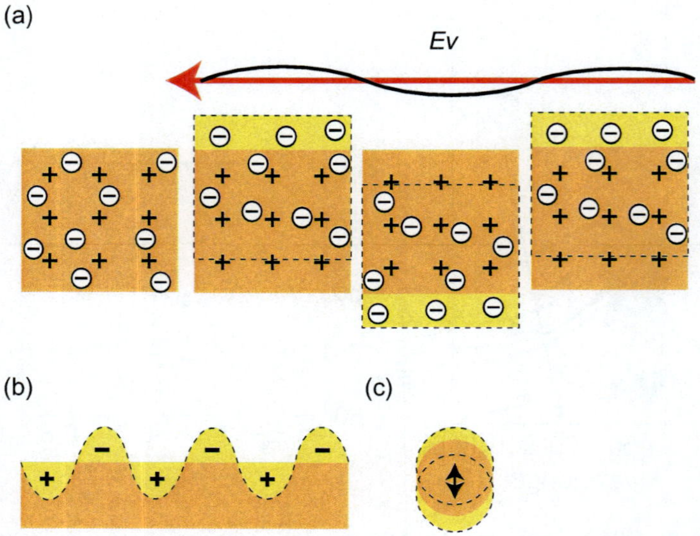

Fig. 6.10 **a** Schematic of surface plasmon resonance (SPR). When an electromagnetic wave of the appropriate frequency is incident, the electrons in the metal vibrate. **b** At the bulk metal surface, the plasma vibration is transmitted as a wave. **c** In gold colloidal particles, the plasmon resonance is localized within the particle (LSPR)

6.4.2 Raman Scattering

The gold colloidal particles have potential applications as substrates useful for Raman spectroscopy. Raman spectroscopy is a technique for analyzing the molecular structure of materials and identifying substances from spectra obtained by spectroscopy of Raman scattered light. The energy of light having frequency n is hn (h is Planck's constant). As discussed in Chap. 1, Rayleigh scattering occurs when light is irradiated onto molecules that are sufficiently smaller in size than the wavelength of the light. The frequency of scattered light by Rayleigh scattering is the same as that of incident light n_0. In Raman scattering, some of the photon energy is lost as vibrational energy in the molecule or added to the light energy. The frequency changes to $n = n_0 \pm \Delta n$ because of the energy transfer. Here, Δn is the contribution of vibrations in the molecule. Raman scattering contains information on the structure of molecules. However, it is very weak, about 10^{-6} times the intensity of Rayleigh scattering, making it difficult to detect.

When a molecule is adsorbed on a metal surface with a roughness ranging from the atomic scale to several hundred nm, the Raman scattering light is enhanced 10^4–10^6 times compared to bulk molecules. That is, it is as easy as detecting the Rayleigh scattered light. This phenomenon is called surface-enhanced Raman scattering (SERS). In particular, strong electric fields are generated at hot spots formed between very closely spaced metal particles. The Raman scattering from molecules is significantly enhanced when molecules are adsorbed on the hot spots (Fig. 6.11a).

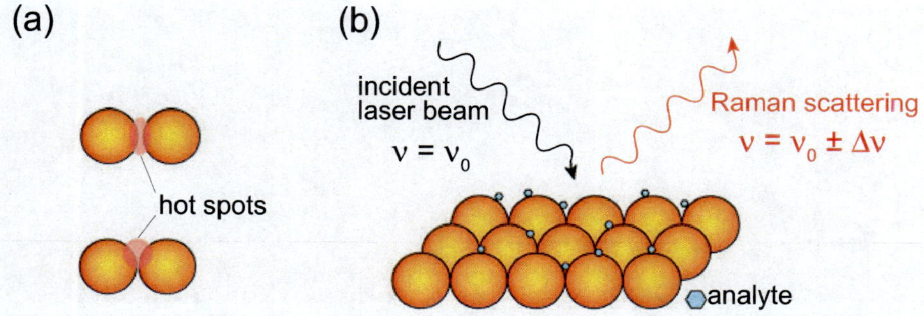

Fig. 6.11 Illustrations of **a** the depletion attraction, **b** hot spots between gold particles, and **c** SERS using the gold colloidal crystal

Because gold colloidal crystals contain many hot spots, they are expected to serve as SERS substrates (Fig. 6.11b). Here, we fabricated gold colloidal crystal structures by depletion attraction and evaluated their performance as SERS substrates.

6.4.3 Crystallization of Gold Colloids and Performance as SERS Substrates

By adding sodium polyacrylate (NaPAA, Mw = 25,000) to the gold colloid, island-like gold crystals with a diameter of several tens of microns were generated due to depletion attraction. Because the gold particles are very bright because of plasmon resonances, a single gold colloidal particle of 80 <nm can be observed with an ordinary optical microscope.

The growth process of the gold colloidal crystals was observed using gold particles ($d = 100$ nm, $\phi = 1.4 \times 10^{-5}$) at $C_p = 0.125$ wt%. Microscopic images of the crystallization process are shown in Fig. 6.12. Measurements were taken from the bottom of the sample cell using an inverted microscope [17].

After the sample was dried, the crystal structure was observed by scanning electron microscopy (SEM) (Fig. 6.13). An SEM image of the crystal observed from an oblique direction by tilting the sample stand is shown in Fig. 6.14. The crystal has an FCC structure consisting of about 20 layers of lattice planes, and its (111) plane was found to be oriented toward the bottom of the cell. The (001) plane was also observed on the crystal side.

A dilute gold dispersion was used as the starting material, but because gold has a high density of about 19.3 g/cm^3, the particle concentration was sufficiently high at the bottom of the cell by sedimentation to crystallize. Furthermore, calculations of the excluded volume show that the depletion attraction between particles and walls is twice as large as the attraction between particles [49]. Thus, due to particle sedimentation and the strong attraction from the walls, gold colloid crystals are formed at the bottom of the sample cell and grew along the bottom in two dimensions.

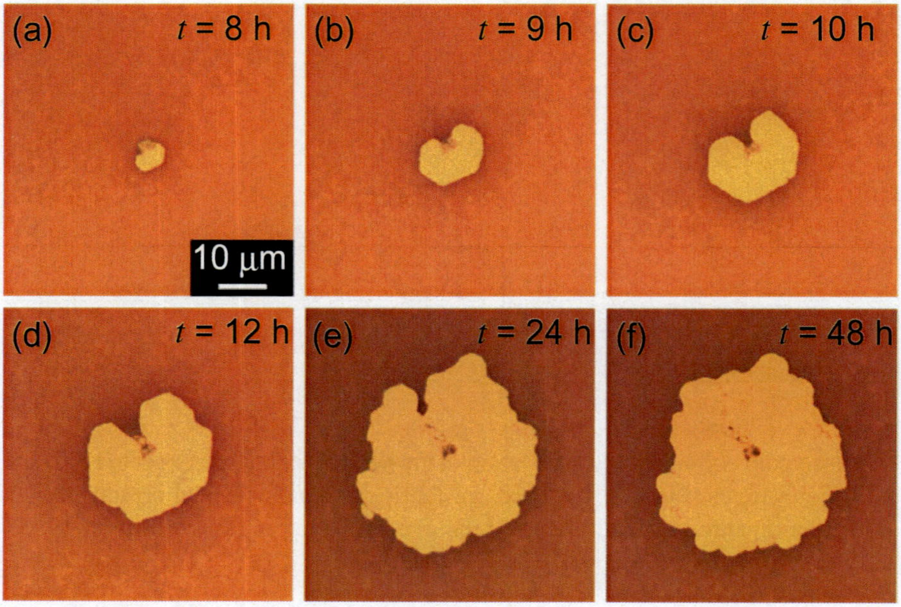

Fig. 6.12 Growth process of the gold colloidal crystals ($d = 100$ nm, $\phi = 1.4 \times 10^{-5}$) at $C_p =$ 0.125 wt%. Reprinted with permission from Ref. [17]. © (2021) American Institute of Physics

Fig. 6.13 SEM images showing particle arrangement in the gold colloidal crystal

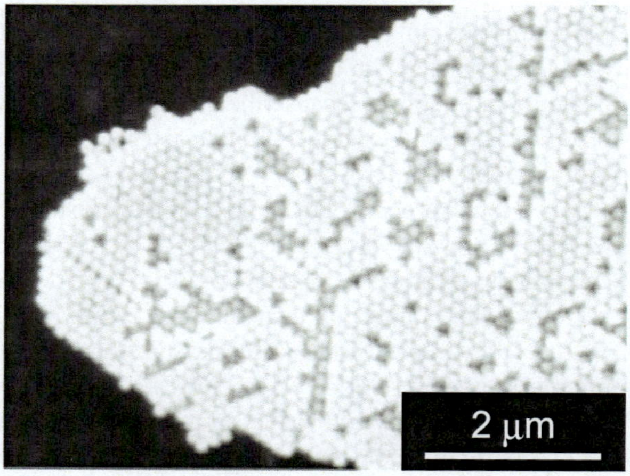

For use as a substrate for spectroscopic measurements, large-area crystals that are not island-shaped are desired. Therefore, as shown in Fig. 6.15a, crystals were grown by placing the sample cell on a tilting table. Figure 6.15b is a micrograph of the edge of the cell at an inclination angle $\theta = 30°$, where an area of several 10 μm from the edge is covered with crystals. Figure 6.15c is micrographs of the crystals at various values of θ ($t = 24$ h). Detailed examination revealed that the tilted bottom of the

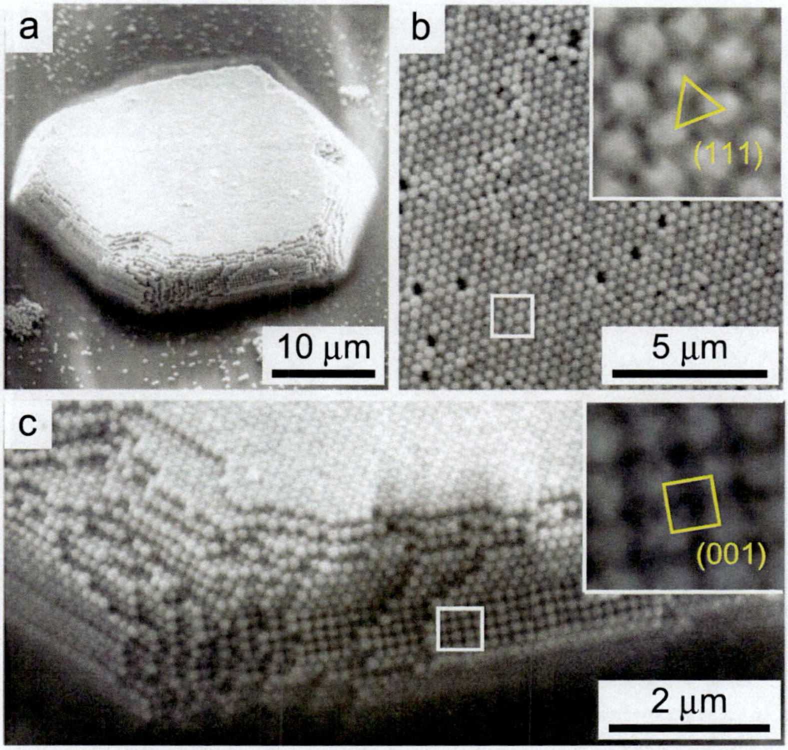

Fig. 6.14 **a** SEM images of the gold crystals observed from an oblique angle using a tilted sample stand. **b** FCC (111) planes oriented to the cell bottom. **c** FCC (001) planes on the crystal sides. Reprinted with permission from Ref. [17]. © (2021) American Institute of Physics

cell efficiently supplied gold colloidal particles to the growing crystals, resulting in the formation of large crystals.

The performance of gold colloidal crystals as SERS substrates was evaluated. Colloidal crystals were prepared on a glass plate using gold colloidal particles ($d =$ 150 nm) at $\phi = 1.3 \times 10^{-5}$ and $C_p = 0.07$ wt% and then dried. The Raman spectra obtained by a fiber Raman spectrometer are shown in Fig. 6.16. Measurements were taken on a sea-island structure and on a substrate with the crystals grown at $\theta =$ 15° (entire bottom surface covered), respectively. These results indicate that the Raman scattering light is significantly enhanced by the gold colloidal crystals. In the spectrum, the peaks with wavenumbers are those attributed to the rhodamine molecule.

2D Colloidal Crystals

The 2D crystals with a gap between the particles are obtained by electrostatically adsorbing 3D-charged colloidal crystals onto a flat substrate charged with an opposite

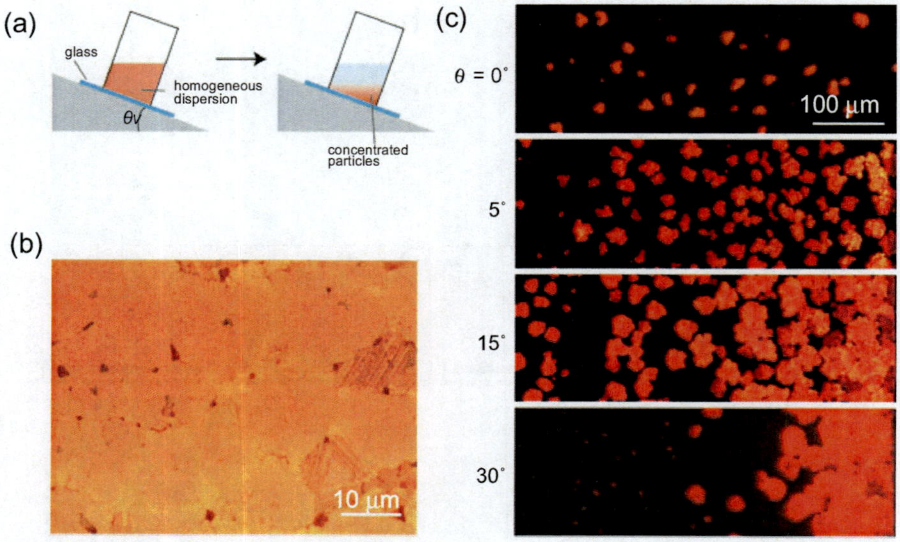

Fig. 6.15 **a** Illustration of the crystal growth on a tilting table. **b** A micrograph of the edge of the cell at an inclination angle $\theta = 30°$. **c** Crystal structures at various inclination angles. Reprinted with permission from Ref. [17]. © (2021) American Institute of Physics

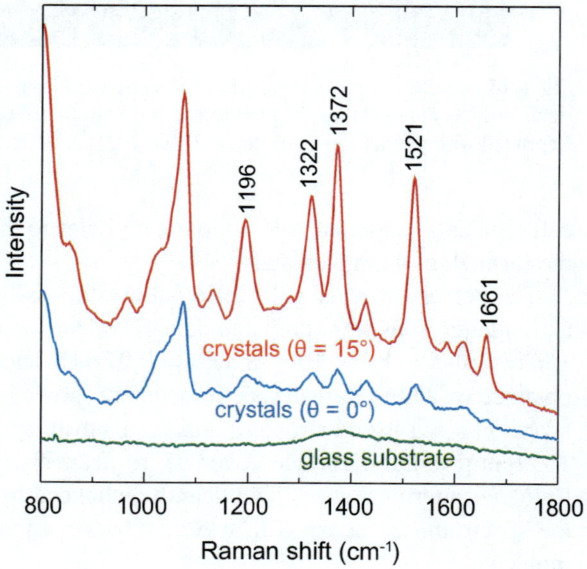

Fig. 6.16 Raman spectra of R6G on glass substrates and gold colloidal crystals (d = 150 nm, $\phi = 1.3 \times 10^{-5}$, and $C_p = 0.07$ wt%). Island structure and on a substrate with the crystals grown at θ = 15°. Reprinted with permission from Ref. [17]. © (2021) American Institute of Physics

sign [28]. The mechanism is schematically illustrated in Fig. 6.17. After adsorption, the non-adsorbed particles were washed with water to obtain 2D crystals.

An example of 2D colloidal crystals of PS particles ($d = 500$, $Z = 5290$, $\phi = 0.07$) is shown in Fig. 6.18. The 3D crystals of charged colloids were prepared in

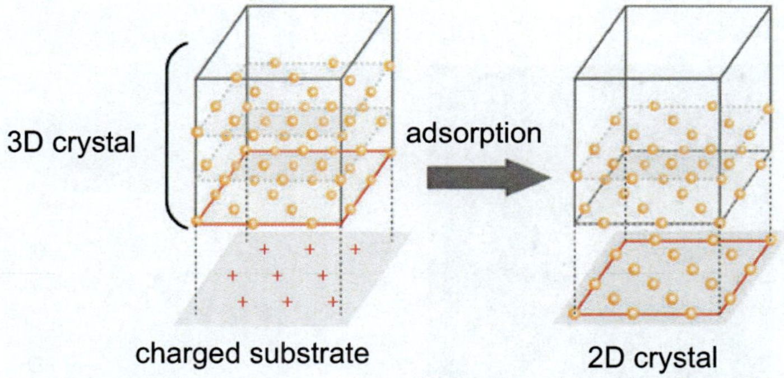

Fig. 6.17 Illustration of the fabrication method of the 2D colloidal crystal by the electrostatic adsorption of the three-dimensional (3D) charged colloidal crystal on an oppositely charged substrate. Reprinted with permission from Ref. [28]. © (2019) American Chemical Society

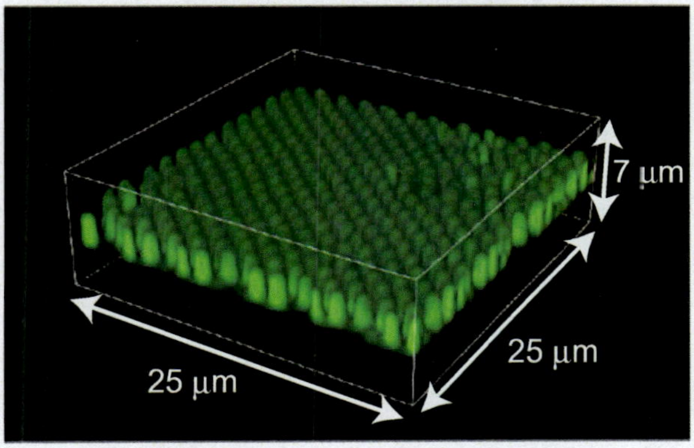

Fig. 6.18 LSM image of the 2D crystal of polystyrene particles ($\phi = 0.07$). Reprinted with permission from Ref. [28]. © (2019) American Chemical Society

water and adsorbed on a glass surface that had been previously positively charged with a cationic silane coupling reagent, aminopropyltriethoxysilane (APTES). The lowest layer of 3D crystals oriented parallel to the substrate adsorbed to the substrate surface, forming 2D crystals [28].

Figure 6.19a shows an optical micrograph of a 3D crystal of silica particles (510 nm), showing a six-fold symmetric pattern attributed to the FCC (111) plane, oriented parallel to the cover glass surface. Figure 6.19b presents the radial distribution function g(r) (r is the distance between particle centers) obtained from the image processing. The g(r) of the 2D crystal is shown in Fig. 6.19b as a red curve, which is almost identical to that of the 3D crystal, indicating that the crystal lattice was adsorbed without disturbing the crystal structure.

The glass substrates modified with APTES have an isoelectric point around pH = 8 and are positively charged on the acidic side and negatively charged on the alkaline side. Using negatively charged particles, 3D colloidal crystals were prepared under alkaline conditions and dropped onto the APTES-modified glass substrate. Under alkaline conditions, the APTES-modified glass substrates are negatively charged.

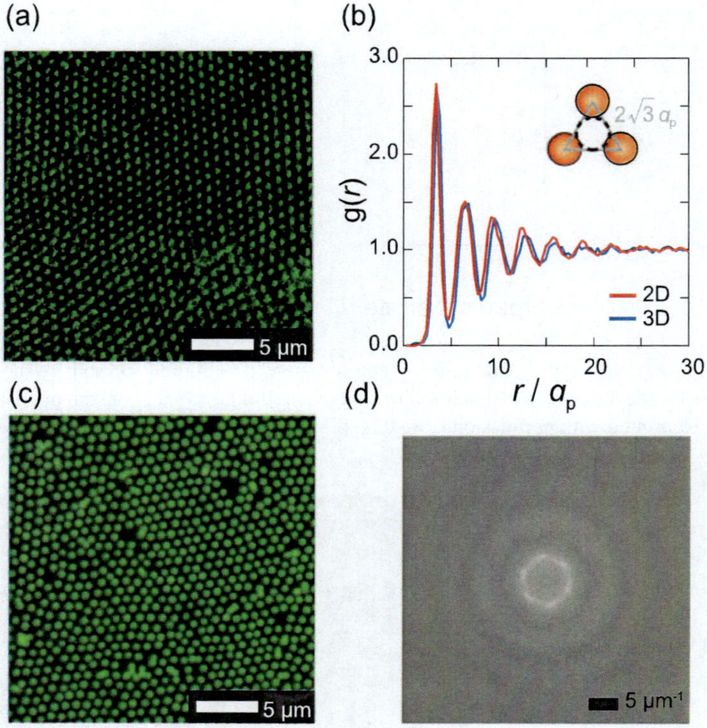

Fig. 6.19 **a** Micrograph of the 3D colloidal crystal. **b** Radial distribution function g(r) of the 3D and 2D crystals. **c** Microscopy image of the 2D colloidal crystal. **d** Fourier spectrum of the image in **c**. Sample, colloidal silica (d = 510 nm). Reprinted with permission from Ref. [28]. © (2019) American Chemical Society

Subsequent deionization with an ion exchange resin changed the charge of the substrate to positive and allowed the negatively charged particles to be adsorbed. Furthermore, by combining the unidirectional crystallization of silica colloids under a base concentration gradient (Chap. 5) and the unidirectional adsorption of APTES, we also succeeded in creating large cm-sized 2D silica crystals.

6.5 Space Experiments on Colloidal Self-Assembly

The ISS is a huge manned experimental facility built in orbit about 400 km from the ground (Fig. 6.20). Here, we describe two projects on colloidal self-assembly conducted in collaboration with the Japan Aerospace Exploration Agency (JAXA) and a number of other organizations.

Fig. 6.20 International Space Station (ISS). © JAXA

6.5.1 The Three-Dimensional Photonic Crystal (3DPC) Project

First, we describe the 3D Photonic Crystal Formation Project (3DPC Project). The Japanese Experiment Module "Kibo" on the ISS was completed in 2009 after more than 20 years of work. Prior to the full-scale use of the space environment by Kibo, space experiments were conducted at the Japan Aerospace Exploration Agency (JAXA) using Russian service modules and other equipment. The 3DPC project was one of the space missions.

The 3DPC project involved various experts from industry, academia, and government in colloid science, optical materials, and particulate synthesis, as well as experts in space experiments and experimental equipment development. The goal of the 3DPC project was to create a material with colloidal crystals fixed in a gel in space. The overview of the project is illustrated in Fig. 6.21. Generating 3DPC with excellent optical properties requires the use of fine particles with a high refractive index. Still in general, materials with an optical refractive index have a high density and, therefore, a high specific gravity (density). As the density of the fine particles increases, gravity distorts the crystal structure, making it impossible to produce large uniform crystals. Therefore, the project aimed to generate large 3DPCs using fine particles with a high refractive index (high density) by taking advantage of the microgravity environment in space. As a method for growing colloidal crystals, we used unidirectional crystal growth described in Chap. 5. Figure 6.22 illustrates the space-experimental cell. A

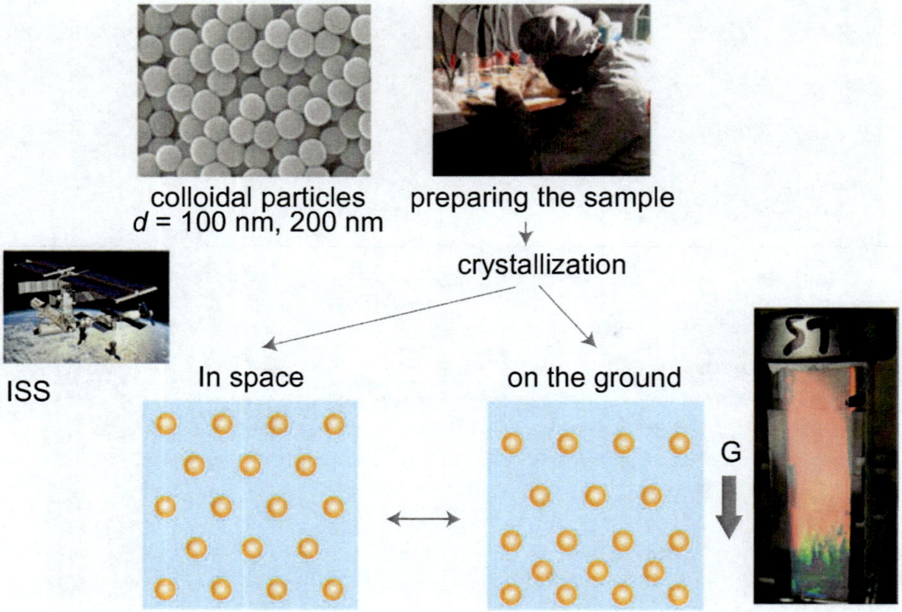

Fig. 6.21 Overview of the 3DPC project. © JAXA/NASA

semi-permeable gel membrane is placed in the center of the cell. About half of this cell is filled with an aqueous solution of base, and the other half with a colloidal dispersion of silica or other material. As the base diffuses, silica colloidal crystals grow. A gelling agent is dissolved in the sample, and after the crystals have grown, they are irradiated with ultraviolet light to fix the crystal structure.

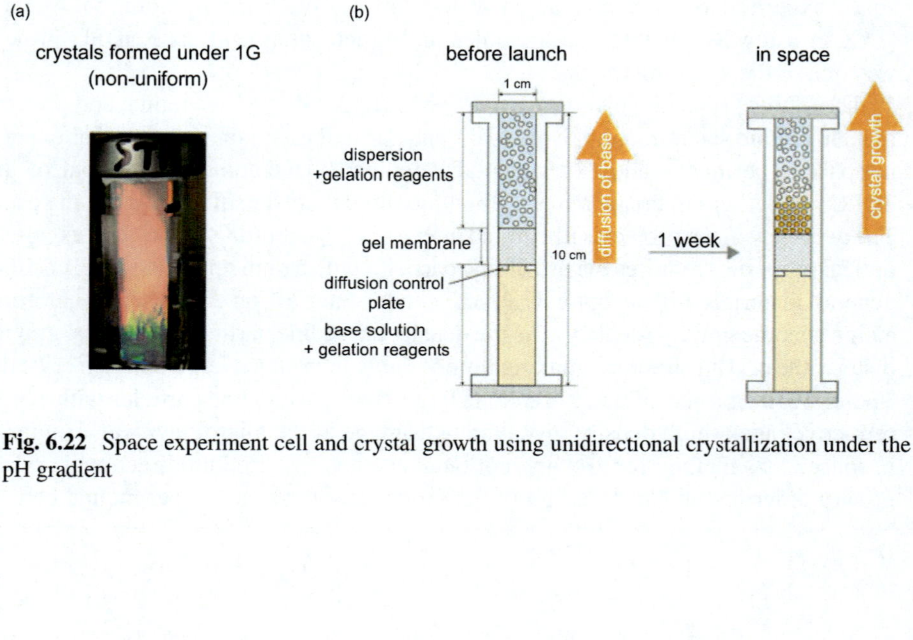

Fig. 6.22 Space experiment cell and crystal growth using unidirectional crystallization under the pH gradient

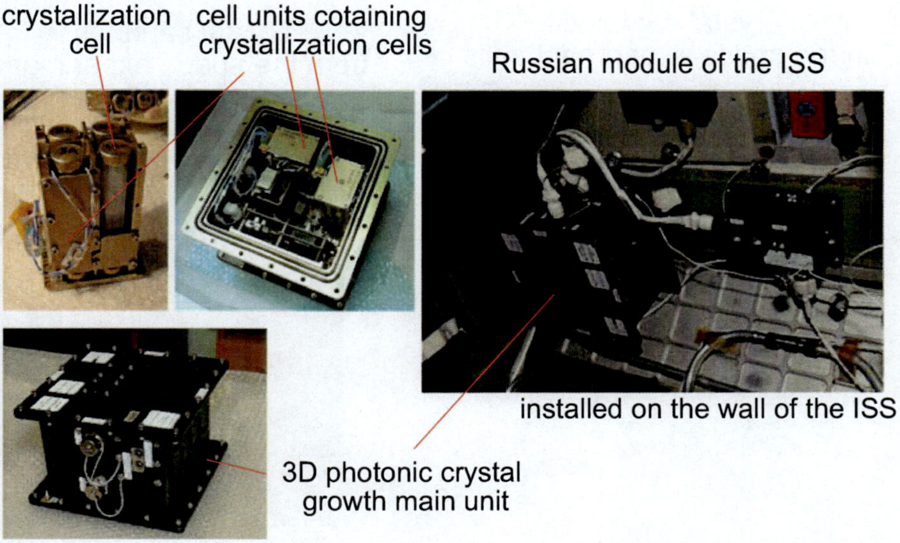

Fig. 6.23 Space experiment equipment used in the 3DPC project. © JAXA/NASA

After more than five years of preparation, the space experiments were conducted in 2005 and 2007. Figure 6.23 shows the apparatus used in the space experiments. A crystal growth cell is placed in a metal container of about 30 cm on a side, and a UV-LED is attached near the cell. The photo on the right shows the actual experimental apparatus installed on the wall of the ISS. The details of the experimental setup were reported in Ref. [50].

The gelled sample returned safely from the ISS. Figure 6.24 shows examples of gel-immobilized colloidal crystals fabricated in space. This experiment demonstrated for the first time in the world a technique for growing colloidal crystals in space, fixing them with gel, and returning them to the ground. This made it possible to evaluate in detail on the ground the characteristics of crystals formed in space as photonic crystals. The experiment also demonstrated the improved uniformity of crystals compared to those on the ground, as well as the success in creating mm-sized crystals using silica particles 200 nm in diameter, which had previously been difficult to achieve on the ground.

6.5.2 The Colloidal Clusters Project

Clusters of a small number of atoms and molecules are widely observed in nature, from the initial process of crystal growth to the formation of protein complexes in cells. When positively and negatively charged particles are mixed, clusters of various association numbers are formed (Fig. 6.25). In particular, the tetrahedral

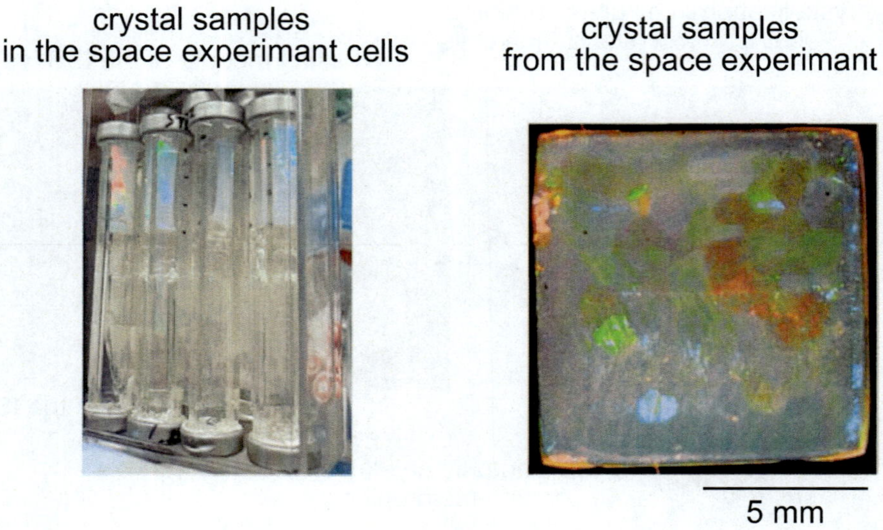

Fig. 6.24 **Left** crystal samples in the space experiment cells **Right** crystal samples obtained from the space experiment

cluster is the building block of the diamond crystal, which has attracted attention as a high-performance photonic crystal (Fig. 6.26).

We have conducted space experiments on the clustering of colloidal particles in the ISS [50]. Figure 6.27 shows the sample bags used in the experiments. Dilute dispersions of positively or negatively charged colloidal particles were introduced into each of the two rooms of the plastic sample bag, which are divided by a separator. The sample solution also contains gelation reagents. The sample bags are kept under microgravity for two days after the space crew press and squash the bags on the ISS to break the separators. The photograph on the right shows the samples launched into space. They are connected with strings to prevent them from falling apart under the microgravity inside the ISS. The samples are then gelled by UV-LED irradiation and returned to the ground.

The samples were launched by a rocket from SpaceX of the U.S.A. (Fig. 6.28), which carried space crews and materials to the ISS. The experiment was conducted in July 2020. Figure 6.29 shows the space crew working on the ISS. The solutions in the two rooms of the sample bag were mixed by pushing and squeezing the bag after the separator is broken. Figure 6.30 shows the sample bags placed in the experimental apparatus. After setting, they were kept in microgravity for 2 days to form clusters. Two days later, the samples are gelled. After the gelation, the space crew removed the samples from the device and put them in a container for collection (Fig. 6.31).

Figure 6.32 is a gel-immobilized sample returned from space. No supernatant or precipitate was observed in the sample obtained in a sedimentation-free microgravity environment, even in the high-density Titania samples. Samples were cut for microscope observation. The cross-section of the gel was observed under an optical

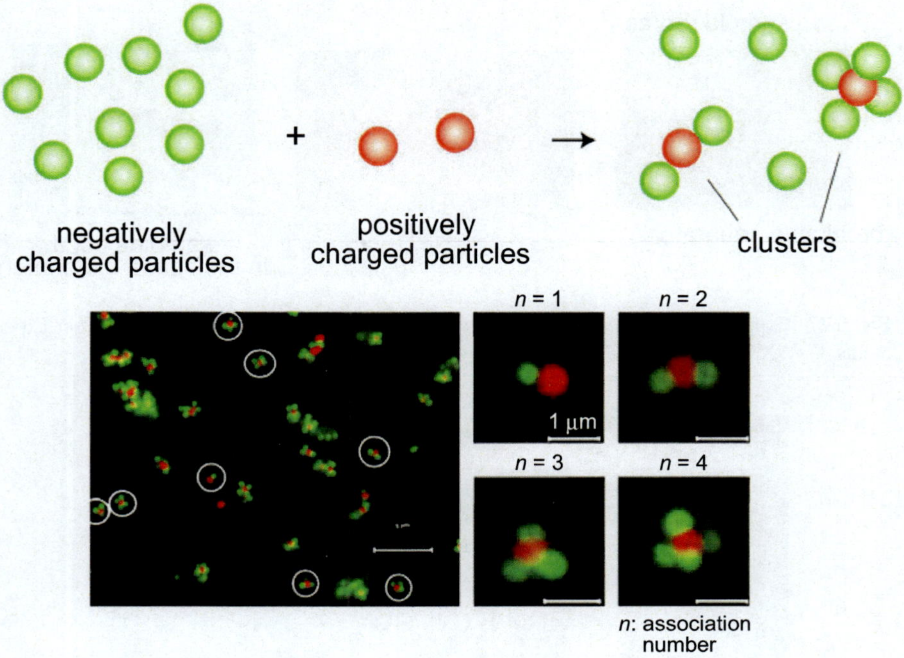

Fig. 6.25 Colloidal clusters in binary mixture of oppositely charged polystyrene particles

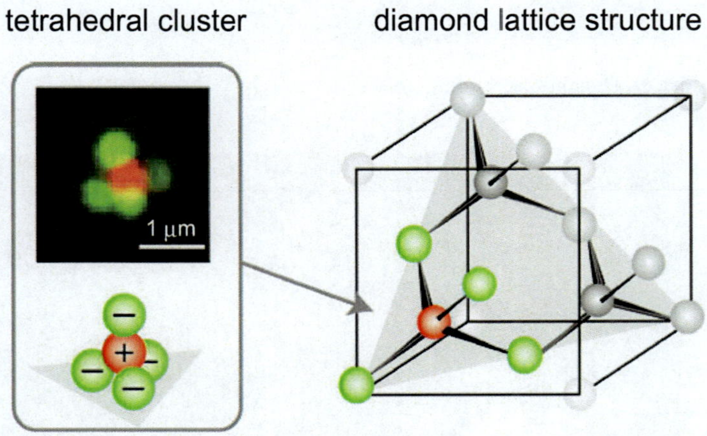

Fig. 6.26 Tetrahedral clusters and diamond lattice structure

microscope. Figure 6.33 shows an example of a micrograph. The sample is a binary colloid of positively and negatively charged PS particles, colored with fluorescent dyes [PS(+) (red), $d = 789$ nm, $\zeta = +38.7$ mV, $\phi = 2 \times 10^{-5}$; PS(−) (green), $d = 1025$ nm, $\zeta = -50.1$ mV, $\phi = 10^{-3}$]. The clusters with various associations were formed under microgravity.

Fig. 6.27 Experimental apparatus equipped with UV-LEDs used in the colloidal cluster experiments. © JAXA/NASA

Fig. 6.28 SpaceX Dragon spacecraft. © JAXA

Fig. 6.29 Snapshot of the space crew performing the cluster experiments. © JAXA/NASA

Fig. 6.30 Sample bugs in the experimental apparatus

Fig. 6.31 Snapshot of the space crew taking out the sample bags after the gelation. © JAXA/NASA

The structures of the clusters formed in the space were analyzed by optical microscopic observation, as well as by small-angle and ultra-small-angle neutron scattering experiments [51]. Even for PS particles with low density, the average number of clusters was found to be about 50% higher in the space sample than in the ground control. In addition, for clusters having association numbers of 4 and 3, the space sample showed greater structural symmetry than the ground sample. These findings suggest that slight settling and convection effects on the ground influence the association of micrometer-sized colloidal particles.

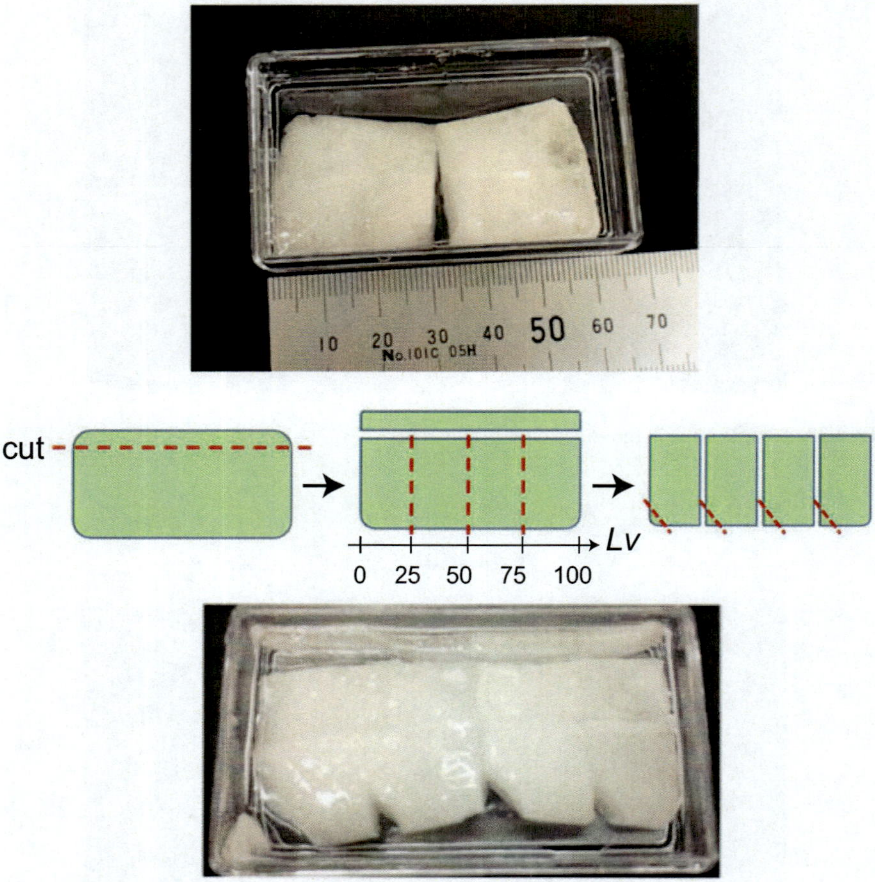

Fig. 6.32 Space experiment sample (Titania particles) taken out of the sample bag after returning to the ground. No precipitation or supernatant was observed, and the particles were almost uniformly distributed. ©JAXA/NASA

Furthermore, tetrahedral clusters of Titania particles were obtained for the first time in space. As will be discussed in Chap. 7, tetrahedral clusters are the building blocks of the diamond lattice, which works as photonic crystals. We expect that the findings from this space experiment will be useful for the development of photonic materials.

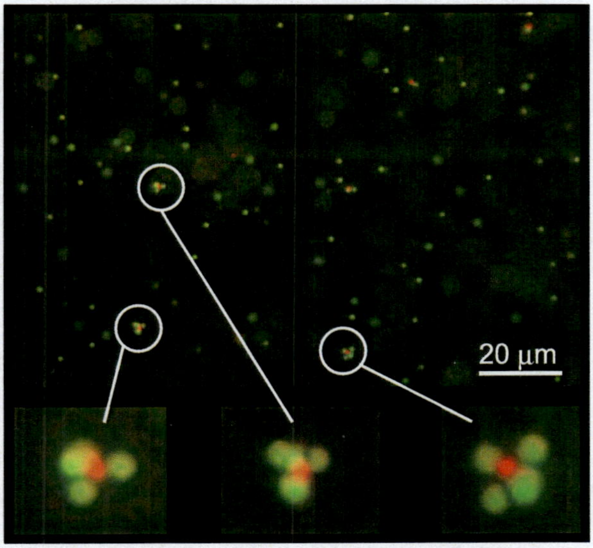

Fig. 6.33 Clusters of oppositely charged PS particle obtained by the space experiments. ©JAXA/NASA

References

1. Joannopoulos JD, Winn JN (2008) Photonic crystals: molding the flow of light, 2nd edn. Princeton University Press, New Jersey
2. Moon JH, Yang S (2010) Chem Rev 110:547–574
3. Li F, Josephson DP, Stein A (2011) Angew Chem Int Ed 50:360–388
4. Lawrence JR, Ying Y, Jiang P, Fougler SH (2006) Adv Mater 18:300–303
5. Furumi S, Sakka Y (2006) Adv Mater 18:775–780
6. Holtz JH, Asher SA (1997) Nature 389:829–832
7. Kamenetzky EA, Magliocco LG, Panzer HP (1994) Science 263:207–210
8. Lee K, Asher SA (2000) J Am Chem Soc 122:9534–9537
9. Asher SA, Kimble KW, Walker JP (2008) Chem Mater 20:7501–7509
10. Asher SA, Alexeev VL, Goponenko AV, Sharma AC, Lednev IK, Wilcox CS, Finegold DN (2003) J Am Chem Soc 125:3322–3329
11. Iwayama Y, Yamanaka J, Takiguchi Y, Takasaka M, Ito K, Shinohara T, Sawada T, Yonese M (2003) Langmuir 19:977–980
12. Fudouji H, Sawada T (2006) Langmuir 22:1365–1368
13. Fernandez-Nieves A, Wyss H, Mattsson J, Weitz DA (eds) (2011) Microgel suspensions. Wiley-VCH, Berlin
14. Alsayed AM, Islam MF, Zhang J, Collings PJ, Yodh AG (2005) Science 309:1207–1210
15. Mohanty PS, Nöjd S, van Gruijthuijsen K, Crassous JJ, Obiols-Rabasa M, Schweins R, Stradner A, Schurtenberger P (2017) Sci Rep 7:1487
16. Sato Y, Toyotama A, Okuzono T, Yamanaka J (2019) Chem Lett 48:1319–1321
17. Ioka M, Toyotama A, Yamaguchi M, Nozawa J, Uda S, Okuzono T, Yoshimura M, Yamanaka J (2021) J Chem Phys 154:234901
18. Pelton M, Bryant G (2013) Introduction to metal-nanoparticle plasmonics. Wiley, New Jersey
19. Maier SA (2007) Plasmonics: fundamentals and applications. Springer, New York
20. Mayer KM, Hafner JH (2011) Chem Rev 111:3828–3857
21. Halas NJ, Lal S, Chang W-S, Link S, Nordlander P (2011) Chem Rev 111:3913–3961
22. Stewart ME, Anderton CR, Thompson LB, Maria J, Gray SK, Rogers JA, Nuzzo RG (2008) Chem Rev 108:494–521

23. Pieranski P (1980) Phys Rev Lett 45:569–572
24. Goldenberg LM, Wagner J, Stumpe J, Paulke B-R, Görnitz E (2002) Langmuir 18:5627–5629
25. Isa L, Kumar K, Müller M, Grolig J, Textor M, Reimhult E (2010) ACS Nano 4:5665–5670
26. Trau M, Saville DA, Aksay IA (1996) Science 272:706–709
27. Vogel N, Weiss CK, Landfester K (2012) Soft Matter 8:4044–4061
28. Aoyama Y, Toyotama A, Okuzono T, Yamanaka J (2019) Langmuir 35:9194–9201
29. Russel WB, Saville DA, Schowalter WR (1989) Colloidal dispersions. Cambridge University Press, Cambridge
30. Hiemenz PC, Rajagopalan R (1997) Principles of colloid and surface chemistry, 3rd edn. CRC Press, Boca Raton
31. Berg JC (2010) An introduction to interfaces and colloids, the bridge to nanoscience. World Scientific, Singapore
32. Davis KE, Russel WR, Glantschnig WJ (1989) Science 245:507–510
33. Davis KE, Russel WR, Glantschnig WJ (1991) J Chem Soc Faraday Trans 87:411–424
34. Davis KE, Russel WB (1989) Phys Fluids A 1:82–100
35. Crandall RS, Williams R (1977) Science 198:293–295
36. Shinohara T, Yamada H, Sogami IS, Ise N, Yoshiyama T (2004) Langmuir 20:5141–5144
37. Piazza R, Bellini T, Degiorgio V (1993) Phys Rev Lett 71:4267–4270
38. Ras M, Philipse AP (2004) Nature 429:857–860
39. van Roij R (2003) J Phys Condens Matter 15:S3569
40. Murai M, Okuzono T, Yamamoto M, Toyotama A (2012) and. J Yamanaka J Colloid Interf Sci 370:39–45
41. Zhu J et al (1997) Nature 387:883–885
42. Russel WB, Chaikin PM, Zhu J, Meyer WV, Rogers R (1997) Langmuir 13:3871–3881
43. Cheng Z, Chaikin PM, Zhu J, Russel WB, Meyer WV (2001) Phys Rev Lett 88:015501
44. Schöpe HJ, Wette P (2011) Phys Rev E 83:051405
45. Tomita Y et al (2018) Int. J Microgravity Sci Appl 35:350303
46. Yamanaka J, Murai M, Iwayama Y, Yonese M, Ito K, Sawada T (2004) J Am Chem Soc 126:7156–7157
47. Murai M, Yamada H, Yamanaka J, Onda S, Yonese M, Ito K, Sawada T, Uchida F, Ohki Y (2007) Langmuir 23:7510–7517
48. Toyotama A, Yamanaka J (2011) Langmuir 27:1569–1572
49. Lekkerkerker HNW, Tuinier R (2011) Colloids and the depletion interaction. Springer, Amsterdam
50. Ohki Y, Ikeda T, Itoh A, Yamanaka J, Ito K, Watanabe Y, Yoda E, Sawada T, Takiguchi Y (2009) Trans JSASS Space Tech Jpn 7:Th21–Th26
51. Miki H, Ishigami T, Yamanaka J, Okuzono T, Toyotama A, Mata J et al (2023) npj Microgravity 9:33

Chapter 7
Summary of the Book and Future Perspective

Abstract In the final chapter, after summarizing the contents of this book, we briefly review recent work on colloidal self-assembly, which offers some perspectives for future research. In particular, the current status of the field with respect to the formation of more complex structures using anisotropically interacting particles and multicomponent colloids is presented. Related topics including the self-assembly of liquid crystals and the effects of electric fields will also be briefly discussed. Also, the study of active matter, one of the hot topics in colloid science, will be briefly introduced.

Keywords Anisotropic interaction · Patchy particles · Janus particles · Multicomponent colloids · Liquid crystal · Electric field · Active matter

7.1 Summary of the Book

We wrote this book for those who wish to begin the study of colloidal self-assembly. First, in Chap. 1, the definition and examples of colloids were introduced and the significance of studying colloidal self-assembly was discussed. Some representative self-assembled structures, including colloidal crystals, are briefly described.

In Chap. 2, we described the basic knowledge necessary to understand the contents of the subsequent chapters. In particular, the concepts of stability and stabilization of colloidal dispersions, which are fundamental to colloid science, were explained from the basics. Next, several important interparticle interactions in colloidal systems, including rigid sphere repulsion, vdW force, electrostatic interaction, and depletion attraction, were discussed based on the interparticle potentials. Furthermore, colloidal crystallization caused by these interactions was described.

Chapter 3 summarized useful methodology for experimental studies of colloidal self-assembly. First, we described the synthesis of colloidal particles having a narrow size distribution. Then, techniques for the characterization of the particles, including concentration, size, and surface charge number, were explained. We also presented examples of the fabrication of colloidal crystals using various interactions. Finally, spectroscopy, optical microscopy, and scattering methods for analyzing the association structures were described with examples.

© The Author(s), under exclusive license to Springer Nature Singapore Pte Ltd. 2023

J. Yamanaka et al., *Colloidal Self-Assembly*, Lecture Notes in Chemistry 108,
https://doi.org/10.1007/978-981-99-5052-2_7

In Chap. 4, we described the basic concepts of numerical modeling, the algorithms, and methods for analyzing numerical data obtained by computer simulation of colloidal systems. First, molecular dynamics methods were briefly introduced, followed by some other methods that are also effective for simulating colloidal systems. Next, the Brownian dynamics and Monte Carlo methods were described as numerical simulation methods for colloidal systems. Finally, pseudocodes for the main algorithms and some images obtained from the simulations were presented.

In Chap. 5, we presented several methods for controlling crystallization to prepare large, high-quality charged colloidal crystals. They included the unidirectional crystal growth under pH and salt concentration gradients. Then, we discussed the temperature dependence of the phase behavior of colloids and described crystallization using temperature variation. When small amounts of different types of particles are added to a single-component colloidal system, these behave as “impurities”. In Chap. 5, we also reported the processes for removing various impurities from colloidal crystals.

In Chap. 6, we reported several applied studies on colloidal crystals and self-assembly. First, we described the preparation of self-standing materials by fixing the crystal structure with polymer gels. Then, the crystallizations of microgels and gold colloids using depletion attraction are described. We also reported recent space experiments on colloidal self-assembly in the microgravity environment of the International Space Station.

Chapters 1 through 4 will help the readers understand the basic concepts of colloid structure formation. For a more in-depth study of colloid science and various related fields (interface and surface science, soft matter science, crystal growth science), many excellent textbooks and review articles are available. Please refer to the references listed in Chapter 1 and other relevant chapters.

A video entitled “Visual Guide to Study Colloidal Self-Assembly” was published by Springer-Nature in 2022, which presents videos of the various phenomena described in this book. Please refer to this video for a further understanding of the contents of this book.

7.2 Perspective of the Research on Colloidal Self-Assembly

In the last part of this book, we present some recent trends in the study of more complex structures by colloidal self-assembly. Thus far, we have mainly discussed the self-assembly of isotropically interacting, single-component colloids. On the other hand, the fabrication of more complex structures using anisotropic interactions and multicomponent systems is currently a major research trend in this field and is expected to develop further in the future. For example, particles with non-uniform surfaces, collectively referred to as “patch particles”, have been synthesized by chemical control of surface states [1–4]. These novel particles can form anisotropic association structures by self-assembly. These particles are often called “colloidal molecules”, as colloidal systems in which isotropic interactions are models of atoms [5, 6].

In the following, we will present the current state of the field with regard to the formation of more complex structures. We will also discuss diamond lattice structures as an example of complex structures, which have received particular attention in recent years. In addition, "active matter", particles that move autonomously, will be introduced. Active matter is attracting attention not only in colloids but also in biology.

Although not discussed in detail in this book, the formation of complex structures inspired by biological structures has also attracted much attention [7]. Recent studies on structure formation using external fields are introduced in Chap. 2, and other interesting studies using various external fields, including magnetic fields and light, are also reported [8, 9]. There are excellent review articles on each of these topics, and the reader is referred to the references cited throughout.

7.2.1 Anisotropic Interaction

Patchy Particles [1–4, 10, 11]

Patchy particles are colloidal particles that are anisotropically patterned, either by modification of the particle surface chemistry. A variety of patchy particles with non-uniform shapes and physicochemical properties have been synthesized (Fig. 7.1).

The simplest example is a spherical particle, consisting of two hemispheres with different properties. Such particles, named Janus particles after the Roman god Janus with two faces, have been actively studied; the differences in properties of the two hemispheres are, for example, hydrophobic/hydrophilic, metallic/polymeric, charged/uncharged, organic/inorganic, and so on. Janus particles were introduced by Casagrande et al. in 1989 [12]. And they were brought to the attention of researchers worldwide when de Gennes, a well-known researcher in polymer physics and soft condensed matter physics, mentioned Janus particles in his 1991 Nobel lecture [13]. It should be noted that in his Nobel lecture, de Gennes used the term "soft matter" to describe substances such as liquid crystals, polymers, surfactants, and colloids that

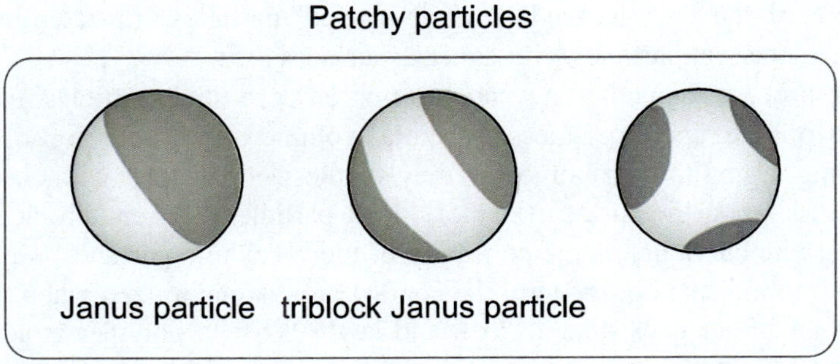

Fig. 7.1 Various patch particles

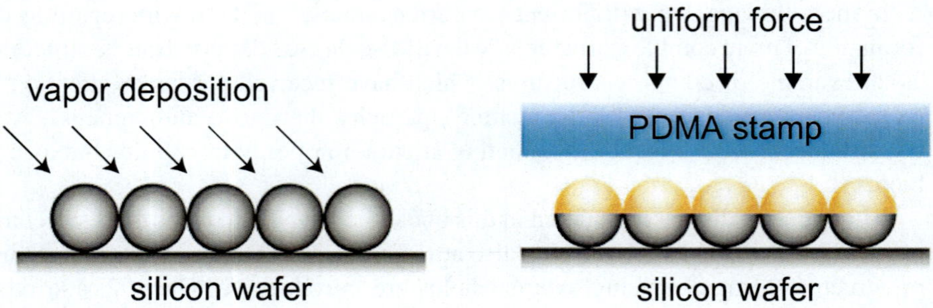

Fig. 7.2 Method for synthesizing Janus particles using metal vapor deposition

are easily deformed by external stresses [13]. This term has spread worldwide since then.

Many kinds of Janus particles have thus far been developed and implemented in new technologies such as emulsion stabilizers, electronic paper, and drug carriers. One of the simplest methods to obtain the Janus particles is shown in Fig. 7.2. First, a tightly packed monolayer colloidal crystal of spherical particles (e.g., polystyrene particles) was made and then a vapor of metal, such as gold, is directly physically deposited on top of it. This modifies half the surface of the particles with gold. Next, the particles are stamped with silicone rubber (polydimethylsiloxane, PDMS) to adsorb and remove the particles.

The group of Granick created "triblock Janus" particles [11, 14] (Fig. 7.1). These Janus spheres have a structure in which both poles of the particle are hydrophobic and the equatorial part is charged. By tuning the electrostatic repulsive interactions between the particles by adding NaCl in water, the particles could be self-assembled by short-range hydrophobic attraction. When appropriate conditions are chosen, a "Kagome lattice" is formed at equilibrium. A unit structure of the lattice is shown in Fig. 7.3. The synthesis and self-assembly behavior, physical properties, and applications of Janus particles are reviewed in detail, for example, in Refs. [2, 3].

Lock and Key Particles [15, 16]

Sacanna et al. have synthesized "keyhole-shaped" particles illustrated in Fig. 7.4a, which are spherical particles with concave surfaces [15].

In the process of synthesizing polymer-based core–shell particles, they discovered that polymerization reduces the molar volume of the core portion, resulting in buckling. Using this phenomenon, they synthesized particles with concavities of controlled size on the surface (keyhole, lock particles). When spherical particles with the same curvature as the concavity of these keyhole particles (key particles) are mixed in a dispersion system, the contact area is maximized when the two are in contact without gaps, like a "lock and key". When a polymer is added to the dispersion and depletion attraction is applied, the potential between lock and key is proportional to the overlap of the excluded volume. In other words, it is maximum

Fig. 7.3 Structural unit of two-dimensional Kagome lattice in which triblock Janus particles are assembled (Ref. [14])

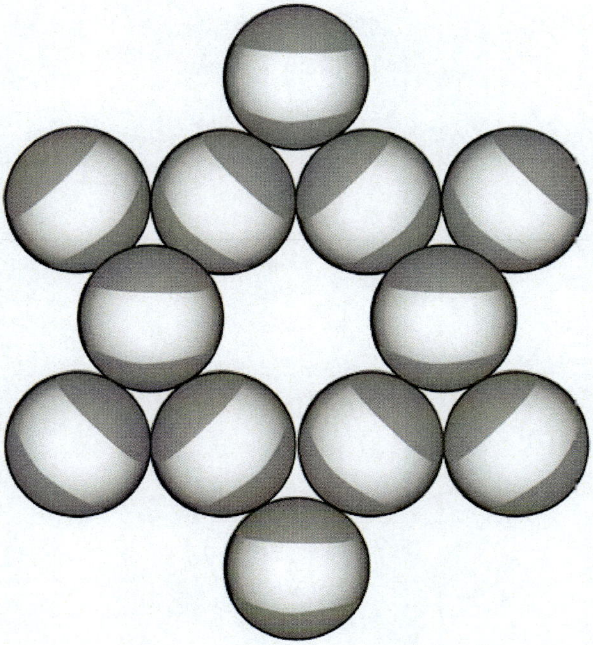

in configurations, where the key fits exactly into a spherical cavity depending on the size and position of the key, making it easy to achieve a "lock and key" (Fig. 7.4b).

Since there is no irreversible chemical bonding between the components in a lock–key coupling, it can move freely as a ball-in-socket joint. Various "colloidal molecules", in various linear and branched forms, have been fabricated using this technique. The same group has also constructed three-dimensional "lock and key" particles by placing multiple concavities on the particle surface, yielding controlled assembly [16].

DNA-Modified Particles

Nanotechnology using deoxyribonucleic acid (DNA) molecules has reached applications in chemistry, materials science, and nanotechnology. The binding of colloidal particles, via DNA molecules modified on the particle surfaces, has been vigorously investigated [17, 18]. DNA is a copolymer of four different nucleotides. Each nucleotide has a base molecule, adenine (A), thymine (T), cytosine (C), and guanine (G). A and C selectively form base pairs with T and G, respectively, through hydrogen bonds. DNA with a certain base sequence (for example, ACTG…) binds strongly to its complementary base pair (TGAC…), and this property can be used for selective binding. Using this property, a "DNA origami" technique has been developed to create DNA assembly structures [19]. This technique is a programmed structure using DNA sequence information. In 1996, the organization of gold nanoparticles using DNA oligonucleotides was reported [20]. In addition, particles can be selectively bound by modifying the surface of submicron to micron-sized colloidal particles

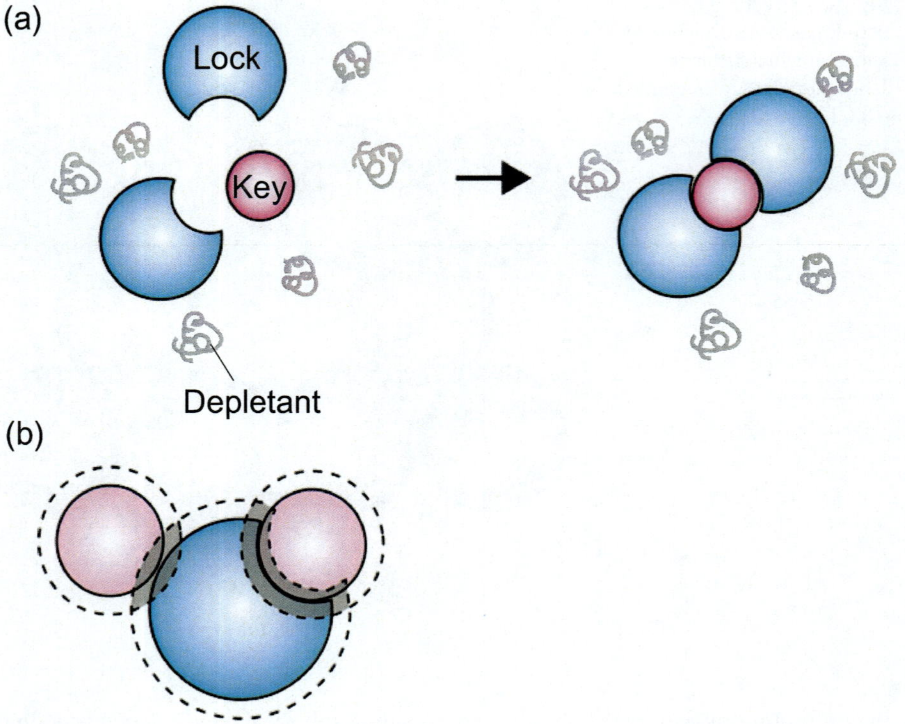

Fig. 7.4 **a** Schematic diagram of a method for synthesizing keyhole-shaped particles. **b** Exclusion volume between particles. When the key particle fits exactly into the cavity of the keyhole-shaped particle, the excluded volume is maximized and the depletion attraction is maximized, thus "key and keyhole" is easily realized (Ref. [15])

with DNA molecules (Fig. 7.5). This technique has been applied to the formation of complex structures such as diamond lattices, as described in Fig. 7.5.

Many other novel particles have been synthesized to date. Hueckel et al. reported a review of the field on the synthesis of various particles [4].

7.2.2 *Novel Self-Assembly Structures*

In recent years, the formation of complex self-organizing structures has been reported using ingenious methods. In this section, we introduce several topics on this issue. Many other interesting colloidal self-assembly has been reported as well, including organization on patterned substrates [21–23] and structure formation at interfaces [24, 25]. Please refer to recent review articles for details [26, 27].

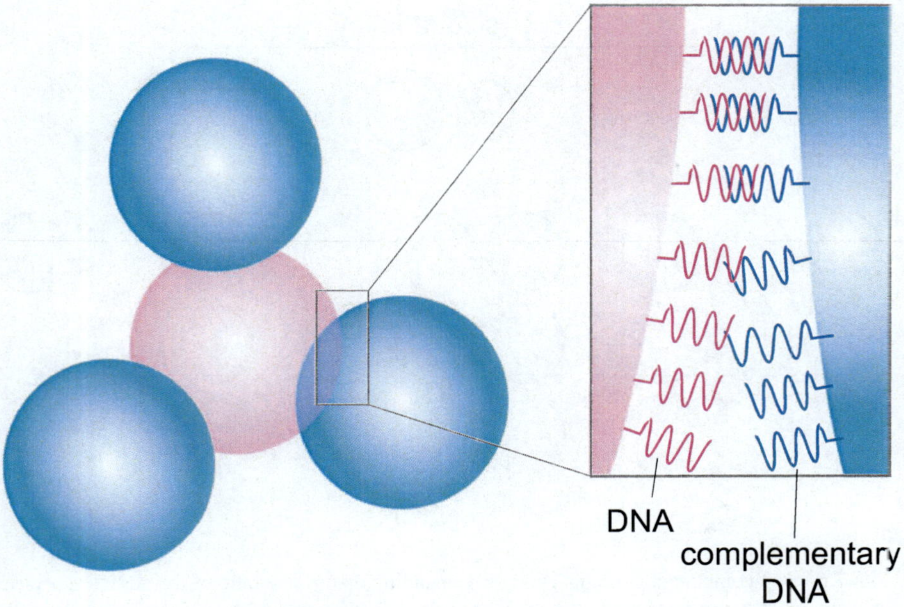

Fig. 7.5 Binding between colloidal particles whose surfaces are modified with DNA molecules and complementary DNA molecules

Electrostatic Interaction in Binary Systems

When particles with positive and negative charges are mixed, particles form clusters due to electrostatic attraction, as described in Chap. 6. If the attraction is too strong, random aggregation may occur. However, regular self-assembled structures are formed under appropriate conditions by modulating electrostatic interactions. Bartlett and Campbell [28] reported that two-component colloids with opposite charges form binary crystals in equilibrium. They used poly(methyl methacrylate) spheres of approximately the same size, positively and negatively charged, and tuned the charge number of the particles by using the adsorption of bromide ions. Face-centered-cubic, cesium chloride, and sodium chloride-type crystals were formed, depending on the charge numbers. The phase transition in the crystal structure was explained by the competing effects of entropy and Coulomb attraction.

Leunissen et al. [29] studied the structure of the oppositely charged binary colloid by adjusting the size ratio of the particles in addition to the magnitude of the electrostatic interactions. They point out that, unlike atomic systems, the sum of the surface charges of the positive and negative particles need not be neutral in a two-component colloidal crystal. This is because the charged colloidal particles become electrically neutral due to the presence of counterions. Since there is no constraint that the particle charges are electrically neutral in total, structures are observed in colloidal systems that are not found in atomic systems. As shown in Fig. 7.6, the crystal structure changed depending on the size ratio of the particles [29, 30].

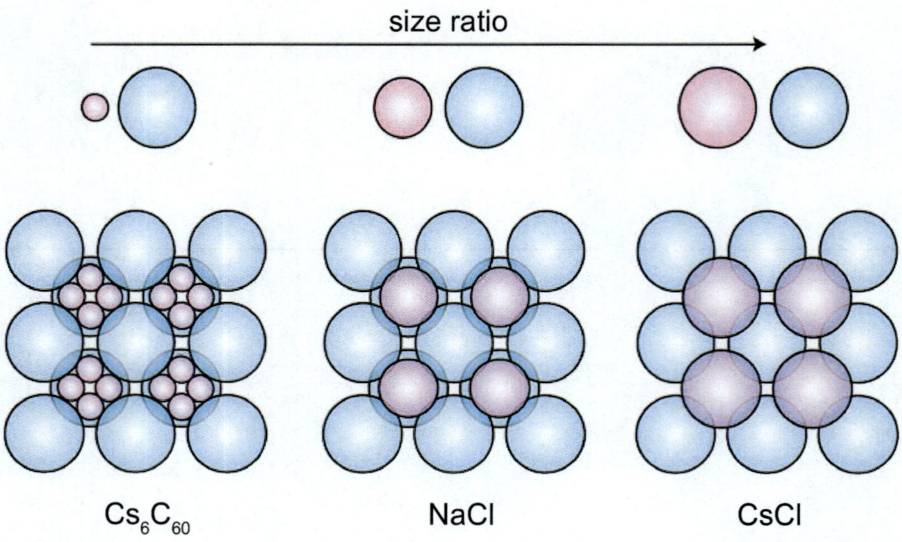

Fig. 7.6 Two-component colloidal crystals. Depending on the size ratio of the particles, a variety of different structures are produced, such as cesium chloride and sodium chloride crystals (Ref. [28])

Hückel et al. have reported that macroscopic colloidal single crystals spontaneously formed in an oppositely charged two-component system [31]. They modified the surface of charged particles with a neutral polymer to fine-tune the interparticle potential.

Effects of the External Field

Force Due to the External Fields

Here, we describe the structure formation by applying external fields. A *field* is a physical quantity that has a value for each point in space and time. The field for electrical forces is the electric field, and an object is subjected to a force from the electric field (see Appendix A). The self-assembly of colloids described so far was caused by particle–particle interactions, due to the fields formed inside the dispersions. In addition to these forces due to the *internal fields*, forces can be applied to the colloidal particles by *external fields.* For example, applying an electric field moves the particles and changes the structure they form.

As mentioned in the section on vdW forces (Column 2–1), most of the interactions we see in our daily lives can be attributed to electromagnetic interactions. Therefore, electric and magnetic fields are fundamentally important external fields. In the following, studies on the effects of applied electric fields are presented in some detail. Many excellent studies on the effects of other external fields have been reported so far. Please refer to the review articles [26, 32–34]. The effect of particle size on phase behavior is also discussed in Ref. [32].

Structure Formation Under the Electric Field

Ions and charged particles are driven by electric fields. As will be discussed in detail in Chap. 3, the potential and number of charges on the surface of particles can be determined from measurements of the electrophoretic velocity and electrical conductivity. On the other hand, the association structure of particles can also be controlled by electric fields. Since charged particles and counterions have opposite sign charges, they will receive the forces in the opposite direction from the electric field. This results in an electric dipole, which can also affect the electrophoretic velocity (Fig. 7.7a). Also, as mentioned in the description of the vdW force, matter polarizes under an electric field. Thus, even if the particles are not charged, the colloidal particles themselves can act as dipoles (Fig. 7.7b). Such electric dipoles may interact with other dipoles to form structures.

Electric fields include both a direct current electric, DC, electric field and an alternating current, AC, electric field. On applying a DC field, the charged particles will accumulate on the electrode of the oppositely charged. This phenomenon is sometimes used as a method of particle accumulation. However, once (non-metallic) particles cover the electrodes, no more current will flow (Because of this reason, many electrical conductometers use AC electric fields). Also, ions may move in one direction under a DC field, which sometimes causes a macroscopic flow of the medium and disturbs the self-assembled structure of colloids (electroosmosis effect). Therefore, AC fields have often been used in many studies of the structure formation

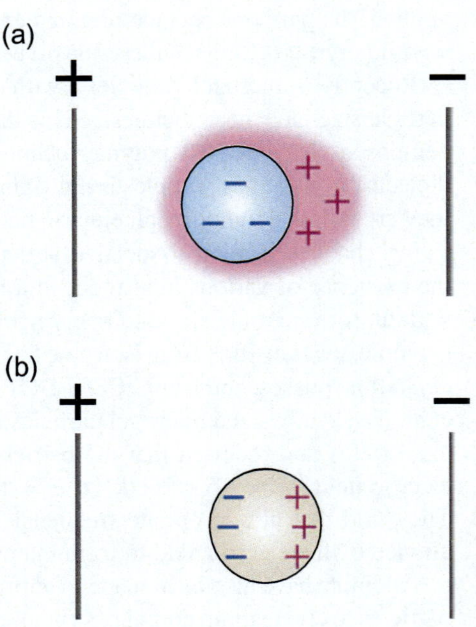

Fig. 7.7 Influence of external electric field. **a** Polarization of counterions and surface charges of particles. **b** Polarization of dielectric particles

of colloids. On applying AC electric fields, instantaneous dipoles (Fig. 7.7a) are induced in the particles, and their orientation changes periodically with time. When the frequency f of the AC field is sufficiently small, the motion of the counterions can follow the changes in the electric field, resulting in the dipoles. At high frequencies, they cannot follow, but dipoles can still be produced due to dielectric polarization. We will introduce several studies on the effect of the applied field below. All the studies use AC electric fields.

Zhang and Liu [35] used an electric field to externally control colloidal crystallization, in their experiments as a model for crystal growth. The sample was an aqueous dispersion of charged polystyrene particles with a diameter of 0.99 μm and a ζ potential of − 26 mV. Indium tin oxide (ITO) plates are used as electrodes. Since ITO is a conductor and transparent, it is often used for microscopy under electric fields. Two ITO plates were fixed parallel 0.12 mm apart, and a sample was introduced between them. A uniform AC field ($f = 450$ to 2500 Hz, a magnitude $E =$ 75 to 230 V/cm) was applied. Then, sea-island-like, close-packed 2D microcrystals were generated, which could be observed in situ with an optical microscope from the bottom of the ITO plate. The authors discuss the growth mechanism in terms of crystal growth science.

Liu et al. [36] studied the structural changes induced by an electric field, for dispersions of rod-like charged silica particles (2.3–3.3 μm in length, surface potential − 70 mV) dispersed in an organic solvent, cyclohexyl chloride. In the absence of the electric field, the particles are arranged three-dimensionally in the medium to form a crystalline-like structure. However, unlike the spherical particles, there is no long-range ordering, and the structure is called a "plastic crystal". When an alternating electric field of high frequency ($f = 1$ M Hz, maximum $E = 600$ V/cm) was applied, the particles became oriented and exhibited an ordered structure similar to a liquid crystal (*cf*. next subsection). The structural change was reversible.

Recently, "microgel particles" with a diameter of micrometers and uniform particle size have been synthesized, as discussed in Chap. 6. Microgel particles are composed of about 10% polymer chains and medium, and although their average dielectric constant as particles is not significantly different from that of the medium, they are polarized under high electric fields and have been reported to exhibit interesting changes in their associated structures. Nöjd et al. [37] have demonstrated the existence of various field-induced transitions by synthesizing microgel spheres with uniform particle size and applying an electric field of about $f = 100$ kHz. For example, the transition from isotropic fluid to string and body-centered square (BCT) crystalline phases, and from FCC to BCT phase at medium concentrations. Crassous et al. [38] synthesized microgel particles of rotational ellipsoidal shape with various axial ratios and reported that the particles form aggregate structures when an AC electric field ($f = 1.6$ ~ 160 kHz, $E = 1670$ V/cm) is applied in water dispersion. They find that at appropriate frequencies, ellipsoidal ellipsoids are polarized and aligned by the electric field to form interesting tubular aggregates.

Attempts have also been made to apply an electric field to aggregates of colloidal particles to create more complex structures. Ma et al. [39] have reported the behavior of colloidal particles assembling and of colloidal aggregates arranging themselves in a regular manner by using an AC electric field.

Note that in all the studies mentioned above *uniform* AC electric fields were employed. When a *non-uniform* electric field is applied to dielectric particles, it exerts a force on the particles. This is referred to as dielectrophoresis (DEP) [40]. Hermanson et al. [41] reported that when a non-uniform AC electric field of f = 200 Hz and E = 50–250 V/cm was applied to an aqueous dispersion of gold particles of 15 to 30 nm in diameter, the particles bound together and produced macroscopic fibrous or dendritic chains, more than 5 mm in length.

Liquid Crystal and Colloidal Self-Assembly

Here, we briefly introduce the assembly of colloids in relation to liquid crystals. Liquid crystals are intermediate phases that appear between liquids and crystals. Not all substances have a liquid crystalline state, but it appears in the case of rod- or disk-shaped molecules. In the liquid crystal state, the molecules are more or less regularly oriented, but they do not have perfect symmetry with respect to their position, as in a crystal. Figure 7.8 explains major liquid crystal states. In a *nematic* liquid crystal (a), rod- or disk-shaped molecules are randomly dispersed in an aligned orientation. *Smectic* liquid crystals (b) are usually one-dimensional periodic structures created by rod-shaped molecules. *Columnar* liquid crystals (c) are often composed of disk-shaped molecules and have a two-dimensional periodic structure. In liquid crystal displays used in televisions and computers, light and color switching is achieved by using liquid crystals whose alignment order changes with the application of an electric field. The reviews by Chandrasekhar [42] and de Gennes and Prost [43] are well known in the field of physics of liquid crystals.

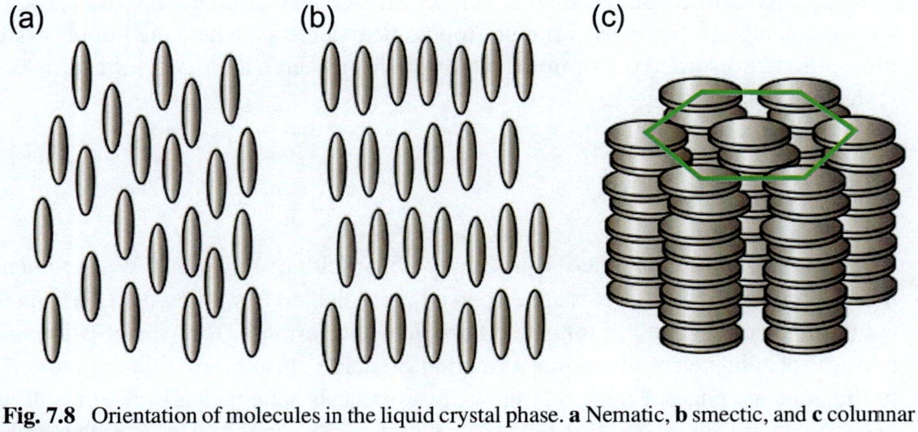

Fig. 7.8 Orientation of molecules in the liquid crystal phase. **a** Nematic, **b** smectic, and **c** columnar liquid crystal

We describe two topics. First is the study of the liquid crystal structures of rod-shaped colloidal particles. The second is the assembly of colloidal particles when they are dispersed in a liquid crystal medium.

The Liquid Crystalline Structure of Rod-Shaped Colloidal Particles

Recently, anisotropic particles including rod-shaped particles have been synthesized. It has been shown that these particles are aligned in a liquid crystalline structure in dispersions. Kuijk et al. [44] have successfully synthesized rod-like silica colloids with a diameter of over 200 nm and a length of 10 μm and an aspect ratio between 1 and 25. Isotropic, paranematic, and smectic phases were observed.

For a comprehensive review of colloids and polymer liquid crystals, see Ref. [45]. Also, a detailed description of structure formation by the depletion attraction of rod-shaped colloidal particles can be found in the textbook by Lekkerkerker and Tuinier [46].

Assembly of Colloidal Particles in Liquid Crystals

Interactions between particles dispersed in liquid crystals have been studied. The presence of particles in a liquid crystal alters the orientation of liquid crystal molecules, due to the affinity to the particles and the geometric factors. The long axis of the molecule is often oriented perpendicular to the spherical particle surface. Therefore, the liquid crystal molecules may be aligned radially from the particle or in the form of an electric dipole field. This orientation causes an apparent force between the particles. Poulin et al. [47] reported that micron-sized water droplets dispersed in a nematic liquid crystal associated to form chain-like structures. Since the medium is an oriented liquid crystal, the elastic energy of the medium is anisotropic. The observed association phenomena of the droplets were discussed in terms of the dynamics of the anisotropic medium. Musevic et al. [48] have reported that spherical colloidal particles confined in a few micrometer thick layer of nematic liquid crystal form a regular two-dimensional crystal structure. The authors assumed that the particles were constrained to regularly arrange "topological defects", where the liquid crystal molecules are disordered. For more details on this research area, see, for example, a review of Stark [49].

Assemblies with Emulsions

Zhang et al. [50] reported the self-assembly of particles in emulsions. When toluene in which polystyrene particles are dispersed is emulsified in water with a surfactant, the particles are trapped in toluene oil droplets and form clusters. Evaporation and removal of toluene results in the formation of stable clusters of colloidal particles in the aqueous phase. Figure 7.9 shows the synthesis schematically. The resulting clusters can be used as building blocks for diamond structures and other higher-order structures.

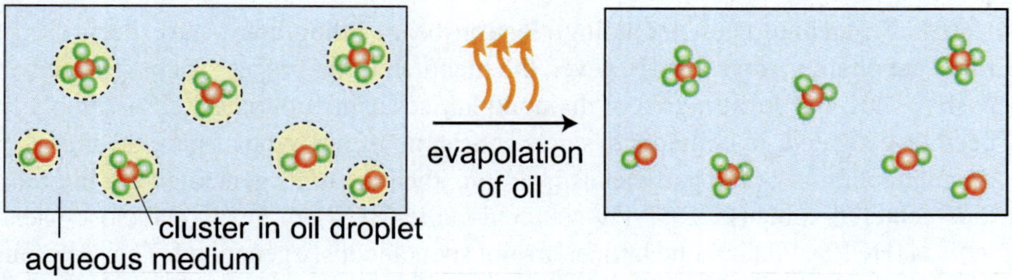

Fig. 7.9 Synthesis of clusters using colloidal particles constrained in an emulsion

7.2.3 Diamond Lattice

An important example of complex structures is the diamond lattice (Fig. 7.10) of colloidal particles. Diamond lattices with periodic structures on the optical wavelength scale are known to function as "perfect photonic crystals" [21, 51, 52], capable of confinement of light [53].

Diamond lattices are composed of regularly assembled tetrahedral structures as building blocks. Recently, it has been reported that "amorphous diamond" structures, in which tetrahedra are randomly assembled, also have band gaps [54]. Diamond lattice structures with a period of about the wavelength of visible light are found in nature, and it has been reported that the coloration of the wings of scarlet macaws, for example, is derived from amorphous diamond structures [55].

Photonic crystals are expected as novel material in the fields of sensing and optical communication and are currently the subject of active research around the world.

Fig. 7.10 Diamond lattice structure

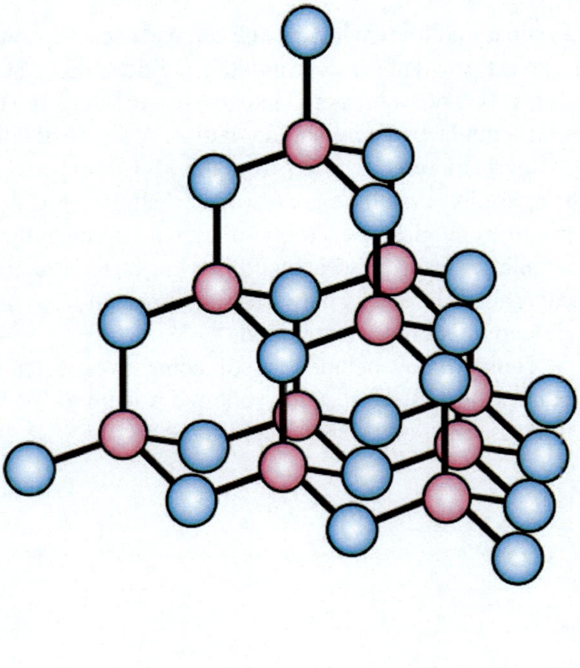

Until now, technologies, including electron beam lithography, have been used to fabricate photonic crystals. However, the manufacturing process is complicated and costly. Thus, the construction of diamond lattices using colloidal self-assembly has been investigated. In colloidal systems consisting of one type of particle and where the interaction between particles is isotropic, the structures generated are limited to face-centered-cubic (FCC), body-centered-cubic (BCC), or hexagonal close-packed lattices (HCP), and diamond lattices are not spontaneously generated. This is because the packing fraction of particles in diamond lattices is as low as 34% (74% for FCC and HCP and 68% for BCC) and structure formation is entropically unfavorable [56].

The following methods have been reported to fabricate the diamond lattices.

Electrostatic Interaction

Because the electrostatic interactions are long range, they may help obtain low-fill lattice structures, like diamond lattices. The formation of large micrometer-sized crystals with a diamond-like lattice structure (ZnS type) has been reported [57] in a two-component system of positively and negatively charged metal nanoparticles of about 5 nm in diameter. However, for particles larger than submicron size, the range of the interaction is shorter than the particle size, and it becomes increasingly difficult to construct a diamond lattice by electrostatic interaction.

Method Using Patchy Particles

Patchy particles with bonding points in the direction of tetrahedral vertices have been synthesized, and their self-assembled structures have been studied. However, the products are mixtures of various structures, and amorphous aggregates are thermodynamically more stable than regular aggregate structures.

Superlattice Crystal Structures

In binary colloids with appropriate particle size ratios, a superlattice of the $MgCu_2$ type is formed under appropriate conditions [56–62]. This superlattice is a combination of two types of crystal lattices, one of which is a diamond lattice, and the other is a structure called a pyrochlore lattice. A diamond lattice was produced by removing the pyrochlore lattice [59] from the $MgCu_2$-type superlattice. However, since there are various crystalline phases other than the $MgCu_2$ type in the binary colloids and the free energies of each phase are not significantly different, it is unclear whether simple self-assembly can yield only the target structure. Therefore, a method using a superlattice of patchy particles modified with DNA molecules [60] has been reported instead of isotropic interaction.

Thus, various methods are still being investigated for the construction of diamond lattices. Pine et al. recently reported a method for fabricating diamond lattices by self-assembly of tetrahedral clusters [42, 43].

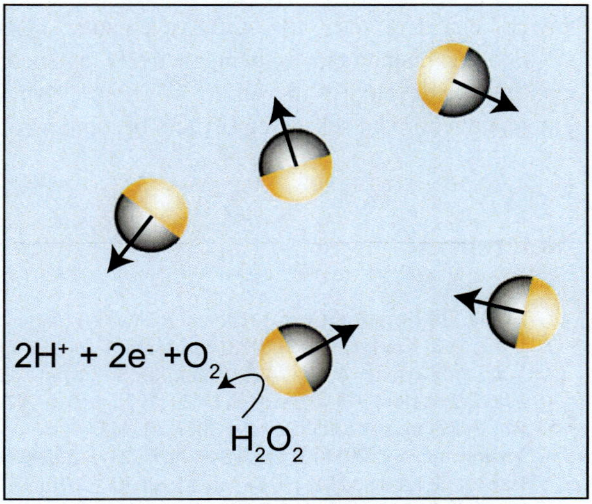

Fig. 7.11 Example of an active particle. Janus particles, whose hemispheres are catalysts, are self-propelled in solution by a chemical reaction

7.2.4 Active Matter

The colloidal particles discussed so far move by Brownian motion, the interaction between particles, or by forces from external fields (such as electric or gravitational fields). On the other hand, in recent years, various particles have been produced that swim autonomously in water, such as those found in microorganisms. These are collectively referred to as "active matter". For example, the self-swimming of particles can be achieved by using a chemical reaction that uses half of a Janus particle as a catalyst to generate an interfacial tension gradient [61, 62] (Fig. 7.11).

Various propulsion mechanisms have been proposed; most of them enable motion by converting locally available chemical energy into mechanical energy. The physical aspects of propulsion mechanisms and spontaneous collective behavior have been studied [61], with a detailed review by Gompper et al. [63]. The effects of osmotic pressure [64] and external fields (electric, magnetic, etc.) [65] have also been investigated. Understanding and controlling the behavior of active matter has attracted considerable attention as an interdisciplinary effort at the interface of biology, chemistry, ecology, engineering, mathematics, and physics. It may also have applications in the pharmaceutical field, such as drug delivery.

7.2.5 Concluding Remarks

As mentioned above, the self-assembly of colloids is of great interest not only for basic research but also for applications, as periodic structures with characteristic lengths in the submicron to micron order. Since this length scale includes the wavelength range of visible light, colloids were originally associated with optical phenomena. Currently, the study of creative anisotropic particles and new self-assembling phenomena is being actively pursued, and this field is expected to continue to develop in the future. We hope that readers of this book will take an interest in colloidal self-assembly and become involved in its study.

References

1. Zhang ZL, Keys AS, Chen T, Glotzer SC (2005) Langmuir 21:11547–11551
2. Pawar AB, Kretzschmar I (2010) Macromol Rapid Commun 31:150–168
3. Walther A, Müller AHE (2013) Chem Rev 113:5194–5261
4. Hueckel T, Hocky GM, Sacanna S (2021) Nature Rev Mater 6:1053–1069
5. van Blaaderen A (2003) Science 301:470–471
6. Manoharan VN (2015) Science 349:1253721-1–1253721-8
7. Isapour G, Lattuada M (2018) Adv Mater 30(1707069):1–36
8. Zerrouki D, Baudry J, Pine D, Chaikin P, Bibette J (2008) Nature 455:380–382
9. Erb RM, Son HS, Samanta B, Rotello VM, Yellen BB (2009) Nature 457:999–1002
10. Zhang G, Wang D, Möhwald H (2005) Angew Chem Int Ed 44:7767–7770
11. Chen Q, Whitmer JK, Jiang S, Luijten E, Granick S (2011) Science 331:199–202
12. Casagrande C, Fabre P, Raphaël E, Veyssié M (1989) Europhy Lett 9:251–255
13. de Gennes P-G (1992) AngewChem Int Ed 31:842–845 (English); Rev Mod Phys 64:645–648
14. Chen Q, Bae SC, Granick S (2011) Nature 469:381–384
15. Sacanna S, Irvine WT, Chaikin PM (2010) Nature 464:575–578
16. Wang Y, Wang Y, Zheng X, Yi G-R, Sacanna S, Pine DJ, Weck M (2014) J Am Chem Soc 136:6866–6869
17. Redl FX, Cho K-S, Murray CB, O'Brien S (2003) Nature 423:968–971
18. Hu Y, Niemeyer CM (2019) Adv Mater 31:1806294-1–1806294-18
19. Xia Y, Gates B, Yin Y, Lu Y (2000) Adv Mater 12:693–713
20. Nelson EC, Braun PV (2007) Science 318:924
21. van Blaaderen A, Ruel R, Wiltzius P (1997) Nature 385:321
22. Lin K-H, Crocker JC, Prasad V, Schofield A, Weitz DA, Lubensky TC, Yodh AG (2000) Phys Rev Lett 85:1770–1773
23. Ozin GA, Yang SM (2001) Adv Funct Mater 11:95–104
24. McGorty R, Fung J, Kaz D, Manoharan VN (2010) Mater Today 13:34–42
25. Li Y, Cai W, Duan G (2008) Chem Mater 20:615–624
26. Li F, Josephson DP, Stein A (2011) Angew Chem Int Ed 50:360–388
27. Li B, Zhou D, Han Y (2016) Nature Rev Mater 1:1–13
28. Bartlett P, Campbell AI (2005) Phys Rev Lett 95:128302-1–128302-4
29. Leunissen ME, Christova CG, Hynninen A-P, Royall CP, Campbell AI, Imhof A, Dijkstra M, van Roij R, van Blaaderen A (2005) Nature 437:235–240
30. Vermolen ECM, Kujik A, Filion LC, Hermes M, Thijssen JHJ, Dijkstra M, van Blaaderen A (2009) Proc Natl Acad Sci USA 106:16063–16067
31. Hückel T, Hocky GM, Palacci J, Sacanna S (2020) Nature 580:487–490
32. Min Y, Akbulut M, Kristiansen K, Golan Y, Israelachivili J (2008) Nat Mater 7:527–538
33. van Blaaderen A, Dijkstra M, van Roij R, Imhof A, Kamp M, Kwaadgras BW, Vissers T, Liu B (2013) Eur Phys J Special Topics 222:2895–2909
34. van Dommelen R, Fanzio P, Sasso L (2018) Adv Colloid Interf Sci 251:97–114
35. Zhang K-Q, Liu XY (2004) Nature 429:739–743
36. Liu B, Besseling TH, Hermes M, Demirörs AF, Imhof A, van Blaaderen A (2014) Nat Commun 5:3092
37. Nöjd S, Mohanty PS, Bagheri P, Yethiraj A, Schurtenberger P (2013) Soft Mater 9:9199–9207

38. Crassous JJ, Mihut AM, Wernersson E, Pfleiderer P, Vermant J, Linse P et al (2014) Nat Commun 5:5516
39. Ma F, Wu DT, Wu N (2013) J Am Chem Soc 135:6–9
40. Sarno B, Heineck D, Heller MJ, Ibsen SD (2021) Electrophoresis 42:539–564
41. Hermanson KD, Lumsdon SO, Williams JP, Kaler EW, Velev OD (2001) Science 294:1082–1086
42. Chandrasekhar FRSS (1992) Liquid crystals, 2nd edn. Cambridge University Press, Cambridge
43. de Gennes PG, Prost J (1993)The physics of liquid crystals, 2nd edn. Clarendon Press, Oxford
44. Kuijk A, van Blaaderen A, Imhof A (2011) J Am Chem Soc 133:2346–2349
45. Vroege GJ, Lekkerkerker HNW (1992) Rep Prog Phys 55:1241–1309
46. Lekkerkerker HNW, Tuinier R (2011) Colloids and the depletion interaction. Springer, Amsterdam
47. Poulin P, Stark H, Lubensky TC, Weitz DA (1997) Science 275:1770–1773
48. Musevic I, Škarabot M, Tkalec U, Ravnik M, Žumer S (2006) Science 313:954–958
49. Colvin VL (2001) MRS Bull 26:637–641
50. Zhang J, Li Y, Zhang X, Yang B (2010) Adv Mater 22:4249–4269
51. Ho KM, Chan CT, Soukoulis CM (1990) Phys Rev Lett 65:3152–3155
52. Edagawa K, Kanoko S, Notomi M (2008) Phys Rev Lett 100:13901-1–13901-4
53. Yin H, Dong B, Liu X, Zhan T, Shi L, Zi J, Yablonovitch E (2012) Proc Natl Acad Sci 109:10798–10801
54. Ducrot E, He M, Yi G, Pine DJ (2017) Nature Mater 16:652–657
55. Kalsin AM, Fialkowski M, Paszewski M, Smoukov SK, Bishop KJM, Grzybowski BA (2006) Science 312:420–424
56. Maldovan M, Thomas EL (2004) Nature Mater 3:593–600
57. Hynninen A-P, Thijssen JHJ, Vermolen ECM, Dijkstra M, van Blaaderen A (2007) Nature Mater 6:202–205
58. Wang Y, Jenkins IC, McGinley JT, Sinno T, Crocker JC (2017) Nature Commun 8:14173-1–14173-8
59. Liu W, Tagawa M, Xin HL, Wang T, Emamy H, Li H, Yager KG, Starr FW, Tkachenko AV, Gang O (2016) Science 351:582–586
60. He M, Gales JP, Ducrot É et al (2020) Nature 585:524–529
61. Paxton WF, Kistler KC, Olmeda CC, Sen A, Angelo SKS, Cao Y, Mallouk TE, Lammert PE, Crespi VH (2004) J Am Chem Soc 126:13424–13431
62. Paxton WF, Sundararajan S, Mallouk TE, Sen A (2006) Angew Chem Int Ed 45:5420–5429
63. Gompper G et al (2020) J Phys Condens Matter 32:193001-1–193001-67
64. Marbach S, Bocquet L (2019) Chem Soc Rev 48:3102–3144
65. Bechinger C, Leonardo RD, Löwen H, Reichhardt C, Volpe G, Volpe G (2016) Rev Mod Phys 88(045006):1–30

Chapter 8
Appendix

Appendix A: Fundamentals of Electrostatics

A.1 Coulomb's Law

Here, we consider that two electric charges, Q_1 and Q_2, are at rest in a vacuum and separated by a distance of r. The unit of electric charge is "Coulomb", denoted by the symbol C. The charge size is assumed to be very small and can be considered a point. They exert force F along the straight line joining them (Fig. 8.1), given by

$$F = Q_1Q_2 / (4\pi\varepsilon_0)r^2 \tag{8.1}$$

where ε_0 is a quantity called the dielectric constant of the vacuum (=8.35 × 10^{-12} F/m). That is, the force between charges is proportional to the product of the two charge quantities and inversely proportional to the square of the distance. This relation was found experimentally by French physicist Charles-Augustin de Coulomb. If the two charges have the same signs, F is positive, implying that the direction of the force is the same as that in which r increases. That is, the force is repulsion. If the charges have different signs, F is negative. Namely, the force is attraction. The relation given by Eq. (8.1) is called Coulomb's law, and the force F is called the Coulomb force. The Coulomb force in non-vacuum media is weaker than that in a vacuum, and F is expressed as

$$F = Q_1Q_2 / (4\pi\varepsilon)r^2 \tag{8.2}$$

Here, $\varepsilon = \varepsilon_0\varepsilon_r$ and $\varepsilon_r(\geq 1)$ is a quantity called the dielectric constant of the medium. The value of ε_r is larger for substances having larger polarity. For air, $\varepsilon_r \sim 1$; for hydrocarbons, $\varepsilon_r \sim 3$; and for water, $\varepsilon_r \sim 80$. Thus, the magnitude of the Coulomb force in water is approximately 1/80 of that in air. Note that (8.1) is an experimental result and is not derived from another more basic equation. Electrostatics is constructed using (8.1) as a starting point. For details, see textbooks on electromagnetism [1, 2].

© The Author(s), under exclusive license to Springer Nature Singapore Pte Ltd. 2023

J. Yamanaka et al., *Colloidal Self-Assembly*, Lecture Notes in Chemistry 108,
https://doi.org/10.1007/978-981-99-5052-2_8

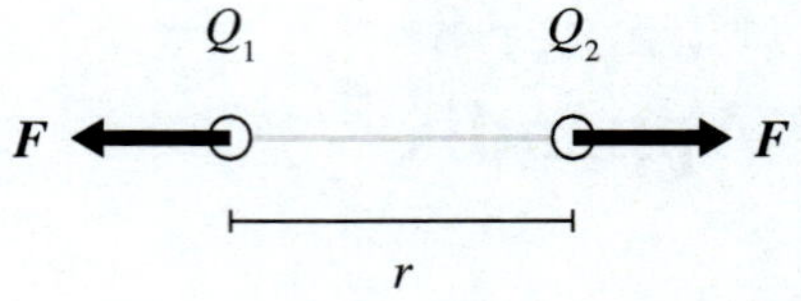

Fig. 8.1 Coulomb force acting between two charges

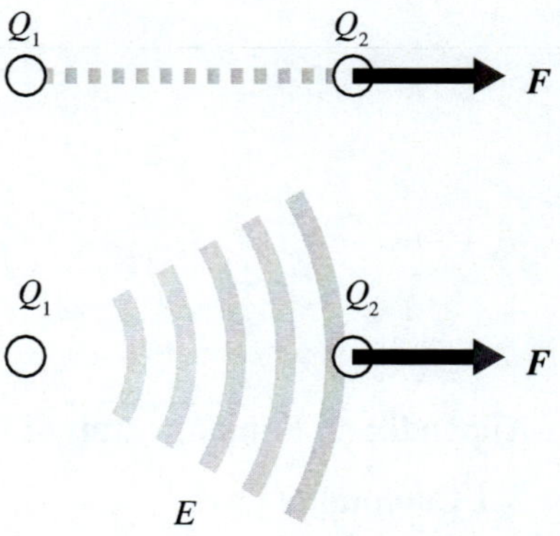

Fig. 8.2 Concept of Coulomb force **a** direct and **b** indirect actions. The latter considers that the presence of an electric charge Q_1 changes the electrical properties of space (electric field) and that space exerts a force on another charge Q_2 placed there

A.2 Electric Field *E*

There are two ways to look at the forces acting between charges described in Sect. 8.1. One is that (1) two charges, Q_1 and Q_2, exert a direct force on each other (direct action, Fig. 8.2, top). Another idea is that (2) the properties of the space are altered by the presence of the charge Q_1, and the charge Q_2 receives a force from that space (indirect action, Fig. 8.2, bottom).

Based on the idea of (2), the physical quantity representing the electrical properties of space is called the electric field $\boldsymbol{E}$. The term "field" comes from the fact that the vector representing the field looks like an ear of wheat in a field.

The electric field equals the force $\boldsymbol{F}$ received when a unit charge (+1 *C*) is placed in that space. That is, $\boldsymbol{F} = \boldsymbol{E}$ for the unit charge. The force acting on a charge Q is expressed as

$$\boldsymbol{F} = Q\boldsymbol{E} \tag{8.3}$$

Comparing (8.2) and (8.3) for $Q_1 = Q$ and $Q_2 = 1$, the magnitude E of the electric field around the charge Q is expressed as

$$E = Q/(4\pi\varepsilon_0\varepsilon_r)r^2 \tag{8.4}$$

Because E is a function of solely r, namely, $E = E(r)$, the direction of the electric field is radial around the point charge Q, outward if $Q > 0$ and inward if $Q < 0$.

A crucial general relationship exists between the charge amount and the electric field's magnitude. Let us consider a closed surface of arbitrary shape S in space and put the total charge in the volume surrounded by S, $Q_1 + Q_2 + \ldots + Q_N = Q$

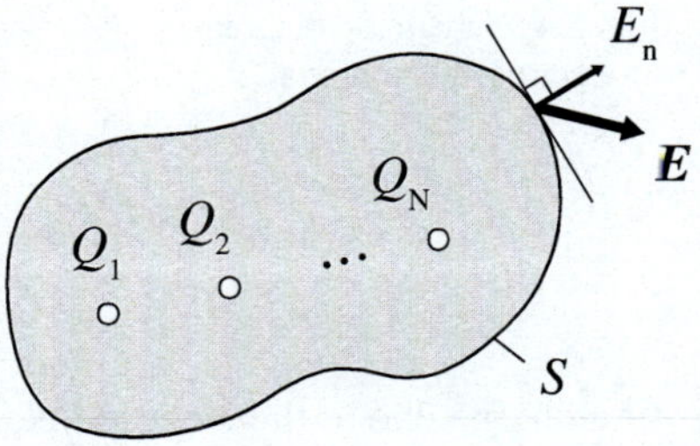

Fig. 8.3 Schematic diagram of Gauss's law

(Fig. 8.3). When the magnitude of the normal component of the electric field on S is given as E_n, then Eq. (8.5) holds.

$$\iint E_n dS = Q/\varepsilon \tag{8.5}$$

Here, dS is a surface element, and the double integral sign means integrating over the surface. The relation (8.5) is called Gauss's law. For a detailed and general derivation of Gauss's law, see textbooks on electromagnetism. Here, let us check this relationship for a point charge located at the center of spherical surface S of radius R. In this case, we have $E_n = E =$ constant and $\iint dS = 4\pi R^2$. Then, Eq. (8.5) gives Eq. (8.4). Gauss's law holds even when the charges are continuously distributed in space.

A.3 Electrostatic Potential φ

When an electric charge is present in an electric field, it has electrostatic potential energy. This potential energy for a unit charge ($q = +1$) is referred to as electrostatic potential φ. As described below, we can calculate $\boldsymbol{E}$ and $\boldsymbol{F}$ by knowing φ.

First, let us review the potential energy U of a mass m due to gravity on the ground. The gravitational force acting on the mass m equals mg, where g is the gravitational acceleration. Let us take the x-axis perpendicularly upward to the ground. U is equal to the work required to carry a mass from a reference position ($x = 0$, usually ground surface) to a certain height ($x = h$) with force $-F = -mg$, which is in the opposite direction of gravity. Because F can be assumed to be constant near the Earth's surface, U is given by

$$U = -\int F dx = -F\int dx = mgh. \tag{8.6}$$

Similarly, let us consider potential energy due to an electric field. Since the electric charge is subjected to a Coulomb force from an electric field, work W is required to move the charge against the force. Therefore, like potential energy in a gravitational field, a charge has electrostatic potential energy in an electric field generated by a charge Q (Fig. 8.4). The magnitude of this potential energy is proportional to the amount of charge. The electrostatic potential φ is defined as the potential energy of a

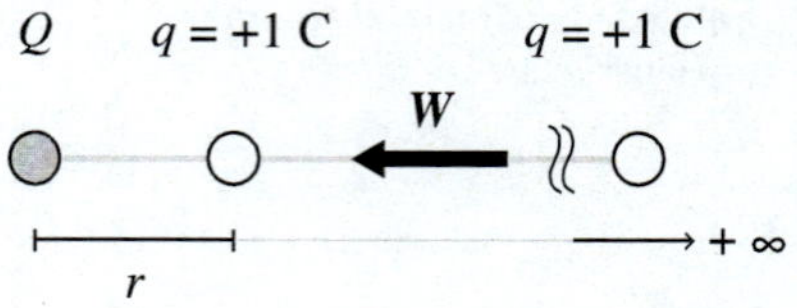

Fig. 8.4 Schematic diagram of electrostatic potential

unit charge due to the electric field. If the distance between two charges is infinitely large, they do not interact. Therefore, $r =$ infinity is usually used as the reference for $\varphi = 0$. Thus, the electrostatic potential is "the work W required to carry a unit charge from infinity to the position r we are considering".

In a medium with dielectric constant ε, we can write

$$\varphi(r) = -\int F dr = -\int E dr = Q/(4\pi \varepsilon r). \tag{8.7}$$

Note that from $\varphi(r) = -\int E(r)\,dr$, we have

$$E(r) = -d\varphi/dr. \tag{8.8}$$

Therefore, the electric field $E(r)$ can be derived from $\varphi(r)$ and $F(r)$ can be calculated from $E(r)$.

References

[1] Purcell EM, Morin DJ (2013) Electricity and magnetism, 3rd edn. Cambridge University Press, Cambridge
[2] Jackson JD (1998) Classical electrodynamics, 3rd edn. Wiley, NY

Appendix B: Debye–Hückel Theory for Strong Electrolyte Solutions

B.1 Overview

As mentioned in Appendix A, the electrical properties of a system are described by the electrostatic potential. Here, let us consider the electric field in an aqueous solution of an electrolyte, such as NaCl solution, consisting of positive and negative ions. Assuming that the ions are point charges, from Eq. (8.7), the electrostatic potential at a distance r from a single ion, $\varphi_0(r)$, is given by

$$\varphi_0(r) = \frac{z_i e_0}{4\pi \in} \frac{1}{r} \tag{8.9}$$

Here, z_i is the valency of the ion, and e_0 the elementary charge (=1.602×10^{-19} C). In the case of electrolyte solutions, because of the large number of positively and negatively charged ions present, the potential around individual ions is greatly affected by their electrostatic interactions with other ions. The Debye–Hückel (D-H) theory [1]

Fig. 8.5 Electrostatic potential around an ion ($z_i = +1$) according to Debye–Hückel theory (red). The Coulomb potential (blue) is also shown (calculated for aqueous solutions at 25°C)

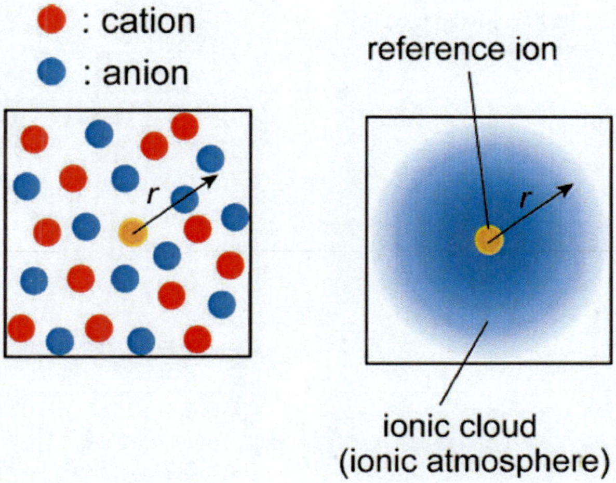

describes the electrostatic potential in an electrolyte solution. For more details on this theory, see, for example, the textbook by Bockris and Reddy [2]. The electrostatic potential by D-H theory, $\varphi_{\mathrm{DH}}(r)$, is given by

$$\varphi_{\mathrm{DH}}(r) = \frac{z_i e_0}{4\pi\varepsilon} \frac{e^{-\kappa r}}{r} \tag{8.10}$$

where κ is a positive constant called the Debye parameter, which is proportional to the square root of the ionic strength I. Comparing (8.9) and (8.10), we see that the contribution of all other ions in the electrolyte solution is represented by $\exp(\kappa r)$. Figure 8.5 illustrates a graph of $\varphi_{\mathrm{DH}}(r)$ *versus* r for $z_i = +1$ at three salt concentrations (Cs) values in water at 25°C A graph of $\varphi_0(r)$ *versus* r for $z_i = +1$ is also shown. Values of both potentials were reduced by thermal energy $k_{\mathrm{B}}T$.

Note that $\exp(-\kappa r)$ is a monotonically decreasing function, which takes 1 for $r = 0$ and 0 for $r \to +\infty$. Therefore, for $r > 0$, $\varphi_0(r) > \varphi_{\mathrm{DH}}(r)$ always holds. Namely, in an electrolyte solution, the electrical force due to ions decreases faster than in the absence of the surrounding ions. This phenomenon is called the electrostatic screening effect. The larger κ is, *i.e.*, the higher the concentration of ions, the faster $\varphi_{\mathrm{DH}}(r)$ decreases. The power of $\exp(-\kappa r)$ is dimensionless, so $1/\kappa$ has the dimension of length, called the Debye screening length. In the following, we will explain the derivation of Eq. (8.10).

B.2 Model of the Debye–Hückel Theory

Consider the electrostatic potential $\varphi(r)$ around one ion, called the reference ion, in an electrolyte solution. $\varphi(r)$ can be calculated by adding up the electrostatic potentials at that position produced by the reference ion plus all the ions in the solution. However, it is not practical to calculate $\varphi(r)$ for individual ions because the number of ions is vast (about the Avogadro constant), and their positions change with time due to thermal motion. Therefore, in the D-H theory, the ions other than the reference ion are considered to have a continuous distribution, called an ionic cloud or ionic

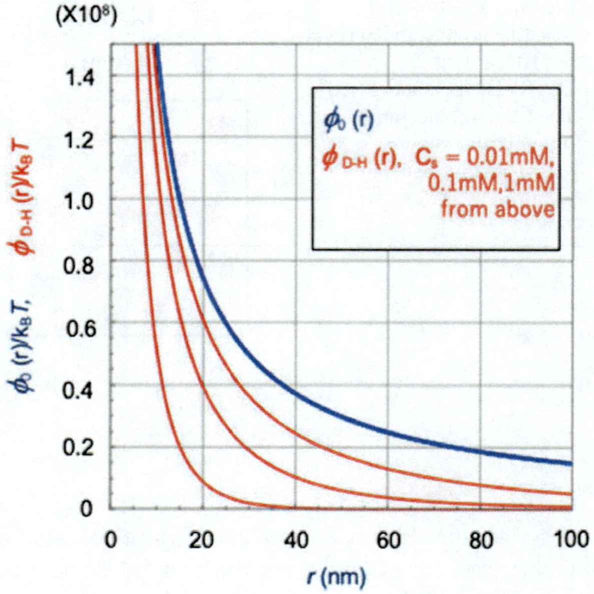

Fig. 8.6 Model used in Debye–Hückel theory

atmosphere (Fig. 8.6). Because the distribution of ions is isotropic, the ionic cloud should be spherically symmetric. In general, the charge density ρ of an electrolyte solution is expressed as $\rho = \Sigma z_i e_o n_i$, where i is the number assigned to each type of ion, z_i is the valence of the i-th ion, which is positive for cations and negative for anions. n_i is the number density of the i-th ion. Because the ionic cloud is spherically symmetric, n_i at a given position r in the cloud is a function of r alone, that is, $n_i = n_i(r)$. Therefore, ρ is a function of solely r, i.e., $\rho = \rho\ (r)$. The ionic cloud contains both positive and negative ions, and when $r \to \infty$, the concentrations of positive and negative ions are equal, namely, $\rho(r) \to 0$. By the law of electroneutrality, the total charge of the electrolyte solution is zero. Therefore, if the charge of the reference ion is $+\ z_i e_0$, the sum of the charges in the ionic cloud is $-\ z_i e_0$.

B.3 Derivation of $\varphi_{DH}(r)$

In the following, we will explain the derivation of the electrostatic potential $\varphi_{DH}(r)$ of the ionic cloud model by Debye and Hückel. The explanation is based on that described in the textbook by Bockris and Reddy [2]. The outline of the derivation is as follows.

(i) First, the relation between $\varphi(r)$ and $\rho(r)$ (Poisson equation) is derived based on the concept of electrostatics.
(ii) Next, we apply the concept of Boltzmann distribution in statistical mechanics to express $\rho(r)$ in terms of $\varphi(r)$.
(iii) From (i) and (ii), we derive the Poisson–Boltzmann (P-B) equation, which is the equation that the ionic cloud $\varphi(r)$ obeys. Furthermore, the Debye–Hückel (D-H) equation is derived by approximating the P-B equation.

(iv) Solve the D-H equation to obtain $\varphi_{\mathrm{DH}}(r)$.

(i) Poisson equation for spherically symmetric systems

First, we derive the relation that $\varphi(r)$ follows in a spherically symmetric system from Gauss's law (8.5) described. As a closed surface *S*, we consider a sphere of radius *r*. Since the electric field is spherically symmetric, $\boldsymbol{E}_\mathrm{n} = \boldsymbol{E}(r)$ and Eq. (8.5) can be written as

$$4\pi r^2 E(r) = Q/\varepsilon. \tag{8.11}$$

Here, $E(r)$ is the magnitude of $\boldsymbol{E}(r)$. By dividing the sphere into spherical shells of radius dr, Q is represented as

$$Q = \frac{1}{\varepsilon}\int_0^r \rho(r) 4\pi r^2 dr. \tag{8.12}$$

Then,

$$E(r) \cdot 4\pi r^2 = \frac{1}{\varepsilon}\int_0^r \rho(r) 4\pi r^2 dr. \tag{8.13}$$

By using (8.8) to express $E(r)$ in terms of $\varphi(r)$, and differentiating both sides by r, we obtain (8.14) that does not include integration.

$$\frac{1}{r^2}\frac{d}{dr}\left(r^2\frac{d\varphi(r)}{dr}\right) = -\frac{1}{\varepsilon}\rho(r). \tag{8.14}$$

Equation (8.14) is referred to as the Poisson equation for a system with spherical symmetry.

We note that in the case of a three-dimensional Cartesian coordinate system, the Poisson equation can be written as

$$\nabla^2\varphi(x, y, z) = -\rho(x, y, z)/\varepsilon. \tag{8.15}$$

Equation (8.14) is also obtainable from Eq. (8.15) by transforming the variables, though it requires a somewhat lengthy calculation.

(ii) Equations for $\rho(r)$ and $\varphi(r)$ based on Boltzmann distribution

Since both $\rho(r)$ and $\varphi(r)$ contained in Eq. (8.14) are unknown quantities, we cannot go any further from electromagnetism alone. Next, using the ideas of statistical mechanics, we express $\rho(r)$ in terms of $\varphi(r)$. $\rho(r)$ is determined by the balance of the Coulomb attraction between the reference ion and the ionic cloud and the thermal motion of ions, which works to make the ionic cloud distribution uniform. In other words, the ratio of electrical energy $U(r)$ to the thermal energy k_BT, $U(r)/k_BT$, determines the ion concentration distribution; here, k_B is the Boltzmann constant, and T is the absolute temperature.

Let us calculate $\rho(r)$. In the following, the type of ions in the solution is denoted by $i = 1, 2, \ldots N$. The charge of an ion is z_ie_0, where z_i is positive for cations and negative for anions. Since $\varphi(r)$ is the potential energy of a unit charge at a location r, we have $U(r) = z_ie_0\varphi(r)$. According to statistical mechanics, $n_i(r)$ follows the Boltzmann distribution and is written as

$$\begin{aligned} n_i(r) &= n_i^0 \exp\left(-\frac{U(r)}{k_BT}\right) \\ &= n_i^0 \exp\left(-\frac{z_ie_0\varphi(r)}{k_BT}\right). \end{aligned} \tag{8.16}$$

Here, n_i^0 is $n_i(r)$ when $\varphi(r) = 0$, that is, the value at $r \to \infty$. Therefore, Eq. (8.16) implies that $n_i(r)$ is larger at lower $\varphi(r)$.

$\rho(r)$ is the sum of the electrical charges of the i-th ion ($=z_i\ e_0\ n_i(r)$) and is expressed as follows:

$$\begin{aligned} \rho(r) &= \sum n_iz_ie_0 \\ &= \sum n_i^0 z_ie_0 \exp\left(-\frac{z_ie_0\varphi(r)}{k_BT}\right). \end{aligned} \tag{8.17}$$

The sum is taken from $i = 1$ to $i = N$. Thus, based on the Boltzmann distribution, $\rho(r)$ can be expressed as $\varphi(r)$.

(iii) Derivation of the Poisson–Boltzmann (P-B) and Debye–Hückel (D-H) equations

Both Eqs. (8.14) and (8.17) give the relationship between $\rho(r)$ and $\varphi(r)$. Substituting Eq. (8.17) into Eq. (8.14) to delete $\rho(r)$, we obtain

$$\frac{1}{r^2}\frac{d}{dr}\left(r^2\frac{d\varphi(r)}{dr}\right) = -\frac{1}{\varepsilon}\sum n_i^0 z_ie_0 \exp\left(-\frac{z_ie_0\varphi(r)}{k_BT}\right). \tag{8.18}$$

Equation (8.18) is called the Poisson–Boltzmann (P-B) equation. Now, we can obtain $\varphi(r)$ by solving the P-B equation. However, Eq. (8.18) is *nonlinear* because of the exponential function on the right side. Nonlinear equations have no general analytical solution. Thus, numerical calculations are used to solve the P-B equation. An analytical solution can be obtained by approximating Eq. (8.18) with a linear expression (under conditions where the approximation is valid).

By using series expansion $\exp(-x) = 1 - x + x^2/2! + \ldots$, we can use an approximation, $\exp(-x) \sim 1 - x$, if x is small enough. Therefore, when $z_i\ e_0\ \varphi(r)/k_BT \ll 1$, (namely, $\varphi(r) \ll k_BT/z_i\ e_0$), we have

$$\exp\left(-\frac{z_i e_0 \varphi(r)}{k_B T}\right) = 1 - \frac{z_i e_0 \varphi(r)}{k_B T}. \tag{8.19}$$

Then, Eq. (8.18) can be written as

$$\begin{aligned}\rho(r) &= \sum n_i^0 z_i e_0 \left(1 - \frac{z_i e_0 \varphi(r)}{k_B T}\right)\\ &= \sum n_i^0 z_i e_0 - \sum \frac{n_i^0 z_i^2 e_0^2 \varphi(r)}{k_B T}.\end{aligned} \tag{8.20}$$

Since the solution is electrically neutral, the first term on the right-hand side is zero. Therefore,

$$\rho(r) = -\sum \frac{n_i^0 z_i^2 e_0^2 \varphi(r)}{k_B T}, \tag{8.21}$$

and by substituting it into Eq. (8.21), we obtain

$$\frac{1}{r^2}\frac{d}{dr}\left(r^2 \frac{d\varphi(r)}{dr}\right) = \frac{1}{\varepsilon k_B T}\sum n_i^0 z_i^2 e_0^2 \varphi(r). \tag{8.22}$$

For simplicity, let us set the coefficient of $\varphi(r)$ on the right-hand side to be κ^2 (the reason for putting it to the second power will be shown later). Thus, we obtain the following relations on $\varphi(r)$ for the electrolyte solution.

$$\frac{1}{\varepsilon k_B T}\sum n_i^0 z_i^2 e_0^2 = \kappa^2, \tag{8.23}$$

$$\frac{1}{r^2}\frac{d}{dr}\left(r^2 \frac{d\varphi(r)}{dr}\right) = \kappa^2 \varphi(r). \tag{8.24}$$

Equations (8.23) and (8.24) are called the linearized Poisson–Boltzmann equation or the Debye–Hückel (D-H) equation.

(iv) Electrostatic potential by the D-H theory

Equation (8.24) can be solved in the following manner. First, we assume

$$\varphi(r) = \frac{\mu(r)}{r}. \tag{8.25}$$

Then, the left-hand side of Eq. (8.25) is

$$\frac{1}{r^2}\frac{d}{dr}\left(r^2 \frac{d\varphi(r)}{dr}\right) = \frac{1}{r^2}\frac{d}{dr}\left(r^2\left(-\frac{\mu}{r^2} + \frac{1}{r}\frac{d\mu}{dr}\right)\right)$$

$$= \frac{1}{r^2}\left(-\frac{d\mu}{dr} + r\frac{d^2\mu}{dr^2} + \frac{d\mu}{dr}\right)$$
$$= \frac{1}{r}\left(\frac{d^2\mu}{dr^2}\right). \tag{8.26}$$

Therefore, we have

$$\frac{d^2\mu}{dr^2} = \kappa^2\mu. \tag{8.27}$$

We can confirm that

$$\mu = e^{\pm\kappa r} \tag{8.28}$$

satisfies Eq. (8.27) by substituting Eq. (8.28) to (8.27). $\mu = e^{+\kappa r}$ and $\mu = e^{-\kappa r}$ are the two essential solutions of Eq. (8.26). The general solution is a linear combination of these basic solutions,

$$\mu = Ae^{-\kappa r} + Be^{+\kappa r} \tag{8.29}$$

That is, $\varphi(r)$ is expressed as follows:

$$\varphi(r) = A\frac{e^{-\kappa r}}{r} + B\frac{e^{+\kappa r}}{r}. \tag{8.30}$$

Here, A and B are constants. The values of A and B can be determined from the boundary conditions. First, by definition, $\varphi(r)$ approaches 0 at $r \to \infty$. For this condition to be satisfied, $B = 0$ must hold. Also, when $\kappa \to 0$, $\varphi(r)$ must match the electrostatic potential in the absence of the ionic cloud.

$$\varphi(r) = \varphi_0(r) = \frac{z_i e_0}{4\pi\varepsilon}\frac{1}{r}.$$

Therefore, we have

$$A = \frac{z_i e_0}{4\pi\varepsilon}. \tag{8.31}$$

That is,

$$\varphi(r) = \varphi_{\mathrm{DH}}(r) == \frac{z_i e_0}{4\pi\varepsilon}\frac{e^{-\kappa r}}{r}. \tag{8.32}$$

References

[1] Debye PW, Hückel E (1923) Phys Z 24:185–206. Translated and typeset by Braus MJ (2019)
[2] Bockris JO'M, Reddy AKN (1973) Modern electrochemistry, vol 1: ionics. Plenum, New York

Appendix C: Electrical Conductivity Titration of Acid Solutions

In electrical conductivity titration, an acid or base of known concentration is added dropwise to a solution sample to be measured, and the change in electrical conductivity of the solution sample is measured to the acid or base concentrations. For example, 3.2.3 of the text described the electrical conductivity measurement when a base solution was added to an acidic polymer solution, in which the dissociation of the polymeric acid and the effect of ion condensation appeared in the titration curve. This section describes titration curves of low-molecular-weight acids, which are the basis for the titration of polymers.

C.1 Strong Acid–Strong Base

When NaOH is added to HCl, the following neutralization reaction occurs.

$$HCl + NaOH \rightarrow NaCl + H_2O$$

HCl and NaOH, which are strong acid and base, dissociate practically 100% in aqueous solutions to give H^+ and OH^- ions and Cl^- and Na^+ ions, respectively. In Fig. 8.7, the conductivity titration curve of 10 mL of HCl solution (concentration $C_{HCl} = 1$ mM) with 0.1 M NaOH (concentration is presented as C_{NaOH}) at 25°C is shown by red symbols. The vertical axis is the sample's electrical conductivity, which is the value obtained by subtracting the conductivity of pure water (approximately 0.5 μS/cm) from the measured conductivity.

In the followings, we present the molar conductivity of ion X at 25°C in the limit of zero ion concentration as $\Lambda_{X.}$ Literature values of $\Lambda_{H}{}^{+}$ and $\Lambda_{Cl}{}^{-}$ are 349.8 and 75.23 Scm²/mol, respectively. ΔK value calculated for $C_{HCl} = 0.001$ M is $\Delta K = (349.82 + 75.23)C_{HCl}/1000 \sim 425$ μS/cm, which is consistent with the experiment. The molar conductivity of the Na^+ ion is $\Lambda_{Na}{}^{+} = 50.10$ Scm²/mol, which is smaller than that of the H^+ ion. The linear decrease of ΔK on additions of NaOH is due to an exchange between H^+ and Na^+ ions. In this region, ΔK, is calculated as $\Delta K = (349.8 - 50.10)C_{NaOH}/1000 \sim 0.300\, C_{NaOH}$ which is in close agreement with the results of Fig. 8.1. After passing the neutralization point, ΔK increases because the added Na^+ and OH^- ions are in excess in the solution. Since $\Lambda_{OH}{}^{-} = 198.3$ Scm²/mol, the slope of the titration curve is $\Delta K = (50.10 + 198.3)C_{NaOH}/1000 \sim 0.25C_{NaOH}$, which is consistent with experimental results.

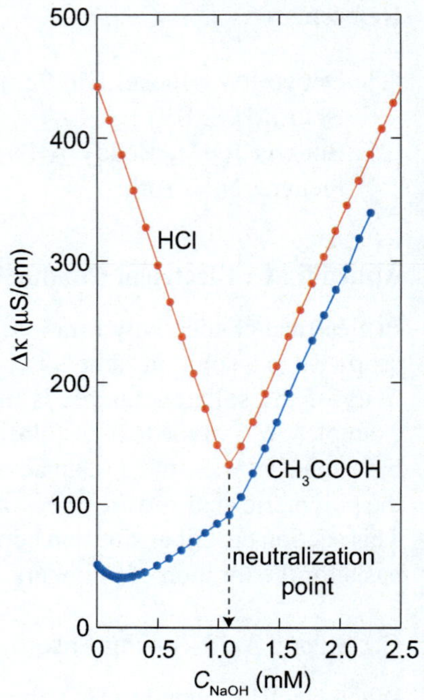

Fig. 8.7 Conductometric titration curves of HCl and CH_3COOH solutions with NaOH

C.2 Weak Acid–Strong Base

The ionization equilibrium of weak acid HA is as follows:

$$HA \rightleftarrows H^+ + A^-.$$

The dissociation is constant of the acid K_a which is expressed as

$$K_a = \frac{[A^-][H^+]}{[HA]} \tag{8.33}$$

Strong acids have larger $[H^+]$ and thus larger values of K_a, while weak acids have smaller values of K_a. The K_a depends on the solvent and varies with temperature. pK_a defined by (8.34) is often used as a measure of acid strength,

$$pK_a = -\log_{10} K_a. \tag{8.34}$$

The molar ratio of dissociated species and total electrolyte, $[A^-]/([A^-] + [HA])$, is called dissociation degree, α. The value α is almost constant at 1 for strong acids, while for weak acids, it varies with electrolyte concentration in the region $\ll 1$. The following neutralization reaction occurs when NaOH is added to the weak acid, for example, acetic acid CH_3COOH.

$$CH_3COOH + NaOH \rightarrow CH_3COONa + H_2O.$$

The α value depends on concentration and is about a ~ 0.15 when [CH_3COOH] = 1 mM. Dissociated acetate ions CH_3COO^- contribute to electrical conductivity, while non-dissociated acetate molecules CH_3COOH do not. In Fig. 8.7, the titration curve of 10 mL CH_3COOH (concentration $C_{CH3COOH} = 1$ mM) with NaOH is shown by blue symbols. Using $\Lambda_{CH3COO^-} = 40.9$ Scm^2/mol, we have $\Delta K = (349.82 + 40.9)C_{CH3COOH}/1000 \sim 390$ μS/cm. At α = 0.15, ΔK value of the acetic acid solution is approximately 58 μS/cm, which roughly agrees with experimental results. On addition of NaOH, the dissociated CH_3COO^- undergoes the counterion exchange (H^+ and Na^+), as in the case of strong acids. Therefore, Δκ decreases as well, but the α value increases on additions of NaOH, resulted in an increase in electrical conductivity. Thus, the titration curve shows a duller bend compared to the case of a strong acid. As more NaOH is added, Δκ increases at a constant rate due to the excess of Na^+ and OH^-, as in strong acids.

Appendix D: Charged Colloids

Charged colloids are systems in which colloidal particles are charged on their surfaces and usually dispersed in a polar solvent such as water. Ions are dissolved in the water, and the system as a whole remains electrically neutral. The ions in water include counter ions resulting from dissociation at the particle surface and ions derived from added salts (strong electrolytes). In the following, for simplicity, we do not distinguish between counter ions and ions due to salts but refer to them as univalent ions. Assume that the particles are negatively charged spheres of radius a, whose (signed) charge number is $Z = 4\pi a^2\sigma / e_0 (< 0)$, where σ is the surface charge density and e_0 is the elementary charge. Consider the case of a charged colloidal system containing N such particles in contact with a reservoir of a strong electrolyte solution of concentration n_r (Fig. 8.8). In the reservoir, the positive and negative ion concentrations are n_r, which in this case is equal to the salt concentration. However, because the charge number Z of a particle is large, counterions are attracted around the particles, resulting in a non-uniform spatial distribution of ions. The so-called electric double layer is formed. In the following, we will look at this situation based on a coarse-grained picture.

Because the solvent molecules and ions are much smaller in size than the colloidal particles and because they are in intense thermal motion, if we look at the spatial scale of colloidal particles (e.g., radius a) and the time scale of their motion (e.g., diffusion time $\tau = a^2 / D_p$; D_p is the diffusion coefficient of the particle), they can be viewed as a continuum that spreads continuously around the particle. In other words, the concentration of ions can be regarded as a continuous function of position $\boldsymbol{r} = (x, y, z)$ and time t. In particular, at equilibrium, there is no dependence on time but only on position. Thus, introducing the local positive and negative ion concentrations at position $\boldsymbol{r}$ as $n_+(\boldsymbol{r})$ and $n_-(\boldsymbol{r})$, respectively, the charge density $\rho(\boldsymbol{r})$ is given as

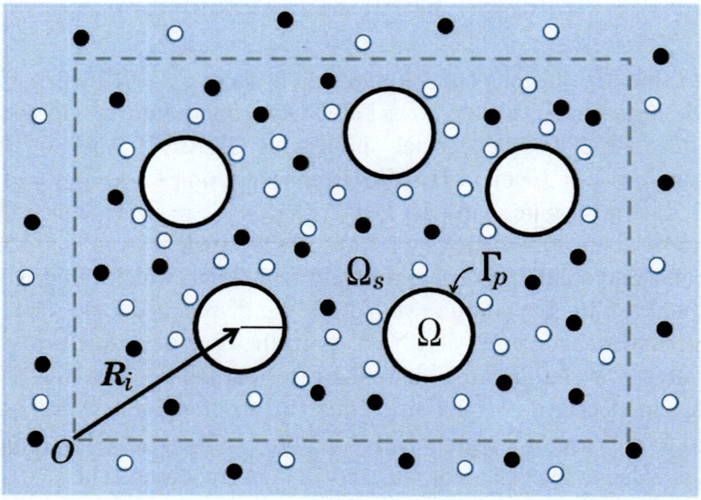

Fig. 8.8 Schematic picture of charged colloidal system. The system which is in contact with a reservoir of an electrolyte solution is the rectangular region bounded by the dashed line. Colloid particles are represented by large white circular regions. Black and white dots represent anions and cations, respectively

$$\rho(\boldsymbol{r}) = e_0 n_+(\boldsymbol{r}) - e_0 n_-(\boldsymbol{r}), \tag{8.35}$$

and the electroneutrality condition of the system is expressed as

$$NZe_0 + \int \rho(r)dr = 0, \tag{8.36}$$

where the integral in the second term is the volume integral over the entire region, excluding the region occupied by particles.

The above system is a strong electrolyte solution in the absence of colloidal particles ($N = 0$), the theory of which was established by Debye and Hückel [1]. From (8.36), the average charge density is zero. However, from a microscopic viewpoint, Coulomb interactions are exerted on each charge, and the electrostatic potential $\psi(\boldsymbol{r})$ fluctuates around 0. The electrostatic potential $\psi(\boldsymbol{r})$ satisfies the equation

$$\nabla^2 \psi = -\frac{\rho(\boldsymbol{r})}{\varepsilon}. \tag{8.37}$$

This is the well-known Poisson equation in electromagnetism. Here, the solvent is considered to be a continuum with dielectric constant ε (using the relative permittivity ε_r and the permittivity of vacuum ε_0, $\varepsilon = \varepsilon_r \varepsilon_0$). Let us introduce the chemical potentials of positive and negative ions, μ_+ and μ_-, respectively, these are given as

$$\begin{aligned} \mu_+ &= \mu_+^\circ + k_B T \ln x_+ + e_0 \psi, \\ \mu_- &= \mu_-^\circ + k_B T \ln x_- - e_0 \psi, \end{aligned} \tag{8.38}$$

where μ_+°, μ_-° are functions of temperature T that determine the origin of the chemical potential, k_B is Boltzmann constant, and x_+, x_- are the mole fractions (or activities) of the positive and negative ions. The last terms $\pm e_0\psi$ represent the electrostatic energy of the ions. From (8.38), $x_+ = \exp\left[\left(\mu_+ - \mu_+^\circ - e_0\psi\right)/k_\mathrm{B}T\right]$, which is equal to the mole fraction x_r of positive ions in the reservoir when $\psi = 0$. Therefore, we can write $x_+ = x_r \exp\left(-e_0\psi/k_\mathrm{B}T\right)$, and ion concentrations are

$$\begin{aligned} n_+ &= n_r \exp\left(-\frac{e_0\psi}{k_\mathrm{B}T}\right), \\ n_- &= n_r \exp\left(\frac{e_0\psi}{k_\mathrm{B}T}\right). \end{aligned} \tag{8.39}$$

Since ψ is, in general, a function of position, n_+ and n_- are also functions of the position, assuming that local equilibrium holds. From (8.35), (8.37), and (8.39), we obtain

$$\nabla^2\psi = \frac{e_0 n_r}{\varepsilon}\left[\exp\left(\frac{e_0\psi}{k_BT}\right) - \exp\left(-\frac{e_0\psi}{k_BT}\right)\right]. \tag{8.40}$$

This is called the Poisson–Boltzmann equation. Note that the Poisson Eq. (8.37) is an expression for finding the electrostatic potential $\psi(\boldsymbol{r})$ for a given charge density distribution $\rho(\boldsymbol{r})$, while in the Poisson–Boltzmann Eq. (8.40), $\rho(\boldsymbol{r})$ varies depending on $\psi(\boldsymbol{r})$ to be found. The electrostatic potential $\psi(\boldsymbol{r})$ is determined self-consistently by (8.40).

Equation (8.40) is a nonlinear partial differential equation and is difficult to deal with. However, if the electrostatic energy is sufficiently smaller than the thermal energy, $\left|e_0\psi/k_BT\right| \ll 1$, then we can expand the exponential function by $e_0\psi/k_BT$ up to the first-order term and obtain the expression of the charge density as

$$\rho = -\frac{2n_r e_0^2\psi}{k_\mathrm{B}T} = -\varepsilon\kappa^2\psi. \tag{8.41}$$

That is, ρ is proportional to ψ. Therefore, (8.40) can be written as follows (Debye–Hückel approximation):

$$\nabla^2\psi - \kappa^2\psi = 0, \tag{8.42}$$

where

$$\kappa^2 \equiv 8\pi l_B n_r, \tag{8.43}$$

and $l_B \equiv e_0^2/4\pi\varepsilon k_BT$ is called the Bjerrum length and is defined as the distance where the Coulomb interaction and thermal energy are balanced ($l_B \simeq 0.7$ nm in water at room temperature). κ is often referred to as the Debye parameter, and its reciprocal is the screening length of the Coulomb interaction (Debye length). Equation (8.42) is a linear equation and can be solved analytically. Using the spherical coordinate system, the solution that is isotropic depends only on the radial coordinate r and has a property $\psi \to 0$ as $r \to \infty$ is

$$\psi = \frac{Ae^{-\kappa r}}{r}, \tag{8.44}$$

where A is a constant. The above solution can be obtained in an elementary manner. Equation (8.42) can be expressed as $\psi'' + 2\psi'/r - \kappa^2\psi = 0$, where $\psi' \equiv d\psi/dr$, $\psi'' \equiv d^2\psi/dr^2$. Letting $u = r\psi$, the above ordinary differential equation can be written as $u'' - \kappa^2 u = 0$. We can easily find the general solution of this equation as $u = Ae^{-\kappa r} + Be^{\kappa r}$ with constants A and B. Since we physically expect that ψ is finite as $r \to \infty$, the constant B must be 0 ($\kappa > 0$), , and we have (8.44). This solution is often called the screened Coulomb potential or Yukawa potential because it shows that the Coulomb potential $(\propto 1/r)$ decays rapidly on the length scale κ^{-1}.

Next, consider the situation where charged colloidal particles are dispersed in a strong electrolyte solution as described above. Colloidal particles are generally huge particles compared to ions, and the charges they carry are also large, making them difficult to treat, as in the Debye–Hückel theory for strong electrolyte solutions. Various attempts [2, 3] have been made regarding the interaction between charged colloidal particles, including density functional theory [4, 5, 6], but we will not go into the details in this book. In this section, we describe how to obtain a practical expression for the particle interaction potential by means of a classical approximation.

Assume that N particles with uniform surface charge density σ are placed at fixed positions $\{\boldsymbol{R}_1, \boldsymbol{R}_2, \ldots, \boldsymbol{R}_N\}$ in a strong electrolyte solution. Let Ω denotes the entire system, the solution part of which is Ω_s and the particle part of which is Ω_p. And let Γ_p denotes the surface of the particle (Fig. 8.8). The boundary condition satisfied by the electrostatic potential ψ on Γ_p with a uniform surface charge density σ is given by

$$-\boldsymbol{n} \cdot \nabla\psi = \sigma/\varepsilon \text{ on } \Gamma_p \tag{8.45}$$

From Gauss's law, if there is no charge inside the particle. In (8.45), $\boldsymbol{n}$ is the outward normal vector on the particle sphere, and ψ is continuous on the sphere and takes a constant value inside the sphere.

In order to incorporate the effect of the particle surface charge into the Debye–Hückel theory, we introduce an "external charge density" $\rho^{\text{ex}}(\boldsymbol{r})$ on the right-hand side of the Poisson Eq. (8.37) as,

$$\nabla^2\psi = -\frac{\rho(\boldsymbol{r}) + \rho^{\text{ex}}(\boldsymbol{r})}{\varepsilon}, \tag{8.46}$$

where $\rho^{\text{ex}}(\boldsymbol{r})$ is the surface charge expressed as a fixed charge density distribution per unit volume. Using the Debye–Hückel approximation as in (8.42), we obtain

$$\nabla^2\psi - \kappa^2\psi = -\frac{\rho^{\text{ex}}(\boldsymbol{r})}{\varepsilon}. \tag{8.47}$$

Fourier transforming (8.47) yields (for Fourier transform, see, e.g., [7])

$$\psi_{\boldsymbol{q}} = \frac{\rho_q^{\mathrm{ex}}}{\varepsilon(q^2+\kappa^2)}, \tag{8.48}$$

where $\psi_{\boldsymbol{q}} = \int \psi(\boldsymbol{r})\mathrm{e}^{i\boldsymbol{q}\cdot\boldsymbol{r}}\mathrm{d}\boldsymbol{r}$ and $\rho_{\boldsymbol{q}}^{\mathrm{ex}} = \int \rho^{\mathrm{ex}}(\boldsymbol{r})\mathrm{e}^{i\boldsymbol{q}\cdot\boldsymbol{r}}\mathrm{d}\boldsymbol{r}$ are the Fourier components of the wavenumber $\boldsymbol{q}$ for $\psi(\boldsymbol{r})$ and $\rho^{\mathrm{ex}}(\boldsymbol{r})$, respectively. Using the inverse Fourier transform of (8.48) and the convolution theorem, we obtain

$$\psi(\boldsymbol{r}) = \frac{1}{4\pi\varepsilon}\int \frac{\mathrm{e}^{-\kappa|\boldsymbol{r}-\boldsymbol{r}'|}}{|\boldsymbol{r}-\boldsymbol{r}'|}\rho^{\mathrm{ex}}(\boldsymbol{r}')\mathrm{d}\boldsymbol{r}'. \tag{8.49}$$

However, it is difficult to obtain specific expressions from this formula when the particles have finite sizes. Hence, we assume that the particle concentration is sufficiently dilute and the particles can be regarded as point particles ($a \to 0$) and express the external charge density, using Dirac's δ-function $\delta(\boldsymbol{r})$ as

$$\rho^{\mathrm{ex}}(\boldsymbol{r}) = Ze_0 \sum_{i=1}^{N} \delta(\boldsymbol{r} - \boldsymbol{R}_i). \tag{8.50}$$

Then, (8.49) becomes

$$\psi(\boldsymbol{r}) = \frac{Ze_0}{4\pi\varepsilon}\sum_{i=1}^{N}\frac{\mathrm{e}^{-\kappa|\boldsymbol{r}-\boldsymbol{R}_i|}}{|\boldsymbol{r}-\boldsymbol{R}_i|}. \tag{8.51}$$

This is a superposition of solutions to the linearized Poisson–Boltzmann Eq. (8.42). That is, the overall electrostatic potential $\psi(\boldsymbol{r})$ can be written as the sum of the potentials $\psi_i(\boldsymbol{r})$ generated by the individual particles,

$$\begin{aligned}\psi(\boldsymbol{r}) &= \sum_{i=1}^{N}\psi_i(\boldsymbol{r}),\\ \psi_i(\boldsymbol{r}) &= \frac{Ze_0}{4\pi\varepsilon}\frac{\mathrm{e}^{-\kappa|\boldsymbol{r}-\boldsymbol{R}_i|}}{|\boldsymbol{r}-\boldsymbol{R}_i|}.\end{aligned} \tag{8.52}$$

Such an approximation is allowed if the overlap of the electric double layer of thickness κ^{-1} formed around each particle is small. However, the size of the particles has not yet been taken into account. Let us consider the case where there is only one spherical particle of radius a with uniform surface charge density σ ($N = 1$). Suppose that the center of the particle is at the origin and the system is isotropic. That is, $\psi(\boldsymbol{r})$ depends only on the distance $r = |\boldsymbol{r}|$ from the origin. Assuming the solution of the form $\psi = A\mathrm{e}^{-\kappa r}/r$ of (8.44), which satisfies the boundary condition (8.45), then

$$-\frac{\mathrm{d}\psi}{\mathrm{d}r} = \frac{\sigma}{\varepsilon} \text{ at } \quad r = a \tag{8.53}$$

is true. Substituting (8.44) into (8.53) and using the relation $Ze_0 = 4\pi a^2\sigma$, the constant A is determined as

$$A = \frac{Ze_0}{4\pi\varepsilon}\left(\frac{e^{\kappa a}}{1+\kappa a}\right). \tag{8.54}$$

If we compare this result with (8.52), we can see that the charge number Z in (8.51) or (8.52) for a point particle should be replaced by the effective charge number

$$Z' = Z\left(\frac{e^{\kappa a}}{1+\kappa a}\right). \tag{8.55}$$

Therefore, (8.51) is now expressed as

$$\begin{aligned}\psi(\boldsymbol{r}) &= \frac{Z'e_0}{4\pi\varepsilon}\sum_{i=1}^{N}\frac{e^{-\kappa|\boldsymbol{r}-\boldsymbol{R}_i|}}{|\boldsymbol{r}-\boldsymbol{R}_i|} \\ &= \frac{Ze_0}{4\pi\varepsilon}\left(\frac{e^{\kappa a}}{1+\kappa a}\right)\sum_{i=1}^{N}\frac{e^{-\kappa|\boldsymbol{r}-\boldsymbol{R}_i|}}{|\boldsymbol{r}-\boldsymbol{R}_i|}.\end{aligned} \tag{8.56}$$

From the above discussion, the expression (8.56) for the electrostatic potential in the N-particle system and the charge density distribution in the whole region, we have

$$\rho(\boldsymbol{r}) = Z'e_0\sum_{i=1}^{N}\delta(\boldsymbol{r}-\boldsymbol{R}_i) - \varepsilon\kappa^2\psi(\boldsymbol{r}). \tag{8.57}$$

The first term on the right-hand side of (8.57) is the distribution of point charges with effective charge $Z'e_0$ (Z in (8.50) is replaced by Z'). It is easy to verify that (8.56) and (8.57) satisfy the electroneutrality condition (8.36). From these equations, the electrostatic interaction energy $U(\boldsymbol{R}_1, \boldsymbol{R}_2, \ldots, \boldsymbol{R}_N)$ for a given particle position is obtained as follows. In general, U can be expressed in terms of the charge density field $\rho(\boldsymbol{r})$ and the resulting electrostatic potential $\psi(\boldsymbol{r})$ as

$$U = \frac{1}{2}\int\rho(\boldsymbol{r})\psi(\boldsymbol{r})d\boldsymbol{r}. \tag{8.58}$$

Substituting (8.56) and (8.57) into this equation, we find the approximation to the first order of $e_0\psi/k_BT$,

$$U(\boldsymbol{R}_1, \boldsymbol{R}_2, \cdots, \boldsymbol{R}_N) \simeq \frac{(Z'e_0)^2}{4\pi\varepsilon}\sum_{i<j}\frac{e^{-\kappa|\boldsymbol{R}_i-\boldsymbol{R}_j|}}{|\boldsymbol{R}_i-\boldsymbol{R}_j|}, \tag{8.59}$$

where the sum of the right-hand side is taken for all pairs (i, j) satisfying $i < j$. Equation (8.59) with the addition of the van der Waals potential is often referred to as the DLVO potential [8] and is the fundamental interaction potential between charged colloidal particles.

In the above discussion, the Debye screening length κ^{-1} is given by (8.43) and, apart from the temperature dependence of l_B, is determined solely by the ionic concentration n_r of the reservoir, i.e., the salt concentration. However, in solution (Ω_s), counterions dissociated from the particle surface are dissolved. Thus, the average values of ρ and ψ in Ω_s are not zero. Then, taking the mean value of ψ as $\overline{\psi}$ and $\psi = \overline{\psi} + (\psi - \overline{\psi})$ in (8.39) for the ion concentration, linearizing with respect to the deviation from the mean value $\psi - \overline{\psi}$, we obtain

$$n_+ = \overline{n}_+\left[1 - \frac{e_0(\psi - \overline{\psi})}{k_B T}\right], n_- = \overline{n}_-\left[1 + \frac{e_0(\psi - \overline{\psi})}{k_B T}\right], \tag{8.60}$$

where

$$\overline{n}_+ = n_r \exp\left(-\frac{e_0\overline{\psi}}{k_B T}\right), \overline{n}_- = n_r \exp\left(\frac{e_0\overline{\psi}}{k_B T}\right), \tag{8.61}$$

are the average concentrations of positive and negative ions, respectively. From these, the linearized Poisson–Boltzmann equation is

$$\nabla^2\psi = \overline{\kappa}^2\left(\psi - \hat{\psi}\right), \tag{8.62}$$

where

$$\overline{\kappa}^2 \equiv \frac{e_0^2}{\varepsilon k_B T}(\overline{n}_+ + \overline{n}_-) = 4\pi l_B(\overline{n}_+ + \overline{n}_-), \tag{8.63}$$

$$\hat{\psi} \equiv \overline{\psi} + \frac{k_B T}{e_0}\left(\frac{\overline{n}_+ - \overline{n}_-}{\overline{n}_+ + \overline{n}_-}\right). \tag{8.64}$$

Here, $\overline{\kappa}^{-1}$ is the screening length incorporating the effects of counterions, and $\overline{n}_+ + \overline{n}_-$ is the concentration of all ions in solution (if the volume of solution is V_s and the salt concentration is c_s, then $\overline{n}_+ + \overline{n}_- = N|Z|/V_s + 2c_s$). Since the electrostatic potential does not change with the addition of any constant if we rewrite $\psi - \tilde{\psi}$ as ψ in (8.62), we obtain $\nabla^2\psi - \overline{\kappa}^2\psi = 0$, which is the same equation with κ replaced by $\overline{\kappa}$ in (8.42). For a more detailed discussion, please refer to Refs. [9, 10].

References

[1] Debye PW, Hückel E (1923) Phys Z 24:185. Translated and typeset by Braus MJ (2019)
[2] van Roij R, Dijkstra M, Hansen J-P (1999) Phys Rev E 59:2010
[3] Hansen J-P, Löwen H (2000) Annu Rev Phys Chem 51:209
[4] Hansen J-P, McDonald IR (2013) Theory of simple liquids: with applications to soft matter, 4th edn. Academic Press, Oxford
[5] van Roij R, Hansen J-P (1997) Phys Rev Lett 79:3082

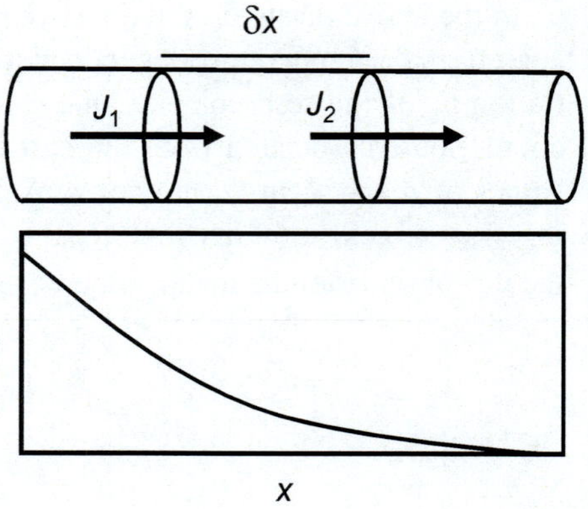

Fig. 8.9 Concentration gradient of solute and flux

[6] Levin Y (2002) Rep Prog Phys 65:1577
[7] Arfken GB, Weber HJ, Harris FE (2013) Mathematical methods for physicists: a comprehensive guide, 7th edn. Academic Press, Oxford
[8] Verwey EJW, Overbeek JTG (1999) Theory of the stability of lyophobic colloids. Dover, NY
[9] Deserno M, von Grünberg H-H (2002) Phys Rev E 66:011401
[10] Zoetekouw B, van Roij R (2006) Phys Rev E 73:021403021403

Appendix E: Fick's Law and Diffusion Equation

E.1 Fick's Law

Thomas Graham, who used the word colloid for the first time, reported his research on the diffusion of gases in 1829 and the diffusion of substances in liquids in 1850. Later, in 1855, a German physician and physiologist, Adolf Eugen Fick proposed Fick's law and defined the diffusion coefficient. Here, we discuss the diffusion of solutes in solution and explain Fick's law.

E.1.1 Fick's First Law

Let us consider the solution where solute concentration C is non-uniform. Here, we consider a one-dimensional case for simplicity. C at a given time is determined by x alone and can be written as $C = C(x)$. Over time, the solute will diffuse from the concentrated region to the dilute region, and then the concentration profile will become uniform. Let us now consider mass transfer through a region with a length dx and a unit cross-sectional area, as shown in Fig. 8.9.

It is convenient to consider "the amount of matter that passes through a unit area in a unit time". This quantity is called the flux J. The flux magnitude accompanying solute diffusion is related to the concentration gradient dC/dx. If there is no concentration gradient ($dC/dx = 0$), no diffusion occurs, i.e., $J = 0$. It is known from experiments that J is proportional to dC/dx. The direction of mass transfer is positive

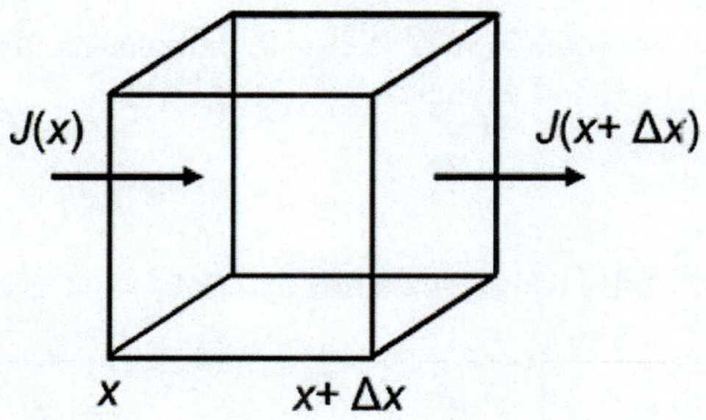

Fig. 8.10 Schematic diagram of Fick's second law

in the right direction (x increases), opposite to the sign of dC/dx. Therefore, we can write

$$J = -D\frac{dC}{dx}. \tag{8.65}$$

Here, D (>0) is referred to as the diffusion coefficient. Though the value of D generally varies with C (i.e., $D = D(C)$) because of the interaction between solute molecules, we can safely assume that D is a constant under dilute conditions. In three-dimensional diffusion, the flux is a vector $\boldsymbol{J} = (J_x, J_y, J_z)$, and Fick's first law can be written as (8.66).

$$\boldsymbol{J} = -D\,\mathrm{grad}C. \tag{8.66}$$

E.1.2. Fick's Second Law

Now, let us consider the change in solute concentration with time t. Again, let us first assume the case of one-dimensional diffusion. Here, C must be regarded as a function of position and time and can be written as s $C = C(x, t)$. As shown in Fig. 8.10, we consider a region with length = Δx and a unit cross-sectional area. The amount of solute in the region is given by a product of concentration and volume = $C\Delta x$, and its time variation is

$$(\partial C\Delta x)/\partial t = (\partial C/\partial t)\Delta x. \tag{8.67}$$

This quantity must equal the difference between the inflow $J(x)$ and outflow $J(x + \Delta x)$. Thus, we have

$$(\partial C/\partial x)\cdot\Delta x = +J(x) - J(x + \Delta x). \tag{8.68}$$

i.e.,

$$\partial C/\partial t = -[J(x + \Delta x) - J(x)]/\Delta x. \tag{8.69}$$

At the limit of $\Delta x \to 0$, the right-hand side of Eq. (8.69) is equal to the derivative of J, and we have

$$\partial C / \partial t = -\partial J / \partial x. \tag{8.70}$$

By using Fick's first law (8.65) and assuming that D is independent of x, we have

$$\frac{\partial C}{\partial t} = D\frac{\partial^2 C}{\partial x^2}. \tag{8.71}$$

The relationship expressed by (8.71) is called Fick's second law, and (8.71) is called the diffusion equation. When D varies with x,

$$\frac{\partial C}{\partial t} = \frac{\partial}{\partial x}\left(D\frac{\partial C}{\partial x}\right) = \frac{\partial D}{\partial x}\frac{\partial C}{\partial x} + D\frac{\partial^2 C}{\partial x^2}. \tag{8.72}$$

Fick's second law in three dimensions can be expressed as Eq. (8.73).

$$\frac{\partial C}{\partial t} = -\operatorname{div}\boldsymbol{J} = D\nabla^2 C \tag{8.73}$$

E.2 Diffusion Equation

The time variation of concentration distribution can be obtained by solving the diffusion equation. However, since the diffusion equation is a second-order differential equation, two additional conditions are needed to find the solution. These are the initial and boundary conditions. Crank's textbook explains in detail how to solve the diffusion equation and its solution under various boundary conditions [1].

As a simple example, let us consider the case where the solute exists only $x = 0$ at $t = 0$ (initial condition), and the diffusion area is infinitely long (boundary condition). This situation is similar to dropping a tiny droplet of ink in water. In this case, the solution of Eq. (8.71) is given by (8.74).

$$C(x, t) = \frac{1}{2\sqrt{\pi Dt}}\exp\left(-\frac{x^2}{4Dt}\right). \tag{8.74}$$

The concentration profiles at various values of t are shown in Fig. 8.11. We note that Eq. (8.74) is the same as the mathematical expression for the Gaussian distribution.

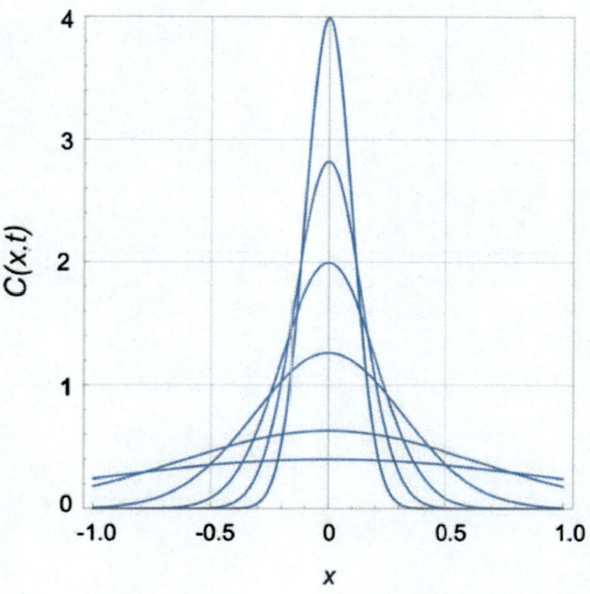

Fig. 8.11 Time variation of concentration profile for diffusion process expressed by (8.74). $D = 10^{-5}$ cm^2/s. $t = 500$, 1000, 2000, 5000. 20,000, and 50,000 from above

Reference

[1] Crank J (1975) The mathematics of diffusion, 2nd edn. Oxford University Press, NY

Index

© The Editor(s) (if applicable) and The Author(s), under exclusive license to Springer Nature Singapore Pte Ltd. 2023
J. Yamanaka et al., *Colloidal Self-Assembly*, Lecture Notes in Chemistry 108,
https://doi.org/10.1007/978-981-99-5052-2